Advances in Underwater Technology, Ocean Science and Offshore Engineering

Volume 14

Submersible Technology: Adapting to Change

ADVANCES IN UNDERWATER TECHNOLOGY, OCEAN SCIENCE AND OFFSHORE ENGINEERING

Vol. 1 Developments in Diving Technology
Vol. 2 Design and Installation of Subsea Systems
Vol. 3 Offshore Site Investigation
Vol. 4 Evaluation, Comparison and Calibration of Oceanographic Instruments
Vol. 5 Submersible Technology
Vol. 6 Oceanology
Vol. 7 Subsea Control and Data Acquisition
Vol. 8 Exclusive Economic Zones
Vol. 9 Stationing and Stability of Semi-submersibles
Vol. 10 Modular Subsea Production Systems
Vol. 11 Underwater Construction: Development and Potential
Vol. 12 Modelling the Offshore Environment
Vol. 13 Economics of Floating Production Systems
Vol. 14 Submersible Technology: Adapting to Change
Vol. 15 Technology Common to Aero and Marine Engineering

CONFERENCE PLANNING COMMITTEE

J. E. Hendrick
T. A. Hollobone
P. Blewett
D. Clark
C. Logan
J. I. M. Morrison
B. G. Redden
A. J. Sandford
Jean Pritchard

Advances in Underwater Technology, Ocean Science and Offshore Engineering

Volume 14

Submersible Technology: Adapting to Change

Proceedings of an international conference ('SUBTECH '87—Adapting to Change') organized jointly by the Association of Offshore Diving Contractors and the Society for Underwater Technology, and held in Aberdeen, UK, 10–12 November 1987

Published by

Graham & Trotman

A member of the Kluwer Academic Publishers Group
LONDON/DORDRECHT/BOSTON

First published in 1988 by

Graham & Trotman Limited
Sterling House
66 Wilton Road
London SW1V 1DE
UK

Graham & Trotman Inc.
101 Philip Drive
Assinippi Park
Norwell, MA 02061
USA

British Library Cataloguing Publication Data

Subtech '87 – Adapting to Change Conference
(1987) Aberdeen, Scotland
Submersible technology
1. Underwater engineering
I. Title II. Association of Offshore
Diving Contractors III. Society for
Underwater Technology
627'. 7

ISBN-13: 978-94-010-7078-2 e-ISBN-13: 978-94-009-1299-1
DOI: 10.1007/978-94-009-1299-1

Library of Congress Cataloging in Publication Data is available.

Softcover reprint of the hardcover 1st edition 1988

Contents

Part III
Saturation Diving

Part IV
Diving Shallower Than 50m

Part V
Developments in Underwater Intervention (Including Robotic Systems)

Part VI
Diversification and Innovation—Adapting to Change

* Late addition from Part I

Part I

An Overview of R & D: Past, Present and Future

1

Sources of Funding for Research and Development in Oil and Gas Technology

R. Ebbers, Offshore Supplies Office,
Department of Energy, UK

The development of new cost-effective techniques for exploration and production is essential for the optimum exploitation of oil and gas resources both on the UK Continental Shelf (UKCS) and worldwide.

The United Kingdom Offshore Operators Association issued a fact sheet in June 1987 in which it was estimated that total annual spending, from all sources, on offshore-related research and development in the UK is in the region of £100–140M. These are significant figures, but when you consider that the annual market for the supplies of offshore goods and services has averaged £2.3–2.4Bn over the last five or six years on the UKCS—it becomes a very worthwhile investment.

In the same fact sheet, the Minister of State for Energy, stated:

> 'As we look forward to the discoveries and developments of the next decade, it is clear that the essential element is R. and D. Only a sustained, determined search for advances in technology which will solve technological problems, reduce costs and maintain close attention to safety, will ensure the full exploitation of our oil and gas resources.'

Development of new technology requires two key ingredients:

(a) the innovative idea;
(b) the funds to undertake the work.

This paper deals with the latter.

There are a number of sources of funding—direct and indirect—which are available worldwide for the development of new products and techniques for use in oil and gas technology. This paper does not purport to cover them all nor can it vouch for the accuracy of all of the text, as the criteria for schemes change from time to time. Whilst every effort has been made to ensure that the information was correct at the time of preparation, companies interested in any source of funding listed are advised to contact the organization listed.

As some new techniques may have a wider application than the oil and gas market, the paper attempts to cover some possible sources of funding which are not specific to oil and gas technology.

Advances in Underwater Technology, Ocean Science and Offshore Engineering, Volume 14: Submersible Technology

R. AND D. SUPPORT

There are two main types of support: indirect and direct. Indirect support is mainly in the form of fiscal relief, while direct sources are much more varied and can be repayable or be in the form of a grant.

UK FISCAL RELIEF

It must be stressed that there is no simple guide to tax relief for offshore and R. and D. projects. Enquiries about whether expenditure specific to R. and D. projects qualifies for tax relief should be made direct to the Oil Taxation Office.

Offshore R. and D. expenditure may qualify for relief for the purposes of:

(a) corporation tax (CT);
(b) petroleum revenue tax (PRT).

Whether research costs are relieved under the CT rules will depend on: (a) the nature of the costs, and (b) in cases involving a subsidiary company, the way in which a group's research is carried out and how the costs are allocated.

As for petroleum revenue tax, relief is given for expenditure incurred in a particular field for one or more of the stated purposes set out in Section 3 of the Oil Taxation Act (1975) and these include searching for oil, ascertaining the characteristics of a field, winning oil from the field, measuring and transporting of the oil and the initial treatment and storage of oil won.

In addition, Section 64 of the Finance Act (1987) allows certain expenditure on oil-related research, which does not at present qualify for PRT relief under Section 3 of the Oil Taxation Act (1975), to be offset against a participating PRT liability. The contact is Mr A. Beacham, Oil Taxation Office, Melbourne House, Aldwych, London WC2B 4LL. Tel: 01-438 7579.

DIRECT FUNDING

Banks and financial institutions are an obvious possible source of funding. The availability and extent of funding will be influenced by market opportunities, the degree of risk involved and the status of the company, etc. Some banks have sections specifically geared to producing the right packages in oil and gas lending. Depending on the size of the project, support can be in the form of an overdraft, short-, medium- or long-term financing, provision of venture capital, loan life-cover/project-cover against cash flow, or a leasing arrangement. Contact should be made with the oil- and gas-project financing group of the bank, if they have one, or alternatively with your local bank manager in the first instance.

OIL COMPANIES

Oil companies are the main clients for the oil and gas technology and are therefore another obvious source as they have the most to gain from the successful development and implementation of innovative, cost-effective techniques. Oil companies do not support R. and D. for its own sake. They must have a clearly identified need—perhaps as a result of a specific field development—or alternatively be convinced that the development of the new product/technique will bring them a net benefit—i.e. be cheaper, or safer, or save time, or a combination of all three.

Oil company spending in the UK covers a wide range of offshore technology, of which a significant portion was directed at subsea techniques over 1985/87. Expenditure on subsea R. and D. is expected to be increased over the next few years with the development of marginal fields and exploration and exploitation in deeper waters.

The Offshore Supplies Office (OSO) might be able to assist companies with ideas in subsea technology, or indeed any other area of oil and gas technology, to identify interested oil companies, but generally speaking, it is better if suppliers contact the oil companies themselves. The contact is: Research and Development Branch, Offshore Supplies Office, Alhambra House, 45 Waterloo Street, Glasgow G2 6AS. Tel: 041-221 8777, ext. 321.

GOVERNMENT FUNDING

Department of Energy

The Offshore Supplies Office of the Department of Energy is the primary focal point within Government for the offshore supplies industry in the UK. Funding is available to contractors on a shared-cost basis for the development of new techniques in oil and gas technology, for high-risk R. and D. projects which otherwise would not go ahead. The funding is repayable only on commercial exploitation of technology via a levy on sales. The levy rate is set to enable repayment of funding (plus interest which accumulates from the date of first exploitation) over a period of approximately five years. Levy agreements run for 15 years, and once repayment of principal and interest has been made, a reduced secondary rate of levy applies on future sales for the remainder of the levy period.

A copy of the guidelines and an application form can be obtained from the following address: Research and Development Branch, Offshore Supplies Office, Alhambra House, 45 Waterloo Street, Glasgow G2 6AF. Tel: 041-221 8777, ext. 521.

The Petroleum Engineering Division of the Department of Energy has the responsibility for oil-field delineation and safety. Research and development in this field provides the material and understanding necessary to enable the Department to fulfil its regulatory role in respect of the UK's hydrocarbon resources and their safe and efficient development. There are three areas which may be of interest to seminar participants:

(a) enhanced oil recovery (EOR);
(b) reservoir simulation;
(c) safety.

The Department's EOR and reservoir simulation programmes are primarily based at the Atomic Energy Research Establishment at Winfrith (with, in the case of EOR, a support programme of underlying research at universities) and are geared to fulfil the Department's statutory requirements.

The objective of the safety programme is to promote the safety and effectiveness of oil- and gas-related activities offshore. Safety R. and D. often identifies a need to change the manner in which offshore systems are designed, constructed or operated. Such research may create a requirement for new equipment and service, and hence the market for these. Alternatively, it will show that existing practices are unnecessarily conservative and may permit a relaxation which will be of economic benefit to UK industry. In some cases the Department funds the activities on its own. Increasingly, however, it shares the cost with industry or trade associations, thereby ensuring that the results are available widely across industry. The contact is Mr D. Carpenter, Petroleum Engineering Division, Department of Energy, Thames House South, London. Tel: 01-211 7083.

Department of Trade and Industry (DTI)

The Department of Trade and Industry offers financial support to encourage firms to undertake innovative research and development. Support for Innovation (SFI) is open to all firms operating in the UK. The support is available in all sectors for projects which would not otherwise go ahead or would be seriously delayed. The main form of support is a grant of up to 25% of eligible costs which is subject to a £5M limit. Collaborative, longer-term applied research projects which are carried out by consortia including more than two organizations may be eligible for a higher rate of grant—normally up to 50%.

There is a minimum of £100 000 eligible costs, below which applications can only be considered from small firms: i.e. in the manufacturing sector, independent firms with less than 200 employees or firms which are part of a group which itself employs less than 200 in total; in the service sector, independent firms with a turnover of less than £1M or firms which are part of a group which itself has a turnover of less than £1M in total.

In addition, the DTI operates a Pre-production Orders Scheme which involves the DTI purchasing modules of equipment embodying new techniques and lending them to potential users for a trial period. However, at the time of preparation of this article, the scheme was not available and was under review. The contact is RTP Division, DTI, Ashdown House, 123 Victoria Street, London SW1 6RB. Tel: 01-212 3203.

Small-firm Merit Award for Research and Technology (SMART)

SMART is competitive to ensure that the best of the very technologically risky projects get a chance to show their potential to private backers such as venture capitalists before a firm commitment is required.

Two SMART competitions have now taken place. The first was to encourage firms to take up dormant instrumentation on biotechnology projects and the second was to encourage biotechnology and advanced materials technology. Two more rounds are planned to start in 1988 and 1989.

The competition takes place in two stages. Stage 1 gives 75% of eligible costs up to £37.5k for a one-year feasibility study. Stage 2 gives 50% of eligible costs up to £100k for a further year, following which projects may receive help through Support for Innovation (SFI). Stage 1 is open to individuals and small firms who wish to start a project in the chosen technologies, but Stage 2 is limited to Stage 1 participants. As an experiment, an open competition is being run in the North-east and South-west regions of Britain to help redress the imbalance of R. and D. being concentrated in the South-east. The contact is DTI, SMART Office, Room 310, Ashdown House, 123 Victoria Street, London SW1 6RB. Tel: 01-212 0183.

Marine Technology and Shipbuilding

The Department of Trade and Industry marine-technology policy is designed with advice from the Maritime Technology Committee to encourage the development and application of advanced technologies relevant to UK maritime industries. Two programmes of collaborative R. and D. have been introduced—the Efficient Ship Programme, and the Resources from the Sea Programme. Individual projects can be funded under the standard Support for Innovation scheme; terms already mentioned. The contact is Mr D. M. Hoddinott, SEE (2) Division, DTI, Room 261, Ashdown House, 123 Victoria Street, London SW1 6RB. Tel: 01-212 5997.

BRITISH TECHNOLOGY GROUP (BTG)

The British Technology Group is a self-financing public organization which licenses new scientific and engineering products to industry and provides finance for the development of new technology. Under an industrial project agreement, BTG will contribute an agreed sum—usually up to 50%—of the cost of a project. British Technology Group shares in the success of the project by means of a negotiated levy on turnover from the resulting product or process. There are no formal limits to the size of BTG's investment, although the majority of projects fall within the range of £50 000 to £1M. The Group also provides equity finance to new companies to assist innovation in industry. The address is British Technology Group, 101 Newington Causeway, London SE1 6BU. Tel: 01-403 6666.

COLLABORATION IN R. AND D. WITH UNIVERSITIES

Marine Technology Directorate

Marine Technology Directorate Limited (MTD Ltd), launched in October 1986, discharges the Science and Engineering Research Council's (SERC) responsibilities in marine technology. The company's purpose is to develop the national capability in ocean-related activities and the longer-term use of the oceans through the support and management of research, and promotion of education and training in marine technology.

Research funds amounting to about £6.6M per year are allocated to universities and polytechnics for research of a strategic nature and, to a lesser extent, for fundamental and applied research and for programme management.

The term 'marine technology' covers a wide range of engineering disciplines (civil, structural, mechanical, electrical and advanced computing methods) in five broad areas: Marine Resources; Physical Ocean Environment; Ocean Structures and Materials; Ships and Floating Structures; Underwater Working.

Many of these engineering disciplines are brought together in large, multi-project research programmes which are coordinated, managed and jointly funded by the sponsors from industry, Government and SERC. During the period July 1985 to July 1987 there were 14 of these managed programmes which accounted for about 60% of the total programme funding. The remaining funds were allocated to individual projects in universities and polytechnics, and may or may not be jointly funded by industry and/or Government.

In the next phase of the programme, beyond 1987, the balance of the funding between managed programmes and individual projects will again be about 2:1. The detailed content and number of the managed programmes to be supported are currently being assessed by the board of MTD Ltd. Nine of the existing managed programmes, as follows:

- Decommissioning;
- Fatigue;
- Welded materials;
- Automated subsea tasks
  hydrodynamics;
  communications and control;
- Defect assessment;
- Structural integrity monitoring;
- Floating production systems;
- Fluid loading;

are being considered for further support. In addition to these, four new priority areas:

- Composite materials;
- Processing and equipment;
- Robotics;
- Subsea completion systems;

are being considered as possible candidates for managed programmes in the future. The contact is Mr A. M. Adye, MTD Ltd, 2 Little Smith Street, Westminster, London SW1P 3DH. Tel: 01-222 7331.

THE LINK RESEARCH INITIATIVE

In December 1986 a £210M government initiative to speed up the development of products and services from scientific and technological research carried out by universities, government departments and industries was announced. Through the initiative called LINK the government will support up to 50% of the cost of collaborative programmes between the scientific community and the industry. The Government's contribution will reach £210M over the next five years. Overall, the initiative will generate expenditure by Government and industry of at least £420M over the next five years, provided industry matches the Government funding.

The LINK initiative will consist of a range of programmes in key strategic areas of science and technology; the areas most likely to be of interest to readers of this book include ocean technology covering composite materials, subsea communications, subsea resource exploration and assessment, and underwater advanced robotics. The address is LINK Secretariat, Department of Trade and

Industry, Room 237, Ashdown House, 123 Victoria Street, London SW1E 6RB. Tel: 01-212 6402.

TEACHING COMPANY SCHEME

This scheme is operated by the Science and Engineering Research Council (SERC) and the Department of Trade and Industry (DTI) and seeks to establish active partnership between higher-education institutions and industry. The fundamental aim of the scheme is to raise the level of 'capability' as distinct from 'knowledge' in both industry and academia in order to:

(a) improve industrial performance through the effective introduction of modern manufacturing and management techniques, promote the development and retraining of company staff and eliminate the under-use of academic expertise and facilities;
(b) provide, for the most able graduates, early career opportunities leading to senior technical and management posts in manufacturing industry;
(c) give Universities and Polytechnics continuous and direct involvement in industry at all levels in order to develop and retrain academic staff, enhance the relevance of teaching and benefit research.

The scheme operates through Teaching Company Programmes which bring together academics, company staff and young, high-quality graduates, ideally with some industrial experience. The graduates, known as Teaching Company Associates, are appointed on two-year contracts to work at the company under the guidance of company and academic staff on tasks within a programme. This can include a research and development project. Each company involved in a programme contributes to the costs within specified guidelines. The balance is funded equally by DTI and SERC. The address is Teaching Company Directorate, Science and Engineering Research Council, Polaris House, North Star Avenue, Swindon, Wilts. Tel: 0793 26222, ext. 2335.

EEC FUNDING

Technology Projects in the Hydrocarbons Sector

Under Regulation EEC No. 3639/85 funding is available for technological development projects directly related to exploration, production, storage or transportation in the hydrocarbons sector. Funding is restricted to EEC member states but is available to any individual organization or group of organizations who can demonstrate their ability to carry out the relevant projects.

Preference is given to projects involving at least two undertakings established in different member states and to projects submitted by small and medium-sized undertakings.

Funding of 140M ECUS was approved for a four-year programme ending 31 December 1989. Application is made within the deadline specified in each of the Commission's invitations for proposals to be submitted and published in the official journal. The deadline for project submission under the most recent invitation was 28 November 1986. To be eligible for support, companies must satisfy the following conditions:

- their project must promise to develop innovatory techniques, processes or products, or exploit a new application of technique;
- their project should offer prospects of industrial, economic and commercial viability;
- they must establish a need for the funding.

The rate of award is up to 40% of the eligible cost of the project. The address is: Research and Development Branch, Offshore Supplies Office, Alhambra House, 45 Waterloo Street, Glasgow G2 6AS. Tel: 041-221 8777, ext. 488, or alternatively: Commission of the European Communities, Directorate-General for Energy-XVII/C-1-TH, Community Technology Projects (Hydrocarbons), Rue de la Loi 200, B.1049, Brussels, Belgium. Tel: 010-322-235-3625.

Eureka

Eureka is a framework for promotion of collaborative projects in fields of advanced technology which was agreed between 18 European countries and the European Commission in 1985. Its aim is to improve European competitiveness in world markets in civil applications of new technology. Eureka projects will be in the fields of advanced technology which include marine technology and materials. It is for companies to identify topics on which they wish to collaborate and to seek prospective partners.

Through Eureka, Governments and EEC can assist the development and commercial success of projects by providing a means for selecting partners and to help overcome market barriers (for instance, by applying pressure in the appropriate form for studies to be adopted). Participants in the UK in Eureka projects will be eligible to apply for financial support under existing UK schemes.

Under DTI's Support for Innovation, firms in the UK participating in Eureka are eligible for support of up to 50% of their share of applied research costs and up to 25% of their share of development costs. The address is Eureka Projects Office, Department of Trade and Industry, Ashdown House, 123 Victoria Street, London SW1E 6RB. Tel: 01-212 0249.

GOVERNMENT FUNDING OF RESEARCH AND DEVELOPMENT IN OIL AND GAS TECHNOLOGY IN OVERSEAS COUNTRIES

It is generally the case that the company must be registered in the country, and the research must be undertaken and exploited there, for the company to be eligible for assistance. Detailed criteria or guidelines are not always readily available and it is suggested that companies get in touch directly with the countries listed for up-to-date information.

Canada

The address is The Programme Secretary, NRC Committee on Industrial

Research Assistance Programme, Montreal Road, Ottawa, Ontario K1A 0RG, Canada. Tel: 010 1 613 993 0331.

Denmark

The address is Mr M. Nielsen, Danish Ministry for Energy, Slotscholmsgrade 1, 1216 Copenhagen—K, Denmark. Tel: 010 451 927500.

France

The address is Director General of Research and Development, Fonds de la Recherche et de la Technologie (FRT), Ministère de la Recherche et de l'Industrie, 1 rue Rescortes, 75231 Paris, Cédex 05. Tel: 010 33 1 634 33 33.

Ireland

The address is Mr J. N. Hughes, The Industrial Development Authority of Ireland, Lansdowne House, Ballsbridge, Dublin 4, Eire. Tel: 001 686 633.

Italy

The address is Italian Industrial Fixed Assets Institution, Viale dell'Arte 25, Rome, Italy. Tel: 010 39 6 54501.

Netherlands

The address is Mr W. Rullens, Ministry of Economic Affairs, Bezvidewhoutseweg 6, The Hague, Netherlands. Tel: 010 317 079 7978.

Norway

The address is Ministry for Industry, POB 8014 Dep. Oslo 1, Norway. Tel: 01047 2 119090.

Sweden

The address is SIND (National Industrial Board Ocean Technology Programme), 11786 Stockholm, Sweden. Tel: 010 468 744 9486.

ADDENDUM

Small Firm Merit Award for Research and Technology (SMART)

This scheme is under review. The two rounds proposed for 1988 and 1989 may not take place.

Marine Technology Directorate

(*a*) The managed programmes to be supported have now been agreed. All but 'Defect Assessment' have been included.

(*b*) 'Subsea completion systems' has been dropped from the list of priority areas.
(*c*) The contact for this scheme is now Dr E. M. Forster.

EEC Funding: Technology Projects in the Hydrocarbons Sector

The deadline for project submission under the most recent invitation is 15th January 1988.

2

The Funding of the Hydra-Lok Swage Pile Connection Development

J. M. Lowes, BUE Hydra-Lok Limited, UK

The sources of funding available to development projects are considerable. Most are linked to government; however, it is possible with an interesting programme and the potential of a 'technical or financial return' to obtain funds through operating companies in the offshore industry.

The development of Hydra-Lok has made use of various sources of funding, and some experience has been gained on the type of projects that are attractive to the various types of funding organizations. The following paper illustrates, through the development of Hydra-Lok, various aspects of funded research and development, including:

(1) how to fund a project, and the levels of funding available;
(2) the way funding was applied to Hydra-Lok;
(3) the advantages and disadvantages of funding a successful project. (The advantages of funding an unsuccessful project should be clear.)

SOURCES AND LEVELS OF FUNDING AVAILABLE

BUE Hydra-Lok have obtained funding for the development of the Hydra-Lok system from a number of sources. We have applied to, and obtained funding at one time or another from, each of the following sources:

(1) UK Department of Energy—OETB (Offshore Supplies Office sponsored);
(2) UK Department of Energy—OETB (PED sponsored);
(3) EEC—Community projects in the hydrocarbons sector;
(4) operators of the oil industry—single- and multiple-sponsored projects.

The availability of funding from these various sources depends upon the specific project, the level of prior development, and clearly the level of interest that can be raised within the sponsoring organization. The levels of funding can vary from as little as 10% to 100% in exceptional circumstances. In general, most sponsoring organizations look most favourably at projects where there is apparent parity between the contribution of the sponsor and sponsored.

It has also been our experience that it is easier to obtain 'government cash' (OETB/EEC), rather than industrial sponsorship at an early stage of the

Advances in Underwater Technology, Ocean Science and Offshore Engineering, Volume 14: Submersible Technology

development, whereas, although this cash is equally available for later 'precommercial development' it may only be available subject to other industrial sponsorship. This is consistent with the opportunities available regarding industrial sponsorship, which is more readily available when the light at the end of the tunnel is becoming brighter. In other words, it can be said that the higher the level of 'R.' in the project the less interest there is from industrial sponsors, whereas the higher the level of 'D.', the more readily obtainable is government funding in conjunction with industrial support.

The ability to obtain funding depends entirely on the ability of the company to prepare a suitable 'proposal', which sells the 'embryonic product or service' to the sponsoring organization. A proposal, in order to be successful, should address the following points:

(1) The concept has a sound basis and has been thoroughly thought out.
(2) The potential pitfalls have been identified and the appropriate investigations proposed.
(3) The development project will produce a 'product or service' that the industry will use and there is a clear technical and commercial advantage to be made, and hence a ready market.
(4) The sponsored company has identified a route to commercialize the product.
(5) The financial implications of the development have been defined, both from the point of view of the cost and the likely return.

The above may seem extremely obvious, but it is necessary to consider each issue when looking for sponsorship. A development proposal contains most issues covered by a business plan.

HOW FUNDING HAS BEEN APPLIED TO HYDRA-LOK

The development of Hydra-Lok was achieved using the above approach. It was recognized from the very beginning that the development would be phased over a number of years. It was also a concept that was unlikely to achieve its ultimate goal without considerable enthusiasm within the industry, and this would be true regardless of the level of cash available either from government sources or from BUEH's own resources. The development was undertaken, therefore, in a number of phases, each being partly funded by sponsorship.

Development of the Hydra-Lok Concept and Preliminary Testing

BUEH had outlined the likely geometry and various other salient parameters prior to embarking on the initial development. At this stage, although the oil industry was interested, it was believed inappropriate to solicit funds other than through the Department of Energy (OETB). The level of funding was therefore limited to 50%.

The early development work optimized all the basic parameters for the connection: geometry; method of forming; design of suitable tools. Preliminary trials were undertaken on model size pile/sleeve connections to establish the likely strength, fatigue properties, corrosion resistance, etc. Final tests were done on 36-inch diameter tubulars which also illustrated the operational capability of the system. Although the industry maintained a watching brief and keen interest, it was only in the latter days of the development that real enthusiasm began to show. This initial phase of development led the industry and ourselves to conclude that we had a viable product—*at least under laboratory conditions.*

The Offshore Trial

Although Hydra-Lok had been demonstrated at 36-inch diameter, and several full-size connections had been made and tested, since it is a technique for subsea application, the industry required an operational trial. This trial was designed to simulate as far as possible an offshore operation. If Hydra-Lok could pass this test, then two oil companies had indicated an intention to use it. At last a real interest from the industry in Hydra-Lok as a pile connector had been realized. It should be noted that Hydra-Lok had been successfully used some two years previously to make a riser connection at a depth of 500 ft, but this procedure did not contain the same operational steps.

It was, at this stage, that oil industry support for the development was first canvassed and successfully obtained. This enabled the offshore trial to be undertaken with both oil industry and the Department of Energy's assistance. The trial included the operational sequences associated with the entire operation of swaging two 42-inch piles, in a dummy template, at 400 ft water depth. Since the system was considered diverless, the operation had to be undertaken without any diver assistance and an ROV was used as the only means of intervention. Success of the trial was to lead to the Balmoral Installation for North Sea Sun Oil and the South-east Forties installation for BP. Failure would mean . . .

The trial was successful and led to BUEH's first commercial application of the system as a pile/template connection and this has since allowed the system to be utilized extensively in the North Sea.

Development of the System to Its Ultimate Goal of Pile/Jacket Connections

Although the system was by now being used extensively offshore, various tests were required to convince the operators and the certifying authorities that Hydra-Lok could be used successfully on large deepwater structures. The industry was now extremely interested; some operators had used the technique and were enthusiastic regarding its potential applications. The opportunity was therefore present to develop the system further with considerable oil-industry support. Later development has therefore been jointly funded between the oil industry, Department of Energy and BUEH. Levels of funding from the third source have been as high as 85%, depending on the specific content of the project.

Development covered within this phase has included development of very large-diameter connections (72 inch) suitable for deepwater jackets, and the further investigation of the properties of the connection, including static strength and strength under cyclic loading.

Investigations have also been undertaken into the material of the pile, both during formation of the joint and in its subsequent service life. The quantity of data generated will enable a full understanding of the connection under all likely service connections and give the industry the confidence to apply the technique to their structures.

As the oil industry becomes involved in these development programmes, the opportunity for applications becomes more visible, and completion of the 72-inch development programme has led to the systems application on BP's SNS development project.

THE ADVANTAGES AND DISADVANTAGES OF FUNDING R. AND D. PROJECTS

There are many advantages of funding successful development projects. The only disadvantage is perhaps the 'pay back'. This 'pay back' is particularly appropriate to governmental support where clear guidelines are laid down as to how development expenditure is to be recovered. The details are, however, an issue for individual discussion and it is not possible to give generalized comment other than to say the intention is to recover the expenditure without prejudice to the commercial success of the system/product.

The benefits of funding are considerable. It could be argued that the financial implications are not the primary consideration. The involvement of the sponsoring organizations generally arises following considerable internal digestion of the 'proposal' and this is a very useful and fairly intense form of advertising.

The involvement in the project requires reporting which is often distributed throughout the sponsor's organization. The responsible engineer in the sponsor's organization, therefore, develops a considerable insight into the system, becoming capable of true assessment of the system for his company's requirements. This creates a favourable atmosphere towards application of the technology.

On balance, therefore, it must be said that aside from the financial benefits of funding, albeit that it may be necessary to pay back, the overall benefits of obtaining funding are considerable and could be regarded as the first phase of eventual acceptance into the commercial world.

3

The Drill Support Tooling Module Project

B. G. Redden, Wharton Williams, UK

This paper describes an ROV-related development project undertaken by the author's company with the assistance of external funding support. It will outline the project aim and discuss the source and level of funding, how application was made, criteria for acceptance and the project monitoring and administration requirements, etc.

THE PROJECT

The project was conceived in 1984 when we recognized that a technical gap existed in the equipment fit and general capability of our work-class ROV systems. To fill this gap a number of component subsystems were identified and it was decided that development of this equipment as a module for use with the Rigworker ROV would be a worthwhile project. Hence the name 'drill support tooling module' (DSTM). In defining the project concept two principal criteria were applied:

(a) that the module would enable a range of drill-support tasks to be accomplished by the vehicle on one dive without need to recover and reconfigure, and
(b) each one of the DSTM components would be designed to stand alone if required.

The DSTM components identified were classed as either subsystems or tooling.

Subsystems included the DSTM control console, an electronics bottle to interface the ROV telemetry system to the DSTM (Fig. 1), a solenoid valve pack (Fig. 2) and a proportional flow and control valve pack (Fig. 3).

Tooling included a range of guidewire and overshot tools (Figs 4, 5 and 6), torque wrenches (Fig. 7), AX ring tools (Fig. 8), HP pump, LP seawater pump, hot line stab-in (Fig. 9) and a grabber.

Figure 10 shows what the DSTM looked like when put together in a sealed-tube chassis.

Advances in Underwater Technology, Ocean Science and Offshore Engineering, Volume 14: Submersible Technology

Fig. 1 Electronics bottle.

Fig. 2 Solenoid valve pack.

Fig. 3 Flow and control valve pack.

Fig. 4 Vetco overshot tool.

Fig. 5 Regan latch removal tool.

Fig. 6 Cameron lockdown tool.

Fig. 7 Torque wrenches.

Fig. 8 AX ring tool.

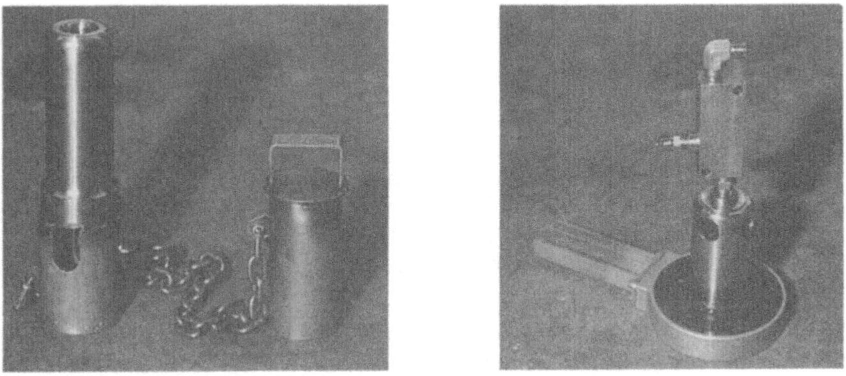

Fig. 9 (*left*) Hot line stab-in probe.
(*right*) Hot line stab-in probe and receptacle and cover.

Fig. 10 DSTM module.

SOURCE OF FUNDING

An outline concept and general arrangement with estimate of project timescale and costings was prepared. This was used in house to convince management the DSTM was worthwhile. However, it also highlighted the cost and technical risk of undertaking such a project, and for this reason we approached the Offshore Supplies Office (OSO) to enquire whether any support schemes were available. After some discussion we were told that funding of up to 50% of the project capital cost was available and were given a ten-page set of explanatory notes describing how to apply for it!

MAKING THE APPLICATION FOR OSO FUNDING

To qualify for OSO funding support a project must first meet one or more of the general requirements set out by the Offshore Energy Technology Board (OETB) as advisors to the Department of Energy. These are that any technological development must lead to:

- improvements in the efficient and economic exploitation of the oil and gas resources on the UK Continental Shelf;
- improvements in the standards laid down by the Department of Energy in pursuance of its statutory responsibilities for safety;
- improvements in the competitiveness of British Industry in the field of offshore oil and gas.

The project must then meet specific criteria which have to be addressed and answered in the proposal for funding application. Briefly I will list these and, as an example, show how the DSTM project conformed.

(1) Relevance

The project must satisfy one of the previously stated objectives. In the case of the DSTM I think we met all three.

(2) Necessity and (3) Novelty

It is necessary to show that the end result could not be achieved as effectively without the expenditure of R. and D. monies and that the product is novel, i.e. not already available. The DSTM complied as there were no comparable drill-support tools or subsystems available from the Rigworker manufacturer or elsewhere on the open market.

(4) Credibility

The objectives of the R. and D. project need to be clearly specified and likely to be achieved in the time proposed and in a cost-effective manner.

This is really an invitation to justify the project and to explain how your organization would set up and manage the works. We obviously managed to do this!

(5) Exploitability

You need to explain how the results of the project could be exploited commercially, outline a philosophy on manufacture and marketing and show that your organization has the standing and resources to implement such a plan.

For the DSTM a manufacture-under-licence scheme was drawn up. As the project has now been completed, these plans are being implemented and the DSTM components should be available on the open market within the next few months.

(6) Dependence on Support

You must show that, for whatever reason, the project is dependent upon Department of Energy support to proceed.

In the case of the DSTM project this was very straightforward in that there were high technical and commercial risks. Without OSO funding as an insurance to lessen the exposure to these risks my management would have decided not to proceed.

(7) UK Competitiveness

This is somewhat similar to Exploitability, with the addition that a benefit to the UK must be demonstrated.

For the DSTM products, we described the international ROV market and showed how this technology could make British ROVs and tooling more attractive and competitive to overseas buyers as well as giving the operators of Rigworker a technical advantage.

(8) Need for Further Work

Finally, it should be stated whether the proposed programme is complete in itself or whether it is in fact the first part of an extended project.

The guidance document goes on to suggest an ideal proposal format and gives more detailed advice on the information to be provided as well as a pro-forma showing the preferred spreadsheet layout. Whilst all of this may seem a little pernickety and bureaucratic, it is in fact quite useful. If followed reasonably closely it saves a lot of time in helping to focus the proposal on the key areas, as seen by the Department of Energy. In our experience this was certainly the case, and in fact our proposal was accepted more or less as originally written.

LEVEL OF FUNDING

Funding of this type is normally up to a maximum contribution of 50% of actual project costs. However, one word of warning. There are strict criteria for what are acceptable costs and what are not and it is possible to incur some costs inadvertently which under normal commercial conditions would be project chargeable, but under the D. of E. contract terms are not! Let me give a simple example: postage, Datapost or special transport of documentation or components.

GENERAL PROJECT ADMINISTRATION

Once the proposal is accepted, you then come to one of the unforeseen problems with OSO funding. This is that most commercial companies like my own have to set up specific accounting routines to handle record keeping and preparation of financial reports to comply with the ministry paperwork system. This will of necessity be a steep learning curve, almost certainly incurring additional, unbudgeted expense. From our own experiences, and with the benefit of hindsight, I would suggest that any company proposing to make an application for government funding would be wise to research carefully the project-management, accounting system, back-up paperwork and claim-submission requirements prior to making the application. This may seem elementary but my point is that the average company's in-house accounting and project-management systems may not be extensive enough to cater for government requirements. Ours was not!

Once the project is underway any changes in the schedule, workscope or spend must be identified and agreed in writing prior to implementation. Providing the laid-down procedures were followed we found that the OSO-appointed project officers were understanding and accommodating in their approach to monitoring the work and authorizing changes, and on many occasions were very helpful in steering us through the system.

CONSTRAINTS

There are obviously constraints put on the development company by the nature, definition and requirements of government type funding, but these are obviously discussed and agreed before execution of the contract.

I believe it is usual for the developing company ultimately, to wholly own the resulting product or knowledge, and certainly in the DSTM project this is the case.

IN CONCLUSION

In conclusion, I must first make reference to the fact that with OSO funding, if the project is successful, all monies must be subsequently repaid in the form of levy and royalties.

Whilst this is, of course, known in advance and does not detract from the value of having this funding source available, it must of necessity make potential developers think twice, and maybe move them towards other sources of non-repayable funds.

However, for specific projects, funding from the Department of Energy is well worth considering and the benefits of this risk-sharing far outweigh the problems in dealing with the inevitable bureaucracy.

Part II

Reassessment of Standards and Operating Practices

4

Review of Standards, Codes, Requirements and Regulations: their Impact on Underwater Projects

T. A. Hollobone, Association of Offshore Diving Contractors, UK

The underwater engineering industry, in terms of its support for offshore oil/gas projects, is young. Although the first underwater projects relating to the exploration for, and exploitation of, hydrocarbon resources took place some 40 to 50 years ago, major underwater work has only been carried out for the past 30 or so years, and as far as the North Sea area is concerned, only since the mid-1960s.

The title of this paper includes the phrase 'their impact on underwater projects'. Having agreed to present it I have come to the conclusion that I can only deal realistically, in the timescale available, with part of the total legislation and associated standards and therefore propose to cover diving, ROV/submersibles and dynamic positioning. The full range of 'underwater projects' would have also required cranes and lifting equipment, medical aspects, life saving and subsea production systems to be covered in similar detail. Perhaps all of that can be reserved for another occasion!

Again, because of time constraints, I have decided in the main to address the subject with particular emphasis on the North Sea area. That was not to chicken out of dealing with the total international scene but simply to keep the paper within manageable proportions. It is appreciated that underwater work is carried out in a supporting role on an international basis and that littoral States of the North-West European Continental Shelf area have no exclusive hold over 'standards, codes, requirements and regulations'.

There are a limited number of international standards which will, however, be addressed, as will particular client requirements, either by way of example or with reference to client groups, such as UKOOA, NIFO, and E. and P. Forum.

The paper will address some specific problem areas, but only very briefly. Much more will be said on those during the paper's presentation.

Brief comments are also included concerning the future, in the sense of 'do we need more or less legislation, rationalization, harmonization, or what?'

Finally I apologise now for errors which this article is bound to contain. The subject is so wide and complex that it has not been possible for me to check, double check and triple check every single reference or to choose, mull over and edit every single phrase and sentence. I do hope, however, that there are no significant errors and those that there are, are of detail only.

Advances in Underwater Technology, Ocean Science and Offshore Engineering, Volume 14: Submersible Technology

NORTH SEA AREA

This section will address in turn the standards, codes, requirements and regulations in relation to underwater projects, with particular emphasis on offshore oil and gas activities, in the following countries: UK; Norway; Denmark; Holland; and Eire.

Whilst the basis of legislation in each of these littoral States varies, the same broad principles apply. Time does not permit descriptions of the way regulations are formed in relation to national law making requirements.

United Kingdom

The principal Acts and Regulations which cover diving activities in UK waters and from UK ships are set out below. The main ones are SI 1981.399—The Diving Operations at Work Regulations, and SI 1975.116—The Merchant Shipping (Diving Operations) Regulations. The latter are being amended to bring them into line with SI 1981.399.

SI 1977 No. 1232
The Health and Safety at Work etc. Act (1974) (Application Outside Great Britain) Order (1977)

SI 1981 No. 399
The Diving Operations at Work Regulations (1981)

SI 1973 No. 1842
The Offshore Installations (Inspectors and Casualties) Regulations (1973)

SI 1976 No. 1019
The Offshore Installations (Operational Safety, Health and Welfare) Regulations (1976)

SRO 1931 No. 679
The Gas Cylinders (Conveyance) Regulations (1931)

SI 1975 No. 116
The Merchant Shipping (Diving Operations) Regulations (1975) and Amendment SI 1975 No. 2062

SI 1976 No. 923
The Submarine Pipelines (Diving Operations) Regulations (1976)

SI 1977 No. 835
The Submarine Pipelines (Inspectors etc.) Regulations (1977)

The Protection of Military Remains Act (1986), ch. 35

The Health and Safety at Work etc. Act (1974) applies through the Application Outside Great Britain Order, SI No. 1977.1232.

Of the nine pieces of legislation listed above, mention should be made of the Protection of Military Remains Act (1986) which came onto the statute book in September of that year but of which, at the time of writing this paper, little more is known.

As far as codes are concerned, and in the context of this article, reference is made to four sets of certification rules, details of which are set out below:

Rules and Regulations for the Construction and Classification of Submersibles and Diving Systems (1987)
Lloyd's Register

Rules for Certification of Diving Systems (1982)
Det norske Veritas

Rules for Building and Classing Underwater Systems and Vehicles (1979)
American Bureau of Shipping

Rules for Underwater Technology
Chapter 1: Diving Systems and Diving Simulators 1986
Chapter 3: Underwater Equipment (in preparation)
Germanischer Lloyd

Last, but certainly by no means least, are the very large number of other codes, guidance notes, guidelines and similar documents. A partial list is set out below:

A Guide to the Diving Operations at Work Regulations (1981)
HS(R)8

Offshore Construction—Health, Safety and Welfare
HS(G)12

The RN Diving Manual (BR.2806), 6th amendment updated 1985
(available to approved applicants)
Ministry of Defence

Guidelines for the Specification and Operation of Dynamically Positioned Diving Support Vessels (1983)
Department of Energy/NPD

The Principles of Safe Diving Practice, UR23 (1984)
UEG, CIRIA

Code of Practice
Association of High Pressure Water Jetting Contractors

Guidance Manual for Material Selection and Inspection of Inert Gas Systems
American Bureau of Shipping

Continental Shelf Operations Notices

21 Action required in the event of a Diving Accident or Diving Fatality (1974)

24 Statutory Controls under the Mineral Workings (Offshore Installations) Act (1971—Diving Operations (revised August 1976)

25 Breathing Gas (Diving) (1974)

38 Diving Safety Memorandum No. 7—Manning Levels and Frequency of Diver Practice (1978)
Department of Energy

M.895: The Application of the International Regulations for Preventing Collisions at Sea (1972) to Dynamically Positioned Vessels and the Dangers to Divers Operating from such Vessels (1979)
M.904: Amendment to M.895
Department of Transport

Diving and Submersible Operations: Guidance Notes for Offshore Installation Managers (1982)
E. and P. Forum

In addition, the Diving Inspectorate of the Department of Energy publishes Diving Safety Memoranda (DSMs) on a whole variety of subjects and have done so since 1975. The total number must now be in excess of one hundred. It is simply not practical to refer to all of them. A number of them, the early ones in particular, have been overtaken by events, and it is hoped that when the new *Guidance Notes to the Diving Operations at Work Regulations* are published, by HSE, it will be possible for the Department of Energy to rationalize their list. The Association of Offshore Diving Contractors (AODC), of course, publishes guidance notes and codes, the total number of which is now just under 50. Some of these are significant, stand-alone documents, such as the *Code of Practice for the Safe Use of Electricity Under Water,* whilst others are very short, one-page documents on specific topics.

The independent Diving Medical Advisory Committee (DMAC) also publishes relevant guidance notes, the total number now being 17. Details of all DMAC and AODC guidance notes are available from the AODC, 28/30 Little Russell Street, London WC1A 2HN.

Reference should also be made to a large number of relevant British Standards, some of which have a direct, whilst others have an indirect, bearing on the industry. At the time of writing this paper, AODC is represented on three BSI Committees: Purity Standards for Breathing Gases; Underwater Breathing Apparatus; and Design, Testing, Examination and Certification Requirements for Offshore Containers.

As far as *manned submersibles* are concerned, there are five specific Regulations as set out below:

SI 1981 No. 1098
The Merchant Shipping (Submersible Craft Construction and Survey) Regulations (1981)

SI 1976 No. 940
The Merchant Shipping (Registration of Submersible Craft) Regulations (1976)

SI 1985 No. 1607
The Merchant Shipping (Fees) Regulations (1985), as amended

SI 1987 No. 306
The Merchant Shipping (Submersible Craft) (Amendment) Regulations (1987)

SI 1987 No. 311
The Merchant Shipping (Submersible Craft Operations) Regulations (1987)

All of these apply to manned submersibles and are enforced by the Marine Division of the UK Department of Transport.

There are four published sets of certification rules for manned submersibles and two further sets in the course of preparation. These are:

Rules for Building and Classing Underwater Systems and Vehicles (1979)
American Bureau of Shipping

Rules and Regulations for the Construction and Classification of Submersibles (1976)
Bureau Veritas

Rules for Certification/Classification of Submersibles (1985)
Det norske Veritas

Rules and Regulations for the Construction and Classification of Submersibles and Diving Systems (1987)
Lloyd's Register

Rules for Underwater Technology
Chapter 2: Submersibles (in preparation)
Germanischer Lloyd

Ships' Rules and Mobile Offshore Unit Rules
Part 6 chapter 1 and part 1 chapter 2. New draft to cover support vessels deploying submersibles August (1984)
Det norske Veritas

As far as guidelines, guidance notes, etc., are concerned, the list is rather less formidable than for diving and there are two main documents, one published by the Association of Offshore Diving Contractors (AODC) and the other by the MTS. These are:

AODC Code of Practice for the Operation of Manned Submersible Craft (1985)

MTS International Safety Standard Guidelines for the Operation of Undersea Vehicles

Contractors complying with the AODC *Code of Practice for the Operation of Manned Submersible Craft* will comply with relevant UK operating legislation.

There is no formal legislation relating specifically to the operation of ROVs, apart from the general duty of care required by other offshore legislation and the Health and Safety at Work etc. Act. AODC through its ROV and Submersible Committee has, however, been active for the past year or more on the preparation of a *Code of Practice for the Safe and Efficient Use of ROVs*.

Regulatory requirements for *Dynamic positioning systems* for diving-support and other vessels can be found in the mass of merchant shipping legislation, although there is no specific legislation *per se*. Set out below, however, are four references in the context of certification rules:

Tentative rules for the Construction and Classification of Dynamic Positioning Systems for Ships and Mobile Offshore Units (1977)
Det norske Veritas

Rules and Regulations for the Classification of Ships:
Part 7 (ch. 4): Rules for the Construction and Classification of Ships and Mobile Offshore Units with Installed Dynamic Positioning System
Lloyd's Register

Regulations for the Construction and Testing of Computers and Computer Systems (1980)
Germanischer Lloyd

Classification of Dynamic Positioning Installations Guidance Note NI 187 (1984)
Bureau Veritas

In 1983, the UK Department of Energy and the Norwegian Petroleum Directorate jointly published *Guidelines for the Specification and Operation of Dynamically Positioned Diving Support Vessels* and it is probably fair to say that that particular document has gone a long way towards improving the standards of DP vessels and their operations, in the diving context. Guidance Note No. 15 from the AODC covers the particular subject of *Surface Orientated Diving from DP Vessels* and gives clear guidance on procedures which should and should not be adopted.

Norway

The pattern of legislation, certification rules, guidelines, guidance notes, codes, etc. in Norway is similar to that in the UK, but with differences in emphasis which reflect each country's legislative regime and related philosophy. In general, in Norway, considerable emphasis is given to the concept of internal control. It is not proposed to address that in any detail in this article, as its principles should be already well known.

As far as diving is concerned, the main Regulations are set out below:

Regulation on Construction of Rescue Craft etc.
4 January 1965
Norwegian Maritime Directorate

Royal Decree of 29 April 1977 concerning Regulations for Safety Delegates and Working Environment Committees
Latest amendment 1 March 1984

Regulations Concerning Launching Arrangements for Life-saving Appliances and for Boats which are Lowered and Hoisted with Persons on Board
23 June 1977
Norwegian Maritime Directorate

Regulation on Life-saving Appliances on Fixed Installations for the Production etc. of Submarine Petroleum Resources
8 February 1978, amended 3 February 1982
Norwegian Maritime Directorate, supervised by Norwegian Petroleum Directorate from 1 July 1985, except as regards seaworthiness

Provisional Regulations for Diving on the Norwegian Continental Shelf
1 July 1978, with later amendments, latest 1 April 1980
Norwegian Petroleum Directorate

Royal Decree of 1 June 1979 concerning Regulations relating to Workers' Protection and Working Environment, etc. in connection with Exploration for and Exploitation of Submarine Petroleum Resources, latest amendment 13 September 1985

Regulations Concerning Medical Examination of Employees on Facilities and Installations in Connection with Production, etc. of Submarine Petroleum Resources
1 August 1980
Ministry of Social Affairs

Regulation on Approval of Survival Suits for Use in Petroleum Activities
10 November 1980
Norwegian Maritime Directorate, supervised by Norwegian Petroleum Directorate from 1 July 1985, except as regards seaworthiness

Regulations on Working Hours for Diving Personnel on Norwegian Ships, Drilling Platforms and Other Mobile Installations in the Sea
26 June 1981
Norwegian Maritime Directorate

Regulations for Life-saving Appliances etc. on Drilling Platforms and Other Mobile Installations in the Sea
3 February 1982
Norwegian Maritime Directorate, supervised by Norwegian Petroleum Directorate from 1 July 1985, except as regards seaworthiness

Regulations for Control of Diving Systems etc. on Norwegian Offshore Units
10 April 1984
Norwegian Maritime Directorate

Act of 8 June 1984 on Permanent Inquiry Commission for Diving Accidents, No. 61

Regulations relating to a Permanent Inquiry Commission for Diving Accidents
25 January 1985

Royal Decree of 28 June 1985 on Regulations Concerning the Licensee's Internal Control in Petroleum Activities on the Norwegian Continental Shelf

Royal Decree of 28 June 1985 on Regulations Concerning Safety, etc. relating to the Act Concerning Petroleum Activities

Royal Decree of 28 June 1985 concerning Regulatory Supervisory Activities with the Safety, etc. in the Petroleum Activities on the Norwegian Continental Shelf.

Further Descriptions of Arrangements for Supervision of Safety, etc. Measures in the Petroleum Industry
1 July 1985
Norwegian Petroleum Directorate

Detailed examination of the above list will show that a number of the references are of greater importance than others, but nevertheless they all apply. The 'Provisional Regulations for Diving on the Norwegian Continental Shelf—1 July 1978, with later amendments, latest 1 April 1980' are the main regulations and these are in the course of being updated with, I understand, a view to their becoming law in a new revised form, probably in 1988.

The certification rules are exactly the same as for the UK and will therefore not be repeated.

The guidelines, codes, guidance notes and associated material of an authoritative form are very similar to the UK list (with the obvious exception of guidance material related to specific UK legislation) but certain specific documents apply to Norway alone. These additional documents are:

Guidelines Concerning First-aid Equipment and Medical Life-saving Kits for

Deep-sea Diving in connection with Exploration and Drilling Operations for Submarine Petroleum Deposits on the Norwegian Continental Shelf
5 November 1979
Department of Social Affairs

Guidelines for Doctors Concerning Medical Examination of Divers
3 March 1980
Directorate of Health

Draft guidelines for minimum performance requirements and standard unmanned test procedures for underwater breathing apparatus 1984
Norwegian Petroleum Directorate

Code of Practice for Offshore Diving on the Norwegian Continental Shelf
October 1986
Association of Offshore Diving Contractors
(with the endorsement of NIFO, and recognized by NPD)

The Norwegian Petroleum Directorate issues Diving Safety Notices on specific topics, and AODC and DMAC guidance material (as mentioned in United Kingdom section) applies in Norway, as elsewhere.

As far as *manned submersibles* are concerned, there are no specific regulations in Norway, although the design and operating characteristics of these vehicles are covered by general legislation. The following certification rules, however, apply:

Rules for Certification/Classification of Submersibles (1985)
Det norske Veritas

Rules and Regulations for the Construction and Classification of Submersibles and Diving Systems (1987)
Lloyd's Register of Shipping

Rules for Building and Classing Underwater Systems and Vehicles (1979)
American Bureau of Shipping

The one principal code cum guidance note is that published by AODC in March 1984—*Code of Practice for the Operation of Manned Submersible Craft.*

There are two particular pieces of legislation dealing with *dynamic positioning systems:*

Royal Decree of 28 June 1985 on Regulations Concerning Safety, etc. Relating to the Act Concerning Petroleum Activities

Further Description of the Arrangements for Supervision of Safety etc. Measures in the Petroleum Industry
1 July 1985
Norwegian Petroleum Directorate

Two specific Certification Rules, both published by Det norske Veritas, are:

Tentative Rules for the Construction and Classification of Dynamic Positioning Systems for Ships and Mobile Offshore Units (1977)
Det norske Veritas

Rules for Classification of Mobile Offshore Units
Part 6,ch. 2: Position Mooring (POSMOOR)
Det norske Veritas

and two specific Guidance Notes:

Guidelines for the Specification and Operation of Dynamically Positioned Diving Support Vessels
1 May 1983
Norwegian Petroleum Directorate/UK Department of Energy

Standard for Insurance Warranty Surveys in Marine Operations
June 1985
Det norske Veritas

Denmark

The principal legislation governing diving operations in Danish waters is included in six documents, as detailed below:

Danish Diving Act No. 214 of 23 May 1979

Order No. 338 of 11 December 1935
On the construction and use of transportable containers of compressed or liquefied gases or gases dissolved under pressure, as amended by Order No. 98 of 25 April 1939, No. 450 of 21 November 1951, No. 147 of 5 May 1954, No. 75 of 19 March 1959, No. 73 of 21 March 1961, No. 345 of 27 September 1963 and Order of 22 December 1971
Ministry of Social Affairs, and Ministry of Industry

Order No. 48 of 7 February 1980
On diving equipment
Ministry of Industry
A draft order has been published which will replace the above order

Order No. 49 of 7 February 1980
On professional diving operations
Ministry of Industry
A draft order has been published which will replace the above order

Order No. 501 of 25 November 1985
On medical examination of professional divers and probational divers
Ministry of Industry

Order No. 502 of 25 November 1985
On training of professional divers
Ministry of Industry

Certification rules, codes, guidelines, guidance notes and similar authoritative material is as generally described above in the sections on United Kingdom and Norway.

Netherlands

Relevant legislation in the Netherlands is as set out below:

Mining Regulations Continental Shelf (MRCS)
Chapter 13 Operations underwater or under pressure

Chapter 16 Article 149. Medical examinations
(1983)
Staatstoezicht op de Mijnen

Special Rules (MRCS)
SR3 Activities under water or under pressure
(1986)
Staatstoezicht op de Mijnen

Guidelines for Incoming Mining Installations
Annex 12 Permit for diving operations
Annex 13 Underwater cutting and welding
(1982)
NOGEPA
Safety Directives for Divers
(1967)
Labour Inspectorate

Certification rules, codes, guidelines, guidance notes and similar authoritative material is as generally described above in the sections on United Kingdom and Norway.

Ireland

In Ireland there are six principal pieces of legislation as follows:

SI No. 422 of 1981
Safety in Industry (Diving Operations) Regulations (1981)

Safety in Industry Act (1955)

Safety in Industry Act (1980)

Factories Act (1955), No. 10

Safety Instructions for Offshore Petroleum Operations (1983)
Section 9, Independent Examination of Pressure Vessels, Lifting Appliances and Gear
Section 40, Diving Operations
Schedule 1, Examination and Testing of Equipment
Department of Labour

SI No. 164 of 1966
The Labour (Transfer of Departmental Administration and Ministerial Functions) Order (1966)

Certificates rules, codes, guidelines and guidance notes and similar authoritative material is as generally described in the sections on United Kingdom and Norway.

At this point, those readers who have got this far, and are more familiar with national legislation and requirements in one of the above countries may howl that one or more of the references is incorrect and that others of equal or greater importance have been left out. I can only say that I have already apologized for any errors or omissions and, if such persons care to give me the specific references, then I will ensure that they are included in any future update of this review, should any organization be so bold as to request it.

INTERNATIONAL STANDARDS

Legislation is, by definition, national. That said, however, good operating practices in relation to underwater activities which apply in the waters of one State are just as relevant in another. Obviously, however, some specific requirements will vary, taking into account environmental and other factors.

Generally the classification society rules as set down in the previous section are applied on an international basis, and although reference has only been made to four classification societies, there are others, whose rules will be applied in particular areas.

The International Maritime Organization (IMO) published its *Code of Safety for Diving Systems* in 1985. This Code covers the design, testing and certification requirements for diving systems which are fixed into diving-support and other vessels. It has already been incorporated into Norwegian legislation, and parts of it are informally recognized by other States.

The International Labour Organization (ILO) has, from time to time, addressed the subject of diving operations in the context of the health and safety of members of the diving team, but no formal documents have been published by that organization. Both the IMO and the ILO are United Nations bodies.

A little closer to home, the European Diving Technology Committee (EDTC), of which Per Rosengren of NPD is the present Chairman, published its *Guidance Notes for Safe Diving* in 1984. The object of that document was to try to harmonize standards, certainly in all European States. More recently the EDTC has addressed a complementary project—Training Standards—dealing with the training requirements for offshore surface-orientated (air) and mixed-gas divers, life-support technicians, and diving and life-support supervisors. The standards contained within this document are based on best current practice, and we look forward to its publication.

CLIENT REQUIREMENTS

In the offshore oil and gas context, this principally means oil operating companies and their relevant national and international Associations. In the North Sea these include UKOOA, NIFO, NOGEPA, NSOC-D, Chambre Syndicale, WEG, and, on the international scene, the E. and P. Forum.

The UKOOA and NIFO have produced checklists for DP diving-support vessels. UKOOA is working, in cooperation with AODC, on a comprehensive checklist for diving systems 'DESIGN' and NIFO has been involved, in particular, in the subject of requirements for deep diving, with particular reference to the Norwegian Continental Shelf.

The international E. and P. Forum have published *Diving and Submersible Operations: Guidance Notes for Offshore Installation Managers* in 1982 as general advice to offshore installation managers having a responsibility for the control of diving and submersible operations.

All the above said, some client companies have their own specific requirements which generally are, or should be, set down in tender and contract documents. It should be said, however, that these 'specifics', although obviously the prerogative of clients anywhere, warrant harmonization in the interests of improved efficiency and safety of operations in an industry which is, after all, international.

SOME SPECIFICS

This section of the paper may be regarded as the 'icing on the cake' and will aim to highlight various specific 'requirements' in relation to particular 'Underwater Projects' which have caused problems in the past. Information is still being collected and it will be more appropriate to summarize a number of cases by way of example, at the Conference, rather than setting them down now. More, therefore, later.

THE FUTURE

This paper has briefly reviewed the various regulations, standards, codes, guidance notes, guidelines and other requirements applying to diving and related aspects of the underwater engineering industry, principally in the North-West European Continental Shelf area. Where do we go from here?

As far as the commercial diver is concerned, he now needs his National Certificate of Competence, Medical Certificate and Offshore Survival Training Certificate. In the future he may also require a specific first aid certificate and refresher training certificate. If he wants to work as an inspection diver he will require 3.1U and 3.2U certificates. If he is a mixed-gas diver and wants to work as an LST he will require his AODC LST Certificate and if he aspires to become a supervisor he will also need his Air and Mixed-Gas Supervisor's Certificates. In addition, he also needs a full CV listing all his dives, with some detail of work undertaken. Each contractor has, of course, to grapple with all of this in addition to the volumes of paper he has to prepare prior to the contract being awarded, let alone to his undertaking the work.

In the UK discussions are still in hand on possible/probable 'First Aid for Diving Regulations'. Work is also in hand on three relevant British Standards (OFFCON, Gas purity, Underwater breathing apparatus). An organization has been set up to consider the certification of inspection bodies but this is still at an embryonic stage. The Department of Transport will be drafting a revised version of the merchant shipping diving regulations.

The International Maritime Organization will be addressing possible requirements for hyperbaric evacuation systems at meetings from mid-1988.

These are just a few of the new documents in the pipeline, which bear on this paper. The above have only been given by way of example and may or may not be mirrored by a similar level of activity in the other littoral States. There seems no doubt that the flood of regulatory and client-type documents affecting the underwater industry continues at a similar rate as in the past and one has to ask the question 'Where will it all end?'

Surely now is the time for consolidation and harmonization, taking into account the lessons learned over the last 20 years, the international nature of the industry and the current relative recession.

ACKNOWLEDGEMENTS

Before closing it is right and proper that I pay full acknowledgement to the Council and Members of AODC with whom I have worked closely over the past

11 years. Also to acknowledge the close working relationship which AODC fortunately has with UKOOA and NIFO and with relevant Government Departments in all of the littoral States mentioned in this article—these are all very much appreciated. Our relations with the various classification societies, the EDTC and the IMO, are also acknowledged.

Finally, but certainly by no means least, I very gratefully acknowledge the help given to me during the preparation of this article by my colleague Judith Mirzoeff, M.A. (Oxon.), who has provided much of the specific information, legislative and other references. Any errors in the paper are mine and not hers.

5

Standardization—Norwegian Experience by NIFO—Norsk Industriforening for Oljeselskap (Norwegian Industry Association for Oil Companies)

W. B. Norton, Norske Shell, Norway

In the development of offshore oil and gas deposits, coordination of effort between those involved in all phases of the activity has generally been a dynamic process. Depending on the location, several groupings of the main parties have evolved, usually consisting of Government, operators and contractors. These groupings eventually stabilize and establish particular functions with clearly defined areas of responsibility and influence. In the USA, for example, the various government agencies have established positions, as have the operators through API and diving contractors with ADC, while in the UK the positions of the Departments of Energy and of transport, the UK OOA and the Association of Offshore Diving Contractors (AODC) are well known. International bodies such as the E. and P. Forum, IADC, the International Maritime Organization (IMO) and the International Labour Organization (ILO) are also involved in various areas of offshore oil- and gas-related diving. In general, the main course of these industrial and governmental groups has been towards the development of standards, whether through regulations, guidelines, approved lists or other means. In Norway, a similar pattern has evolved with four main units representing Government, operators, contractors and trades union. As in most other offshore areas, this evolution has been gradual, but has been based on Norway's well-established broad industrial and social-relations principles and the previous experience of operators and contractors. In the area of subsea operations, liaison with the UK has been of particular importance in this period. As a part of this evolutionary process, a number of recent developments initiated further and more detailed efforts to improve overall efficiency and safety in all phases of offshore activity. These were:

(*a*) revision of the petroleum regulatory system;
(*b*) revised structure of NIFO (Norwegian Industry Association for Oil Operators);
(*c*) changing economics.

CURRENT STATUS

With the introduction of the new Petroleum Act in 1985, the Norwegian

Advances in Underwater Technology, Ocean Science and Offshore Engineering, Volume 14: Submersible Technology

authorities have placed even greater emphasis on the field operators' responsibility for safety, but have left the initiation of detailed actions to meet the overall requirements, to individual operators. While this has enabled the implementation of safety to be tailored to specific field/installation/operational needs, interpretation of the legal requirements has become less predictable. In addition, the new law has required operators to demonstrate their compliance by an internal control system, referenced to a Norwegian Standard, NS 1501. This document gives guidance on quality assurance by internal procedures and is heavily procedurally oriented. Although originally intended as a cost-saving exercise, the implementation of this requirement led to the QA/QC industry expanding very rapidly and a vast amount of documentation being generated, together with the expansion of the necessary personnel to carry out audits, confirm compliance and to document results. The general impact of this is well known, and while increased safety has been achieved, the resultant costs of this 'internal control' policy have been significant.

The Norwegian operators' associations, NIFO, upgraded its function in 1984 and, among many other tasks, recognized the need to review the internal-control aspects of offshore work actively with the authorities on regulatory development, with a view to increasing the efficiency of QA in particular. This was given greater emphasis by the worsening economic situation throughout the industry. The initial NIFO review indicated a need to evaluate further the following areas on a priority basis:

- compatible and unambiguous rules;
- defined responsibility;
- recognition of Standards and codes.

Internally, through various committees, NIFO also began to develop standards, where such a need was identifiable. Areas where this has been considered essential are as follows:

- vessels;
- equipment;
- personnel;
- documentation.

Emphasis has been placed, however, on the use of existing documents wherever possible in order to utilize experience already gained in similar conditions elsewhere.

Other parties concerned, principally AODC and the diving unions, recognized that problems could arise from the change of approach in the new regulations as follows:

The Association of Offshore Diving Contractors were concerned that each NIFO member would attempt to set up his own operating standards for diving operations and that a situation would prevail in which contractors would be expected to meet many different standards.

The unions were concerned at the apparent withdrawal of NPD from their perceived front-line monitoring role and the possible consequences this could have on safety.

Shortly after the issuing of the revised regulations it was agreed between NIFO and AODC that one particular approach under the new law to ease the above concerns could be the introduction of an industry code of safe practice for diving on the Norwegian Shelf, and that the best body to compile this document would naturally be the AODC.

It was recognized from the beginning that, due to variations in practical experience and preferences in methods, each NIFO operating company might have additional requirements in certain areas covered by the code. The code, however, would be proposed as the backbone of the operating standards for diving operations on the Norwegian Shelf.

Contact was made jointly by NIFO and AODC with NPD, with a recommendation that a code should be introduced and a request that this be given 'recognized' status by NPD upon its introduction. This was eventually accepted by NPD, although some concern was expressed over the lack of involvement of the divers' unions in the preparation of the code. In order to expedite the development of the document, however, it was decided to bring the code to a suitable draft form before discussions with other interested groups.

Members of AODC then undertook to compile the new code. This required considerable research into existing recognized standards which included diving, with the NPD regulations as a base line, together with safety memos issued by NPD, other standards such as IMO codes, AODC safety memos, European standards, etc. The task was considerable, taking some 12 months to produce the first draft for discussion, with NIFO. After agreement this was then discussed with the divers' unions and NPD. Some two years after the decision to produce the code, NPD gave approval and NIFO formally endorsed the code as suitable for use by its members.

EXPERIENCE

In an attempt to develop standardization, the implementation of the activities has been a somewhat protracted exercise, as has also been the case in other geographical areas. The main factors influencing the progress are:

- communication;
- expertise;
- national characteristics.

Our experience has been that, with part-time NIFO representatives in various committees and working parties, on governmental and industrial spheres, communication is a major factor. Within NIFO this aspect has been considerably improved by the appointment of a small, permanent, active secretariat who provide continuity and links between individual groups, and by a QA subcommittee. This approach is highly commended to others in similar circumstances.

In addition to communication, there has also been the factor of applying the appropriate level of expertise to particular activities. This is not a new phenomenon, as anyone involved in committee work well understands. In Norway, the development of national expertise in the subsea area of oil-related work has been encouraged by all parties concerned, initially on a general technical level. The development of management expertise in the subsea sector has, however, been a more gradual process and consequently there is a very high demand for the services of management skills on the various committees and working groups.

Another main influence has been the differences in the national characteristics of the participants, and the multinational nature of the industry. This has necessitated a rather lengthy digestion period and a need for more and wider-reaching discussion than may be the case elsewhere. In addition, the wide geographical spread of facilities and operating areas must also be taken into consideration.

With regard to the diving code, it is interesting, some two and a half years later, to see whether or not the original intent behind this code has been achieved and how valuable similar exercises in other activities might be in future, in Norway and also in other areas in which the internal-control philosophy is being used.

The first effect of the code has been to delay the introduction of new legislation from NPD. Since the code incorporates all the requirements set down in the 1978 rules, the *status quo* has, as a minimum, been maintained in respect of operating standards. This has now been accepted by the unions involving diving personnel such that the pressure for new rules has been reduced.

The second most notable effect of the work on the code has been to improve relations between AODC and NIFO, and mutual confidence on safety considerations has been established.

Most operating companies now call for use of the code as an operating standard in all offshore diving operations, which is indicative of the value of the document. On the service-industry side, all contractors have been advised by AODC that they should audit their operations against the code to ensure compliance.

The code has not entirely achieved its aim of standardizing operations and obviating the need for each operator to set his own rules.

It is noticeable, however, that smaller operating companies have placed more reliance on the code than larger organizations with internal diving expertise, which can expand on the minimum standards of safe operations contained in the code.

Nevertheless, it is generally agreed in the industry that the introduction of the code has been useful in providing a base document which may be updated to include any additional safety requirements regularly set by operators. It is anticipated that, with further experience in its use, operators will have sufficient confidence in the code and in operations conducted in compliance with it, and thereby reduce additional requirements to a minimum level.

On a more general front, the detailed level to which some audits have been conducted has been widely discussed, with the acceptance of certification documentation being a particular problem. Lack of a common inspection standard has also been a source of difficulty and considerable expense.

COSTS

On costs, the present detailed level of the acceptance inspection, and the accompanying documentation, has been expensive for the contractors. It is seen by NPD and NIFO, however, as a passing phase during which operating standards and verification levels applied by the subsea industry will be improved to a level perhaps on a par with those employed in the airline business. If a similar confidence level could be achieved then this would be a highly satisfactory exercise.

Costly though it is, the present process of improvement will continue in Norway.

RESULTS OVERALL

In addition to the extensive work on the Diving Code, standardization/harmonization has also been addressed in the areas previously given priority. The financial and industrial impact of the internal control requirements have been discussed at a high level with the authorities.

Recent discussions have been focussed on the following areas:

- identification of a single authority to administer the regulations regarding offshore oil activity;
- improved communication between various groups involved;
- application of the Working Environment Act;
- assessing impact in marine operations.

As part of an overall rationalization exercise it is also hoped that a principle of adopting already existing, recognized standards and guidelines can be implemented. While this is a far-reaching activity affecting the industry as a whole, work has been progressing at a lower level and specific areas of interest to subsea service operations, as follows:

NIFO	Deep Diving Standards; DPDSV checklist advice on hyperbaric rescue
AODC	Training standards electrical safety
IMO	Diving systems

CONCLUSIONS

Given the correct application of standardization, cost savings should be achievable in the longer term, but further effort will be required to develop confidence in common standards and certification processes for equipment, vessels and personnel. It has been suggested that a review of the way in which this has been achieved by the airline operators could be of benefit. All parties concerned should continue to seek for further areas of improvement, and it is hoped that the current NPD/industry strategical discussions will be of considerable assistance.

Detailed discussion will also continue on future requirements for deep diving, standardization of fittings for and recovery of, hyperbaric rescue vehicles (HRVs) and approval of dynamically positioned diving support vessels (DPSVs). Working liaison with all involved parties is being maintained throughout these activities, and good communication is considered to be essential. On past experience, however, progress will be heavily dependent on goodwill and cooperation throughout. It is hoped, nevertheless, that advances will be made on a steady basis with eventual benefit to all from enhanced safety and operability, with reduced costs.

6

Code of Practice for the Safe and Efficient Operation of Remotely Operated Vehicles

A. J. Sandford, Chairman, Association of Offshore Diving Contractors ROV and Submersibles Committee, AODC, UK

Remotely operated vehicles (ROVs) are now almost commonplace in the offshore industry. This is largely because ROV technology has been developed to such an extent that vehicles can effectively undertake a wide range of underwater tasks and, in many instances, provide an acceptable alternative to divers.

It is perhaps strange that, given the versatility and worldwide pre-eminence of ROVs, there are few rules and regulations specifically governing their operation. Having said this, it is true that many ROV activities are influenced or indeed controlled by other general safeguards designed to cover a wide variety of offshore systems and equipment. In addition, it must also be said that virtually all ROV contractors have developed comprehensive operating practices and standards (usually documented in a Company operations manual) which allow the safe operation of their ROV systems. Likewise, many operators have generated internal guidelines covering the operation of ROVs. However, in spite of all this documentation, the industry still lacks a set of rules or guidelines which contractors and clients alike can refer to in the clear knowledge that the contents of such a document can be generally regarded as the 'industry standard'.

This, therefore, is essentially the justification for producing a Code of Practice to cover the safe and efficient operation of remotely operated vehicles.

Briefly, the Code of Practice document was conceived and compiled as described below.

INTRODUCTION

A working group was formed from the AODC ROV and Submersibles Committee to address the requirement for a Code of Practice for The Safe and Efficient Operation of ROVs.

The main objective in producing the Code of Practice was to provide both clients and contractors with a document offering recommendations and advice for improving safety and efficiency during ROV operations. Acknowledgement

Advances in Underwater Technology, Ocean Science and Offshore Engineering, Volume 14: Submersible Technology

of the Code does not, however, negate the need for ROV operators to maintain their own operations manuals and procedures.

The document briefly describes various types of ROV, tools commonly used, tasks performed, support locations and methods of operation, providing an overview of the inter-relationship of a variety of equipment and work tasks which may be combined on a project. Recommendations are made in areas where both commercial considerations and safety implications must be satisfactorily balanced out, particularly stressing the importance of providing sufficient qualified personnel during all aspects of an operation.

It is hoped that the Code can address future changes in operating practices, but it is recognized that the document will require updating from time to time.

The main body of the document is described as follows.

ROV CLASSIFICATION

For the purpose of the document, ROV types are grouped into five main classifications, as follows:

Class I Pure observation
Class II Observation with payload option
Class III Work-class vehicles
Class IV Towed or bottom-crawling vehicles
Class V Prototype or development vehicles

Examples of each classification are provided within the document.

ROV TASKS

Similarly, for the purpose of the document, ROV tasks are grouped into five categories and restricted to the offshore oil and gas industry, as follows:

(1) observation;
(2) survey;
(3) inspection;
(4) manipulative;
(5) specialist.

It is recognized within the Code that a single workscope may encompass both a survey and inspection if specific areas are designated for inspection. Likewise, specialist tasks may combine any of the above categories.

ROV TOOLS

Varying categories of tools used by ROVs are highlighted in the document. It is stressed that tools fitted to the ROV must be used only after suitable forethought and planning in order that their use will not compromise safety during operations.

ENVIRONMENTAL CONSIDERATIONS

Environmental conditions can seriously affect the safe deployment and operation of ROVs. As these conditions can be dramatically different for any given situation, the Code recommends that it should remain the sole responsibility of the ROV supervisor to decide if it is possible to carry out operations safely. A supervisor must therefore be aware of his Company's insurance warranties with regard to environmental considerations and should never allow external pressure to influence his decision regarding the safety of continuing the operation.

For guidance, the document highlights the following environmental considerations:

(1) water depth;
(2) currents;
(3) sea state and swell;
(4) weather (Beaufort scale included in this section of the document);
(5) water characteristics;
(6) seabed characteristics.

Other considerations include the system's readiness and the nature and urgency of the task.

OPERATIONAL CONSIDERATIONS

First, the Code stresses that ROV operating companies should be firmly satisfied that the chosen ROV can safely and effectively carry out the required task. An overview of operational considerations is made within the document, as follows.

(1) Equipment Location, Installation and Security

Remotely operated vehicles employed offshore should comply with Offshore Installation Regulations which are highlighted in this section of the document. Equipment location, although dependent on available deck space, should be chosen carefully, with security being a major consideration. All ROV equipment must comply with statutory safety regulations, particularly when located in 'hazardous areas', and individual site regulations must also be observed.

(2) Handling Systems

Reference is made to the AODC/Department of Energy guidance note for ROV-handling systems entitled *ROV Handling Systems (UK)—AODC 036* and to parts of the Offshore Installations Regulations, which advise that all lifting gear, including the ROV lifting cable, should have test certificates indicating the safe working load. The Code recommends that a list of safe operating parameters should always be available and that due care and attention should be taken at all times during the operation of handling systems. Also, particularly whilst deploying the vehicle, operating personnel should always be equipped with hard hats and safety boots.

(3) Communications

The Code stresses the importance of effective communications between all parties during an operation, as a breakdown in communications could interrupt the supply of vital resources, thus affecting the safety of the operation.

(4) Support Locations

Support locations for ROV operations can vary, thus affecting the levels of support available to the ROV system and crew and the safety of the operation. The Code recommends that a Mobilization Engineer or the ROV supervisor carefully inspect the site to ascertain the optimum location for the ROV system on that particular installation. The seven different types of support location are listed and discussed within the document. The locations highlighted are:

- (i) small workboat, supply boat or stand-by vessel;
- (ii) small air-range diving-support vessels and larger supply boats;
- (iii) dedicated ROV-support vessels;
- (iv) large monohull diving-support vessels (DSVs);
- (v) fixed platforms;
- (vi) temporarily fixed platforms;
- (vii) specialist locations.

(5) Navigation

Guidance for the use of beacons on ROVs for navigation, positioning and tracking should be obtained according to the operating Company's insurance warranty.

(6) Manuals and Documentation

The Code emphasizes the responsibility of every contractor to supply a comprehensive set of manuals, check-lists and log books with each ROV system. It is recognized that this is a major factor in the safe and efficient operation of ROVs. Additionally, the Code recommends that operating personnel should be familiar with reference documentation, including legislative or advisory publications. Some examples are listed within the document for the reader's information.

(7) Umbilicals

The Code discusses the variety of umbilicals which can be used, depending on vehicle type. It is recommended that ROV operating companies should request test data on each umbilical supplied by a manufacturer. Also, termination of umbilicals undertaken by ROV operators should be in accordance with manufacturers' instructions, and a load test carried out on completion. Routine inspection of the umbilical cable is also recommended to ensure system integrity.

(8) Launch and Recovery

The Code recommends that the ROV supervisor should be responsible for establishing a safe and effective launch and recovery procedure suitable for the particular ROV type, deployment system and support location. Additionally, he should ensure that all crew members are familiar with the procedure.

MAINTENANCE

Regular maintenance is recognized as an important factor in ensuring the safe operation of an ROV system. The document highlights the importance of system checks and of observing recommendations given in manufacturers' manuals. Typical pre- and post-dive checks are described. It is recommended that an adequate serviceable supply of spare parts is provided with each ROV system, particularly items which are essential to continued operation and safety.

PERSONNEL

A large section of the Code is devoted to personnel, as it is recognized that suitably qualified personnel are vital to safe ROV operations.

(1) Qualifications

Qualifications demonstrating the normal industry-acceptable standards for ROV personnel are provided. It is stressed, however, that experience can be a substitute for academic qualifications and that each case should be treated individually. Medical fitness should be in accordance with recommendations for topside personnel. The relevant regulations are itemized within the document.

(2) Functions

Key functions of ROV crew members are highlighted.

(3) Team Size

Factors deciding the correct team size for safe operation of an ROV are discussed, but it is stressed that team size and composition must always be sufficient to enable the operation to be carried out safely and effectively. Recommendations are made as to the generally accepted minimum manning levels for 12-hour operations.

(4) Working Periods

The Code recommends that the maximum number of hours for piloting an ROV should not exceed six hours in every 24-hour period. Suggestions are also made within the document regarding length of off-shift time and working periods for ROV operations with a two-man crew.

(5) Training

Suggested safety and technical-training requirements are highlighted, notably the requirement for personnel to attend recognized courses in survival, first aid and fire fighting, and technical-training courses to gain a sound knowledge in the operation and maintenance of ROVs and associated equipment.

CONCLUSION

In addition to examples of ROV classification, a Sea State Chart is included in Appendix 1 of the Code of Practice document.

It is hoped that the Code of Practice for The Safe and Efficient Operation of Remotely Operated Vehicles will be accepted by the industry as a comprehensive and useful reference document for all aspects of ROV operations in the offshore oil and gas industry.

7

Adapting ROV Operations Procedures to Today's Market

J. Mann, Wharton Williams, UK

ROV operations in today's tough market-place are a complex web of circumstances aimed at satisfying a number of inter-relating parameters. These circumstances and parameters will be discussed in this chapter, as will the impact of the guidelines on the market-place.

OBJECTIVE

The objective of the ROV contracting company is to undertake safe, efficient and cost-effective work programmes for the client and in so doing earn a reasonable return on what can only be described as a substantial capital investment. By accepting the Code of Practice for ROV Operations, the industry is taking a major step forward in achieving the objective of self-regulatory safety policies. That is not to say that without the Code of Practice the industry was operating without due consideration for safety, but with the Code of Practice the industry has set out a basic set of operating principles designed at ensuring the industry's excellent track record on safety is maintained and, if possible, improved.

The main parameters in ROV operations are: Safety, Efficiency and Cost-effectiveness.

The priority level attached to each parameter depends very much on an individual's persuasion in life and his outlook on the ROV operation. It is the sincere hope of the author, who was heavily involved in drafting the Code of Practice, that the publication of the Code will educate a wide cross-section of the industry and make people more aware of some of the implications of decisions reached in isolation.

Safety, efficiency and cost-effectiveness are all buzz words in a troubled market-place, and all three are so intricately inter-related that it is extremely difficult to discuss one without the other. In all responsible operations, safety is the first priority: safety of the ROV personnel from the hazards of their own equipment, safety of the ROV personnel from the hazards of the place of

Advances in Underwater Technology, Ocean Science and Offshore Engineering, Volume 14: Submersible Technology

operations, safety of associated personnel involved directly and indirectly with the ROV operations, and safety of equipment in the operation are only a few of the considerations to be taken account of. If we consider the multiplicity of tasks the average ROV operator is required to undertake in the pursuit of his work and add to this the concerns of the oil company's safety organization, it is easy to reach the conclusion that safety must be the highest priority and this is why the Code of Practice restricts itself to highlighting areas of concern rather than stamping an authoritarian message.

It may be mooted that the Code of Practice should have had more bite in its recommendations but, as highlighted in the introduction, it is not designed to remove the absolute need for the ROV contractor's vigilance to safety.

Efficiency in ROV operations is a much debated point, with neither side ever able to win an argument that is based solely upon time to complete tasks. It is the author's view that with the increasing reliability from ROV systems a client who is more acutely aware of the ROV capabilities, an industry which seeks to operate at greater depths, and a definite move to fund research and development into ROV-friendly subsea technology, the overall efficiency of ROV operations cannot fail to improve. If we use, as a guide for the future, the most widely accepted ROV task of today, that of video inspection and observation, we can see that even the outdated RCV 225 technology can be a reliable and effective tool. Indeed the Wharton Williams RCV 225 fleet and others regularly undertake dives in the North Sea exceeding 250 hours without recovering for maintenance or attention. In fact some of our most extended dives have only terminated because the job has been completed. By the same token, everyone will recollect at least one incident when the introduction of some new technology or task into the ROV arena has been initially decried owing to unreliability but subsequently been widely accepted as a valuable aid or tool to the subsea engineer.

With this in mind a great many companies and agencies are willing to persevere with the introduction of new ideas or technology in ROV operations to increase efficiencies and as a consequence increase the cost-effectiveness.

In a world where accountants always seem to have the last word, cost-effectiveness is an ever-popular phrase when operating in a troubled market-place. Cost-effectiveness in ROV operations takes two major sides: on one side is the oil company, who wants to see the savings to be gained by the use of ROVs, and on the other is the ROV contractor, who wants to see maximum utilization and return for his investment. Both sides pay ever-increasing attention to the bottom line of the balance sheet, and we trust that a happy medium can be achieved in order to keep both sets of accountants satisfied. If costs can be viewed on a higher priority than safety, then the need for, and acceptance of, a document such as the Code of Practice can be of paramount importance to ensure that bad practices, on both sides of the line, are eliminated. For our own part we believe we can work together, with the client, to ensure that all of the criteria for safe, efficient and cost-effective ROV operations are in place and are maintained at all times.

CONCLUSION

The Code of Practice for the Safe and Efficient Operation of Remotely Operated Vehicles is required by, and deserves the recognition of, the subsea industry. It is a major contribution to promoting the safety aspects and highlighting some possible areas for concern in the sphere of ROV operations. The author believes it is of extreme significance that the publication of the Code coincides with the continuance of the tough market place which has made many people seriously consider their costs in relation to their return. We can but hope that cost-cutting measures which can adversely affect the safety of our industry are never implemented purely for the sake of cost-effectiveness. Efficiency and, most importantly, safety must surely take precedence.

8

Using Project Management as a Tool: a Necessity in Today's Underwater Contracting Industry

G. P. Gilbert, The Association of Project Managers, UK

The purpose of this paper is to show how modern project management techniques can be applied to today's underwater contracting industry.

By general usage, project management relates only to large capital engineering or construction contracts. However, project management can apply to in-house work where the norm is either flow-line or batch operation, and the one-off 'special' creates serious difficulties. It is also relevant to development work, the preparation of tenders and to every contracting job undertaken offshore.

The paper sets out first to put project management in perspective, goes on to consider its general management basis and to discuss some special project management techniques, before bringing it together as a plan for action, and concludes with suggestions on how to develop project managers.

The drive for greater efficiency applies very much in the offshore underwater contracting industry at this time, and hence the paper's sub-title. It is hoped that the subject will be of interest to the many companies trying to cope with today's market conditions using yesterday's management techniques, who recognize the need for improved performance, and are intent on adopting simple practices and procedures to that end.

PUTTING PROJECT MANAGEMENT INTO PERSPECTIVE

The theme is that project management (PM) can provide the opportunity to improve the way firms run much of their business and, at the same time, to establish better credentials with clients who themselves already practise it.

WHAT'S SPECIAL ABOUT PROJECT MANAGEMENT?

It is probably simplest to start by describing what PM does. It forces the setting of clear objectives. It develops synergy: team-work by which the combined effort is greater than the sum of individual contributions. It improves management

Advances in Underwater Technology, Ocean Science and Offshore Engineering, Volume 14: Submersible Technology

awareness of the project situation and brings to bear specialized approaches and techniques.

WHAT IS PROJECT MANAGEMENT?

Project management is an add-on to general management, in that it cannot readily exist or be effective without a solid management base. On the other hand, the drive for PM can be a spur to improving the general management of the firm. It is, of course, much more than an add-on: it has its own special methodology to bring to bear. Overall, it provides the means of focusing attention on a specific task or project.

Before considering the general and project management aspects, it is appropriate to set down some definitions.

The definition of 'project' that will be used in the paper is significantly wider than that of large capital projects, such as an offshore oilfield development. It is a 'tangible outcome of a non-routine coordinated programme of work, with predetermined output parameters, such as quality, cost and time'. One may even describe the task of creating such an outcome as a project in its own right, since projects can range from technology-based industrial programmes, to the installation of a computer system, or even to the staging of a conference or exhibition. Their time-scale can vary from a few months to many years.

By management is meant 'working with and through individuals and groups to accomplish organizational goals'. It involves interpersonal relationships, leadership and motivation, in addition to technical and coordinating functions.

This leads to the definition of project management as 'the application of general management and additional specialized concepts in a project situation. The discipline of managing a number of related activities involving the integration of resources to achieve a given objective.'

Project management is *not* the application of computerized planning techniques. Nor is it 'crisis management'. Of course, both may apply to any project, but it is possible to manage many of the projects which the immediate audience will be involved in without either. Good PM allows time away from the routine tasks (which can be planned) for corrective action to be taken on the unexpected.

GENERAL MANAGEMENT AS THE BASIS

Three general management concepts have to be considered: process, systems and skills. The management process includes functions such as setting objectives, planning, organizing, motivating, coordinating, and controlling. A system consists of a number of interacting dynamic elements designed to carry out cooperatively a predetermined function, and it is characterized by a coherent pattern and purpose. Management skills are those attributes required to perform management functions. They may be grouped as *technical*—the use of knowledge, techniques and equipment to perform specific tasks; *conceptual*—the ability to understand the complexities of the overall situation and organization, and to act accordingly; and *interpersonal*—ability and judgement in working with and through people.

A number of aspects of general management are particularly relevant in the project situation. They include the allocation of financial responsibilities and authority, correspondence and other office routines. Typically procedures should include items given below.

Financial Responsibilities and Authority

- Authority for making commitments
- Authorized funds and budgets

Accounting

- Recording and reporting commitments and expenditure by cost centre
- Analysing trends for management
- Preparation of invoices
- Preparation of cash flow projections

Contract Administration

Policy	Competitive bidding, standard clauses (insurance, etc.)
Pre-award stages	Pre-qualifying potential bidders, approved bidders list, pre-bid technical discussions, receipt and evaluation of bids
Post-award stages	Kick-off meeting, progress reports/meetings, control of changes, completion/acceptance tests

Purchasing

Policy	When negotiated bidding is acceptable (otherwise competitive bidding), use of established call-up orders, blanket orders, standard terms and conditions, suppliers list
Pre-order stages	Requisitioning, obtaining/evaluation quotations, approval of bid tabulations
Post-order stages	Expediting, acceptance tests, materials receiving reports

Quality Assurance

Quality assurance has become somewhat of a catch-phrase, but it signifies much more than inspection and quality control. For instance, the quality audit techniques contained in BS 5750 are relatively simple but painstaking to apply, and need a strong management commitment to make effective. Inspection of work in progress and mechanical and functional tests upon completion must all be carried out and recorded diligently. Corrective action has to include feedback to avoid recurrence. What we are really considering is fitness for purpose—or value engineering.

Correspondence

There are a few simple office routines which are surprisingly often not practised in what are otherwise well-run firms. These are mentioned here for completeness.

- Distribution of letters, telexes, faxes, E-mail
- Telephone logging, meeting notes
- Filing system

Two specific project requirements are appropriate to mention at this point:

Reports and Presentations

- Format of weekly and monthly reports
- General requirement for management presentations

Project Administration Manual

Most of the above needs to be tailored to suit the particular project requirements, and then it is convenient to assemble them in the form of a manual. Other subjects to be included are:

- List of organizations, roles and relationships
- Organization charts
- Project execution plan
- Work breakdown structure
- Management strategy
- Contracting strategy
- Limits of authority

SOME SPECIAL PROJECT MANAGEMENT CONCEPTS

A number of significant PM concepts are worth noting:

- A structured team approach to organizing work, people and relationships.
- Division into discrete phases
- Planning in the broadest sense
- Management involvement, review and approval at most phases
- Cost/time relationships

Organizing Work, People and Relationships

Much has been said and written over the years on the subject of various organizational forms and their appropriateness for projects. Fundamentally, it has to be agreed as a starting-point that social attitudes, communications technology and the business environment have led to a complex world full of sophisticated people. Simplistic organizational forms, described as 'functional', 'staff-and-line' or 'project', are no longer entirely valid concepts, even if the situation ever was so simple; and we are continually faced with a matrix of groups, people and their relationships. And this is, of course, the fundamental *raison d'être* of PM and project managers.

A key concept is that of the project team. Teams can be formed from, and built around, a number of individuals trained in their own technologies and (especially if they are permanent staff) well versed in corporate policy. But they develop into a team only by working together towards a common objective. By this means, mutual understanding and appreciation can grow, and the overall achievement then exceeds the sum of the individual contributions.

There are two major roles to describe. Within the project teams is the project manager. His own role depends on the nature of the project, its phase and his company's participation in it. He may be identified as the ultimate client, representing the company who has initiated the project and provides the finance for it and who will use the facilities once built; he may be an architect or consulting engineer acting on behalf of the client; an engineering contractor undertaking the design; or a fabrication/construction contractor. These categories are not the only possibilities and will probably not completely match the roles as perceived by many. A passing observation on the role of the client; considering the above list and possible permutations of client–contractor relationships, it would seem that clients are rather like fleas: 'each having a larger one upon his back to bite him'. The other role to describe is that of the project sponsor, or champion.

Within each firm using PM, and for each project, there must be a senior manager responsible for ensuring that the project is successfully brought to fruition, and to whom the project manager reports. A good indication of the seriousness with which a firm takes PM and views a particular project is the reporting level within the organization of the project manager. The project sponsor deals with top management, public relations and other external groups, so providing an umbrella to enable the project manager to get on with his job of running the project.

Project Phases

A project must pass through a number of phases before it can be implemented effectively. Again, not all of these may apply to every project that is encountered, depending on the role of the firm and its starting-point. A general list includes:

- Preliminary analysis of business needs and objectives
- Potential project and its goals identified and analysed
- Preliminary selection of alternative concepts and possible technical solutions; ideas for possible solutions developed into alternative concepts
- Desirable technical solutions identified and classified
- Feasibility of alternatives with assessment of relative merits; development of concepts into possible solutions, with their advantages and disadvantages, costs and benefits assessed, evaluated and classified
- Scoping studies, taking the most attractive alternative to a specific concept
- Development of the most promising alternative solution with sufficient details to decide to proceed, and to justify funding
- A preliminary scheme for implementation
- Detailed design, procurement
- Materials delivery, fabrication, installation and commissioning
- Post-operational audit

Management decisions and commitment are necessary at the conclusion of each stage before the next is allowed to proceed.

The end-point of each phase is not always easy to determine, because the requirements have not been spelt out. This leads to some confusion, since decisions and plans prepared in one phase affect that succeeding. One group's output is the next's input.

The phase at which a contractor becomes involved in a major project for an external client may be either during the feasibility and scoping studies or at the bidding stage for design or supply. Other types of projects may be internal, which would entail starting at the top of the list.

Planning

Planning entails two distinct processes: first, it involves decision-making in the form of implementation planning; and secondly, it deals with the minutiae of actually getting the work done. Decisions have to be taken by everyone involved in a project right up until its conclusion, but the major decisions should have been taken early on, leaving only lower-level decisions to be taken as the project unfolds. Conversely, work scheduling and resource allocation apply to the project definition stages equally as to implementation, although they are techniques especially applicable to labour-intensive phases, such as detailed engineering and site construction.

Planning is not a job restricted to planners; it involves the entire project team and key participants who will be responsible for major activities and cost centre. People, money and time are all important resources. Schedules of activities and events have to be reconciled with achievable manpower levels, work-loads and budget costs. Hence planning takes time and is essential before implementation begins.

Work preparation is vital. All plans are based on assumptions—about the quality and timing of information inputs from preceding activities, economic crew sizes and mixes, productivity and durations. The area in which planning is least successful is the identification and declaration of assumptions relating to inputs, which are invariably other people's outputs. It is essential to flush these out, to test them with the preceding action parties, and to iron out any discrepancies, *before* the detailed programme for the ensuing period is prepared.

It is necessary to decide to what extent parts of the project may be considered as self-contained, discrete work packages. Technical, financial and schedule responsibility for these should be allocated to a single organization. Only when the work-split cannot be clearly defined does responsibility for coordinating the interface have to be retained centrally, and not delegated.

Implementation Planning

The implementation plan divides the overall scope into manageable parts, and establishes the allocation of these to various organization/individuals. For example, it covers the project organization, contracting strategy, procurement policy, quality assurance and cost/schedule aspects. Terms of reference, organization charts, staffing plans and job descriptions for key individuals are required. Administrative policies and procedures for the implementation of the above have to be prepared.

Planning should start in outline form, with details being added later and checked against the previous outline. Then only if the underlying assumptions are shown to be inadequate by later work will the original plan need to be revised. This concept should be applied to all stages of implementation planning.

Implementation Strategy

Starting from the project objectives and with a knowledge of the basis on which the decision to proceed (tender for, or take an order) has been taken, the company's managers now have to take a careful stock of its capability to undertake the project internally, and consider to what extent they will require to call upon external help. Questions to be addressed include:

- Can the detailed design, manufacture, installation and commissioning be carried out in-house?
- What alternative sources of assistance will be appropriate for this particular project, and what are the factors in deciding the extent of the company involvement?
- If part of the project is to be contracted out, will the company still require help to manage the project in detail?
- What then would the company's involvement be?
- What should be the decision and authority levels within the various organizations participating?
- How does the company ensure that the project meets the objectives?

The Work Breakdown Structure

The total project scope has to be divided into manageable parts, the work breakdown structure (WBS). It is necessary to break down the project into tasks or work packages that are clearly defined, and suitable for use in allocating responsibility and for control. For each task, a work statement should be prepared, to include:

- Work definition and interfaces with other tasks
- Inputs required and other restraints on starting measurable output
- Budget allocation and timing
- The earliest possible start date
- Estimated duration
- Company, department and individual responsible for performance

The selection of the satisfactory WBS is an important step in implementation planning, and warrants very careful consideration, involving potential participants. There are three basics:

- Policies and general rules have to be established centrally, for the entire project
- Implementation responsibility should be assigned for integrated technical, financial and schedule performance
- Decoupling of discrete tasks should be carried out as far as necessary to give manageable tasks, provided that responsibilities are not thereby diluted

When the WBS has been set out, each group involved should record its understanding of the following:

- Work scope and method
- Responsibilities and reporting
- Relationships and dependencies on other groups
- Organization and staffing required
- Job descriptions and terms of reference

One common way of recording part of the above information is a cost–time–resource (CTR) form, but this does not usually show all the information which is required for the purpose indicated above.

Such a recording of commitments is more difficult within the organization than without, since there are no formal contracts to be entered into. Nevertheless, internal undertakings should be made and recorded, as though they were contractual obligations.

Management Reviews and Approvals

Much project work has to proceed under delegated authority. The alternative is the 'gold-fish bowl' syndrome, in which as projects have grown larger and often with profound environmental implications, more and more senior management time and effort has to be devoted to a personal involvement in project details. A system of approval by stages, careful setting out of policies and procedures, and periodic presentations and reports has to be established as the framework in which the project can proceed without continuous management intervention.

Management approvals along these lines should be requested in a formal manner, supported by intermediate managers up to the final authority, and recorded for reference at later stages by all needing to know the basic project decisions and their underlying reasons.

In an earlier section, the stages in the development of a project were identified. At the conclusion of each of these, management approval should be required before the next is begun. Unless management appreciate what they are being asked to commit resources for, and the full implications of a decision or non-decision, either their approval will be difficult to obtain or—even worse—fundamental changes may have to be made to the design at a later and more expensive stage.

Apart from the reviews at the conclusion of each phase, management must have periodic summary reports submitted to them, and presentations made by the project sponsor and manager. Thorough preparation for such presentations is essential. A good maxim is for the project manager to carry a set of vu-graphs in his brief-case.

Cost/time Relationships

Cost control

Everyone involved in a project-type activity must be cost-conscious. Like planning, cost control must start at the top and work down through the organization. The need for cost control and value engineering has to be stressed continually to ensure that it does not get overlooked. There is considerable confusion between the concepts of cost control and cost reporting. Control implies action, following conscious, justifiable, and timely decisions. Reporting is historical by nature, and action based on historical reports tends to produce an amplified disturbance, rather than lead to effective corrective action. Cost control difficulties and appropriation overruns in the petrochemical contracting industry lead to the concept of cost engineering. This embraces estimating and cost reporting. By itself, it does not provide cost control, any more than management accounting does.

The fundamental problem is that those making commitments do not ensure that the corresponding accruals and contingent liabilities get entered into the accounting system. Accounting conventions *can* handle this situation, but only if line managers insist that the accounting systems are set up accordingly. Input is required from the departments issuing purchase requisitions and orders, placing contracts and calling up manpower and materials against standing orders. Once the above point has been accepted, it is not so difficult to obtain meaningful reports which not only include accruals right up to the end of the last cut-off date, but extend the information through to the following period. This provides a cushion for the situation to be analysed and timely decisions taken.

Cost control includes the following aspects:

- Preparation of estimates
- Preparation of budgets and requests for funds
- Transposing approved fund requests into working budgets as a control tool
- Establishing and explaining cost control policies and procedures
- Preparation of cash flow projections
- Preparation of cost reports with trend analyses and other management control information
- Cost input to change reviews
- Accountability and auditability of all project business analysis of costs incurred and transporting these into a cost 'data-bank' for future estimates

It may not be appropriate to consider all the above functions as the responsibility of one group, but within large organizations, handling projects on a regular basis, it will usually be found necessary to allocate staff specifically to them. The type and experience of staff required will vary with the nature and size of project. For oil industry projects then cost engineers will be involved.

Two of the above topics are worth special consideration: the control budget and change reviews.

Control the Budget

The cost estimate breakdown used when submitting a request for authorization for funds will almost certainly not reflect the way in which the project will be managed, the work breakdown structure. Until the cost estimate has been recast in such a format, cost control is in a state of black-out, and it is vital to get this done as soon as possible. Important considerations are the allocation of contingencies, and budget authority. Within the overall cost estimate will be overt or hidden estimating allowances and contingencies, to cover unknown or insufficiently defined factors. Management have to make sure that it knows what these allowances are and to decide to what extent these will either be allocated to the various cost centres, and so be available for expenditure without reference back to itself, or be retained as a central fund to be allocated later as necessary by management. Control over contingency funds is as necessary as control over duration floats, and the only valid reason for management relinquishing either is delegation within well-defined parameters to responsible and trustworthy participants in the projects.

Change Control

Changes in work scope and definition are extremely difficult to price and control. It seems that, no matter how thorough the groundwork has been laid and all the preceding stages gone through, changes are proposed with justifications such as safety, a marketing development, or 'site unknowns'. The cost of change is difficult to assess accurately, but rises significantly above the cost of the individual cost elements as the project advances because of the cumulative disruptive effect throughout the entire project, plus an increased possibility of mistakes being created by the process of change. Changes should not be introduced without some formal deterring management approval routine. The disruptive of change should be allowed for by a multiplication factor on the cost estimate. This factor should be 1 at the time of submitting the initial request for total funds, rising to 5 at the end of the definition stage, and progressively increasing to 10 as the project draws to a conclusion and is approaching completion testing. Sound management discourages late changes, deferring them until after commissioning, to ascertain whether or not the proposed changes are confirmed to be necessary from initial operation.

Cost and Time Relationships

So far we have considered planning and cost control as separate subjects. In reality, they cannot be considered in isolation. It is small consolation to be (say) 10% underspent, yet 20% behind progress. So the two aspects have to be taken together.

Spot results of progress (cost/planning) tell part of the story, but more significant are the underlying trends and a continuing reassessment of the balance-to-go. The management question 'Are we on course to meet the time and budget targets?' can only be answered adequately in terms such as 'This is where we were at the last cut-off, having made P_1 progress and used R_1 resources; this is where we expect to be at the next cut-off, with P_2 progress and the consumption of R_2 resources. On the above basis, we expect to complete the project by time T, having used R_t resources.' Such analyses are meaningful at all stages of a project.

WHAT TO DO NEXT

Assuming that the PM approach described and advocated in this paper has made some favourable impression, there are several steps which should now be considered. Firstly, audit your general management practices and processes, to see that PM could stand a chance of being successful. Then consider in more detail the relevance of PM to your own business situation. It may be quite inappropriate. If you decide to go ahead, this must be a top-level decision, which then needs the resolution to push through the consequent organizational changes.

As a start, set up an overall framework and then introduce some simple PM systems initially, which can be added to gradually.

Finally consider how to find and develop your own project managers.

FINDING AND GROOMING YOUR PROJECT MANAGERS

The role of the project manager will obviously vary considerably with the nature and scope of the task. However, the following activities must always be included in his job description:

- Obtain management approval of the task objective
- Break down the task into components; set objectives
- Establish plans
- Organize staff and other resources
- Set up controls and reporting system
- Issue policy directives and other instructions
- Motivate team and communicate with other parties
- Apply innovation and flexibility in seeking alternative courses of action when appropriate

An often overlooked but important end-product for the professional project manager is a critique of the project execution, so that plans and procedures can be improved if necessary for future projects.

The job of a project manager is complex and demanding. Many successful managers of large projects achieve results based on extensive experience, combined with a strong personality. However, both they and others who aspire to follow them, or even make worthwhile contributions to project teams, could improve their performance by greater training in the techniques and skills of management in general and project management in particular.

Skills and Attributes

The project management role varies considerably with the size and nature of the project. Leading his team and dealing with many people not reporting to him and even outside his company are normal aspects of the job, so the ability to work with and through people is paramount. A clear understanding of the company ethos and policies is important. A sound appreciation of the technical features of the project are advisable, provided that the project manager has already learnt that detailed knowledge from an earlier (non-project) existence does not make him still an authority in the field.

Education and Qualifications

Many budding project managers will be engineers, probably chartered. Hence it is probably appropriate to consider the final part of their initial training. The Institutions of Mechanical and Civil Engineers have a joint Diploma in Engineering Management, which is a postgraduate, post-experience qualification. Part 1 deals with fundamental management subjects—organization, managing people, finance, and legislation. Part 2 covers analysis and decision-making in a core subject (quantitative methods and information systems), and a range of functional options, including PM. The PM syllabus covers project organization, project selection, contract strategy and planning and control. Part 3 is a project report or thesis applying the previous work in parts 1 and 2.

The British Institute of Management Diploma in Management Studies covers much the same ground as parts 1 and 2, and is a good basis for going on to PM.

There are a range of first degree and postgraduate courses in Project Management and related subjects offered by universities and polytechnics.

Career Development

No one should be under the illusion that obtaining a degree and then becoming a corporate member of a professional body is the end of his career development. Continuing education, or formation, is increasingly recognized as the way for individuals and industry to stay ahead.

The Association of Project Managers (APM) is one professional body which aims to foster the efficiency and well-being of PM. It provides an environment in which management techniques and experience can be developed, expressed and shared. It also seeks to improve the standards of management theory and practice, by direct and indirect participation in the training of its members. APM is hosting the ninth World Congress on Project Management to be held on 4–9 September, 1988, in Glasgow.

9

The CSWIP Scheme: a Personal View from the Chairman of the Board

T. J. H. Webborn, Haydn Webborn Consultancies Limited, UK

I welcome this opportunity to give an overview of the *Certification for Weldment Inspection Personnel* (CSWIP) scheme in the area of Subsea Inspection Certification, which is the seventh phase of examination categories, brought in by CSWIP. Unlike its forebears, which relate to particular inspection techniques, i.e. Phase 1 Ultrasonics or Phase 2 Radiography, Phase 7 dealt with the perceived need of the Offshore Industry in 1979 to negotiate the complete package of techniques/systems in general use for subsea inspection.

As CSWIP had been preparing examinations and administering them since 1969, it is reasonable to claim that the management board and the industry working parties responsible for bringing the exams on stream had developed considerable expertise and, through consensus agreement, a quality system related to current standards and practice of the day should be the dominating principle for all phases.

At the same time, John Young, a man imbued with high ideals about achieving excellence, was appointed to be the Secretary. He was also responsible for producing the documentation that the scheme would require, and his skills produced a mould into which the outpourings of the Working Parties, who were setting the technical requirements, were injected. This disciplined approach to having a commonality documentation has allowed industry's needs in new areas of certification to be introduced quickly and in familiar format.

It became a matter of routine for the Secretariat, when new techniques/areas for certification were required, to put together a broadly based working party of experts to specify the technical requirements, i.e. experience, mandatory training, types of specimen and numbers. Each new examination is proved by a trial examination where participants in the working party put up candidates with different levels of skills and experience, so that the chief examiner, Monty Frost, who is unaware of the calibre of the candidates, has an opportunity to assess the efficacy of the marking system and the practicability of the exam.

In this overall look at the CSWIP system it is important for you to know that

Advances in Underwater Technology, Ocean Science and Offshore Engineering, Volume 14: Submersible Technology

the management board provides a channel for feedback on problems, introducing new techniques or technology, compliance with new standards, calls for update of specification, and from time to time picking up market-place needs for new certification areas.

In 1979 the Secretary of CSWIP called a meeting of interested parties from owners, certifying authorities, Government and industry to test the requirement for a full underwater examination. The result was a mandate for 'Diving CSWIP'.

The CSWIP is a flexible scheme and has always reacted quickly to this demand, so that when offshore platforms had been in service for a period, a BSC requirement for plate testers (ultrasonic thickness measurement of wall thickness and MPI of crack specimens) was 'translated' into a 'dry CSWIP qualification' for early diver inspectors as an interim measure.

The CSWIP Management Board (Chairman Dr Roy Nichols) followed this up by quickly agreeing to the creation of a new phase of CSWIP, DIV-7 and asked Haydn Webborn to 'chair the working party that would write the examination, choose a suitable contractor to provide facilities in sea water, with not less than 25 ft over the diver's head at all states of the tide, and have the examination available in six months'. Nichols' choice was somewhat restricted as only board members could chair working parties. Webborn, who was the Chairman of the Ultrasonic WP, had had a continuous involvement with CSWIP since its inception, and had worked as an Inspector Diver, was thus given the job of first Chairman of DIV-7.

There was enthusiastic support from the Association of Offshore Diving Contractors (AODC), BInstNDT,DEn Lloyds Register and DNV. Regrettably, the operators did not take membership, which had serious consequences later on.

The CSWIP 3.1D examination commenced in September 1980 with practical tests in ultrasonic thickness measurement and crack detection of welded pieces, together with a written paper on the theory and application of these two NDT techniques.

Three diving schools applied to CSWIP for their courses to be approved. The establishments were physically vetted and the courses at Bovisand, Fort William/Scottish School of NDT, and Prodive were approved. (CSWIP requires demonstrations from all candidates that they have satisfactorily completed an 'approved course'.)

However, in January 1983, the Chairman of DIV-7 became aware that BP and Shell were extremely dissatisfied with the scope of the examination.

A meeting was held in Aberdeen, on January 1983 between Shell, BP and Britoil with the Chairman of DIV-7. Senior management made it quite clear that changes had to be made, but as is common there were many misunderstandings about what CSWIP is and is not. The oil companies had opted out of the first working party and so had made no contribution to the formulation of the examination requirements of which they now complained. A wide-ranging discussion on all topics took place and it was evident that the operators found the examination *much too heavily biased towards NDT*.

The management board approved a further development of Phase 7 and the addition of representatives from BP, Britoil and Shell (2) to join the working party.

This working party met frequently and produced a new examination, 3.1U, to be taken as a preliminary to the existing 3.1U, now renumbered 3.2U. The examination covered extensive practical assessments in CCTV, still photography, damage and corrosion investigation and measurement, and cathodic-protection assessment.

Heavy bias in marking is given to the candidate's ability to place proper emphasis on the nature/size of defects and to demonstrate his understanding of the underlying philosophy of offshore inspection in his verbal commentary.

The new examination was tried out on five 'guinea pigs' who volunteered from Comex and Oceaneering. The trial was witnessed by H. Webborn (CSWIP), S. Freeman (BP) and L. Porter (Lloyds Register). A five-page report was issued, with recommendations for change.

At the time Freeman and Mills (Oceaneering), for the CSWIP Approval of Courses Panel, vetted the diving schools for the new examination. Freeman contacted Webborn to compare notes and reported that the schools had serious limitations in that they were not teaching current offshore inspection principles or philosophy.

The trial examination also revealed two serious flaws:

1. The divers' commentaries were judged to be inadequate, e.g. they did not demonstrate diverse awareness of the need to refer to their close examination for cracking at the toes of welds and showed little awareness of understanding the importance of their role as the key men in the whole inspection.
2. The CSWIP examiners displayed little confidence in handling divers under water and did not establish the relationship necessary between themselves and the divers.

At this time the Department of Energy, Petroleum Engineering Division had been in contact with the DIV-7 Chairman to see if CSWIP could be the route for introducing new technology into the diving schools, following V. S. Davey's report 'Effectiveness of Underwater NDT' and other similar trial work currently going on in the Department. As a result an investigation was funded, by DEn, into the training of inspector divers, which resulted in the sponsoring of a transfer of technology from the operators to the dive schools and CSWIP examiners.

The CSWIP examiners were tutored by BP and Shell at Abington, and both examiners and school instructors were taken offshore, on an on-going basis, to become familiar with practical day-to-day inspection in the real world. A further result of this was that a committee to advise the Department on subsea inspection matters was set up. Thus was formed the Operators' Advisory Committee (OAC) to DEn, which was entitled to representation on the CSWIP management board and DIV-7.

Graham Mills, in his excellent paper 'Underwater Inspection Personnel Certification' IRM 1986, has comprehensively covered all the aspects of the CSWIP examination of structures, and this should be regarded as a source document on Phase 7.

NEW DEVELOPMENTS FOR 1987

Since the Mills paper a year ago, we are expecting the 3.3U Pilot/Observer and 3.4U Inspection Controller examinations to come on stream, so that these new approvals will be available for the 1988 inspection programme.

Inspection of Concrete Platforms

The owners of these structures felt that an equivalent of the steel structure should be available, and if a concrete-examination endorsement was to be introduced then it was reasonable for training facilities of a comparable nature to be constructed for the schools.

As a result of OAC advice to DEn, slip-formed structures in the form of boxes 6 m × 8 m × 4 m, incorporating realistic defects, have been made available to CSWIP and the diving schools.

The concrete endorsement to 3.1U will become available in August 1987.

SUMMARY

I believe it necessary to have set the scene on CSWIP's DIV-7 origin and performance overall as a worldwide subsea inspection qualification. The CSWIP has demonstrated that the consensus view of the industry was reached by reasoned argument by the major parties and also shows that the system has inherent flexibility when and where change is necessary, e.g. the special divers' appeals panel set up at one meeting.

REASSESSMENT OF STANDARDS AND OPERATING PRACTICES OF CSWIP

I can categorically state that the standards demanded by CSWIP examinations are equal to any that would be demanded of inspections done in air. I do believe that in many cases, because of the difference in cost/time/quality ratio, inspector divers do perform better under water in the MPI role than on shore. There is also another factor that leads to better performance standards because of the 'team role' of the diver and inspector/controller working together in close harmony.

Specific Points for 3.1U and 3.2U

The CSWIP has issued more than 16 000 certificates across all its phases. There has only ever been one technical complaint from a plant owner against a CSWIP-approved technician. To date 1000+ certificates have been issued to divers. The (theoretical) failure areas in the two-and-a-half-year diver resits are seen to be high enough to justify the re-examination. We are now approaching the five-year re-examination. If DIV-7 follows the CSWIP pattern it is my judgement that failure rates could be significant.

Specific points for 3.3U and 3.4U

These standards have been set by the OAC and AODC operating in concert. The CSWIP recognizes both these bodies as being authoritative, especially as government and certifying authorities have endorsed the examinations.

OBJECTIVES FOR EXAMINATIONS

In 1982, my working party set objectives for CSWIP, applying to all examinations, i.e.:

- a syllabus based on state-of-the-art requirements;
- fairness of test specimens;
- fair, unambiguous questions;
- objective marking;
- practical tests at defined sensitivity where appropriate;
- standards of defect sizing and POD that relate to the real state of the art.

I consider that these are generally achieved in the conduct of the examination.

CONCLUSION

The CSWIP DIV-7 Committee through its annual audits has refined its examination techniques and ensures that the examiner, dive supervisor and support staff provide a fair examination.

From time to time aberrations do occur but the reporting system is such that these surface quickly and appropriate measures are taken.

The major weakness that exists is that the steel-examination structure does not have sufficient fatigue defects, so that parallel conductor techniques cannot be used. However, a request has been made to DEn for assistance in making a new structure.

I submit that the objectives for conduct of the examinations are honestly met.

All employers and candidates have recourse to appeal to the management board and I would encourage its use by any party that feels aggrieved. Above all, CSWIP is your system, and its requirements are specified by the offshore industry.

APPENDIX

Examples of topics with which a diver should be familiar for the 3.1U examination are as follows

Introduction

- The need for inspection.
- Basic terminology of steel and concrete structures.
- Outline of modes of failure and deterioration experienced in steel and concrete structures.
- Appreciation of how operator's inspection programme attempts to detect and assess such failure and deterioration by the use of the various inspection techniques covered by the Grade I and Grade II syllabus.
- The importance of documentation and record keeping.

Underwater visual inspection

- Cleaning methods, safety aspects.

- The standards of surface finish.
- Types of defects in concrete platforms.
- Types of defects in steel platforms.
- Weldment terminology and weld measurements.
- Identification of visible weld defects.
- Appreciation of likely weld-defect locations.
- Marine growth, types, effects, reasons for removal.
- Dimensional checking of welds, and measurements underwater; principles employed in engineering practice.

Recording methods

- Photography, to include knowledge of the meaning of and relationship between film speed, aperture, shutter speed, focus, depth of field.
- The need for optimum light placement.
- Size references, record keeping.
- Use of CCTV in hand modes.
- Narrative commentary by the diver and dialogue with topside inspection supervision during inspection work.
- The need for fixed, repeatable reference points.
- Structural marking methods.

Photogrammetry

- Principles.
- Requirement for care in use and deployment of equipment.

Corrosion protection systems

- Principles of corrosion.
- Typical instances and causes of visible corrosion.
- How corrosion is prevented or reduced.
- Visual inspection of protective coatings, including monel wraps.
- Modes of deterioration and typical inspection requirements for sacrificial-anode and impressed-current systems.
- Potential measurement methods, typical values obtained, calibration of instruments.
- Effects of debris.
- Safety aspects of impressed-current systems.
- Measurement of wall thickness using digital wall-thickness meters, necessary surface preparation, calibration, appreciation of the shortcomings of these instruments.

Non-destructive testing techniques

An overview of the methods, capabilities and limitations of magnetic-particle, ultrasonic, radiographic and eddy-current inspection techniques.

Reporting

Ability to produce a fluent verbal description during the above inspection activities; ability to produce post-inspection written reports.

REFERENCES

Mills, R. G. Underwater Inspection Personnel Certification.
Mills, R. G. Requirements for the Certification of Underwater Inspectors. No. CSWIP DIV-7-83.
Mills, R. G. Syllabus for Candidates for the Pilot Observer Inspector and Inspection Controller.
Young, J. G. CSWIP. The first fifteen years, Presentation to 3rd European Conference, Florence 1984.
Young, J. G. CSWIP information sheets, (published from time to time).

Part III

Saturation Diving

10

Higher Efficiency through Improved Diver Transportation

A. Haegglin, Chalmers University of Technology, Sweden

The quest for new oil and gas fields has taken the oil companies into areas of great depths and hostile environments. To make the exploitation of these findings possible, there is a need for new and advanced technology.

The distance between the surface support and the diver is increasing. This adds to the technical difficulties and drastically lessens the margins for the diver life-support systems. This calls for new philosophies in many of the problem areas and perhaps totally new concepts for their solution.

Attempts to obtain increased surface independence have led to the development of so-called flying bells. These are diving bells free from guide wires and equipped with thrusters in order to give horizontal manoeuvrability. In the near future we will also meet the large autonomous submarine 'Saga', a mix between the old concept of a stationary habitat and a lock-out submersible as it occurred in the early 1970s.

This chapter intends to describe the pros and cons of these systems and to compare them with the conventional diver transportation systems. This will not be done by a precise statistical method, but rather by analysis of a systematically collected interview material. We shall try to show the potential of this kind of analysis and hopefully be able to continue to make similar studies of offshore diving.

SYSTEM DESCRIPTION

For the comparative analysis of the different transportation systems, a system outline, a performance table and a theoretical bell-run efficiency diagram have been prepared.

System Outline

The outline is roughly divided into structure, main supply systems and diver-worn components. From the beginning our intention was to compare system components of different diver support systems to establish their influence on the total efficiency of the system. We found no such differences of any significance, although some system components, mutual to all systems, proved to be in need of improvement (Fig. 1).

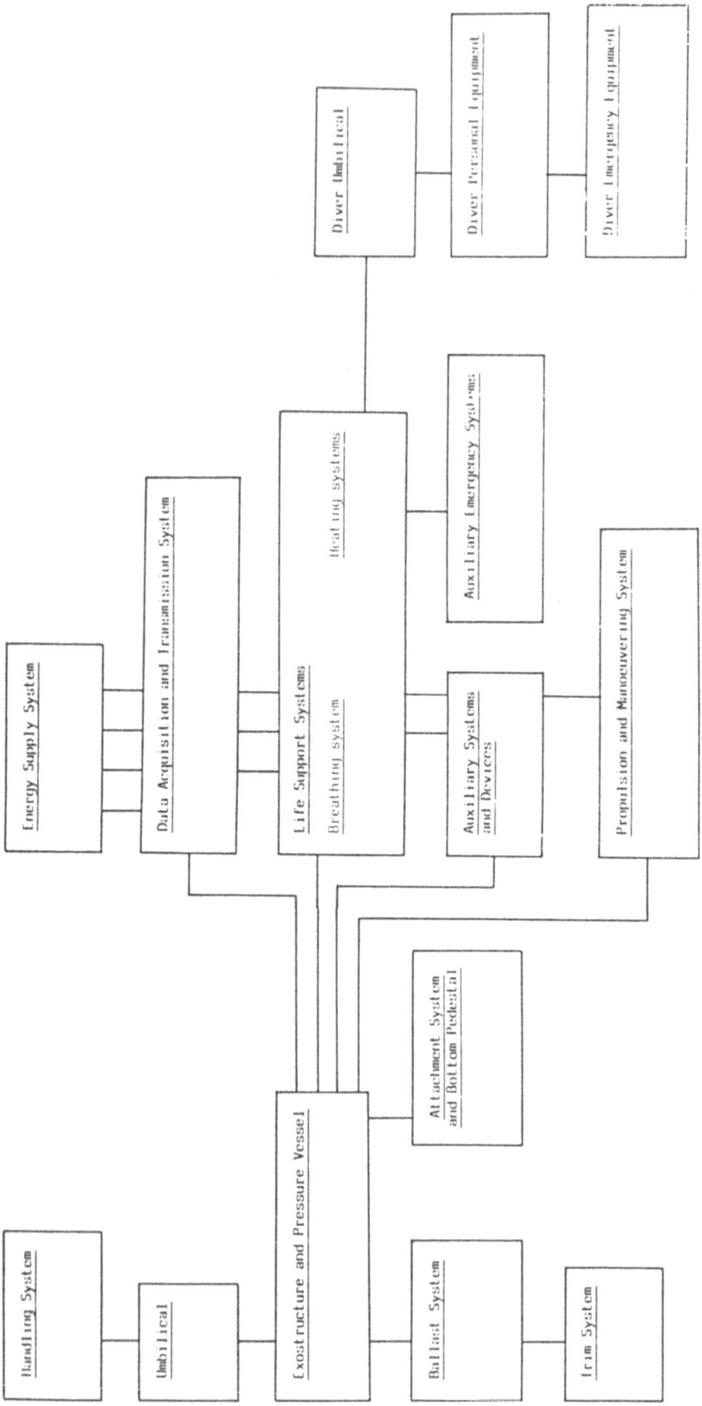

Fig. 1 Principal outline of diver transportation system.

Performance Table

A number of system characteristics have been listed. The information is mostly gathered from commercial brochures (Fig. 2).

Note the following:

MDU: Emergency time: This is most likely 24 hours as defined in classification rules.

FB and MDU: Turnaround time: It is not likely that the FB closes one bell-run and starts the next within one hour. A more credible estimate is two hours. That figure is a rough estimate based on the turnaround time for the SDC and the launch to lock-out time for the MDU.

Flying bell: Launch to lock-out: No information is given concerning this operation. It is, however, very likely that the time for the operation will be close to that of the MDU.

MDU and DLO: Costs: The costs are estimated from old information and the need for a special support vessel.

	length (m)	width (m)	height (m)	weight (tonnes)	payload (tonnes)	number of divers	energy storage (kWh, MWh°/h, days°)	power (kW)	gas storage (SCM)	emergency duration (h)	range (m, nm°)	speed (knots)	saturation interval on 200m (m)	turn around time (h)	launch to lock-out (min.)	costs (k$/day)
SDC/OME	-	2,7	3,0	8,5	-	3	-	NA	90	24	30	-	75	1	5	25
FB/Bruker	-	3,2	3,7	12,0	0,3	3	-	60	144	24	400	2,5	75	1*	NA*	40
MDU/Perry	-	3,7	4,6	20,0	0,2	3	-	10	NA	NA*	185	2	75	NA*	40	40*
DLO/Bruker	8,0	2,9	3,2	19,5	1,0	3	52 / 12	10	132	72	20°	2	40	6	NA	40*
SAGA/Comex	28,1	7,4	8,5	NA	3,0	6	10° / 14°	500	1568 He / 256 O_2	NA	200°	5	40	-	-	NA

Fig. 2 Performance table.

Theoretical Bell-Run Efficiency

Figure 3 shows the theoretical ability to keep a diver at the worksite. The necessary ineffective bell-run time is estimated from interview material. In practice, the influence from various systems of different technical maturity will cause further delays in turnaround time.

Assumptions made for the graph are:

Resting time between dives:	16 hours	
Number of divers in transport:	SDC/FB/MDU:	2 divers
	DLO:	4 divers
	SAGA:	6 divers
Possible working time per diver:	200 metres:	4 hours
	450 metres:	2 hours

The ineffective bell-run time for the SDC, for example, is defined as the crew change-out time. That is, the time which elapses from the moment when diver

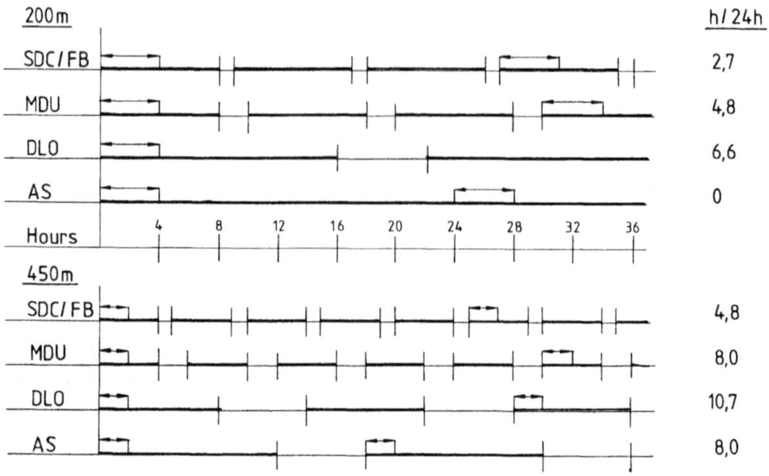

Fig. 3 Ineffective bell-run time due to diver transportation. ←→ Diver number one, first and second bell-run after 16 h rest.

number two of a team leaves the worksite to the moment when diver number one of the next team reaches it. The same is applicable for the flying bell, the MDU and the DLO.

The autonomous submarine does not have ineffective bell-run time as long as the six divers are working back to back. The depth in which the number of divers is insufficient for a continuous operation is probably in practice less than 200 metres. In theory, however, it is somewhat more than 200 metres.

The ratio is calculated as (ineffective bell-run time/total bell-run time) × 24 hours.

Summary:	200	450	
SDC	2.7	4.8	h/24h
Flying Bell	2.7*	4.8*	
MDU	4.8	8.0	
DLO	6.6	10.3	
SAGA	0	8.0	

*This figure is somewhat higher in practice.

MAIN TASK AND ASSUMPTIONS

During the initial stage of our study, we tried to compare system performance and characteristics to a specific task. A main task and a few environmental and technical assumptions were decided on. The main task chosen was a cleaning and non-destructive testing (NDT) operation.

Environmental Assumptions

The installation of this study is located on the 62nd parallel in the North Sea. The weather is average autumn weather with a sea state between 5 and 6. The current at depth is 0.5 knots, the temperature is 2°C and the visibility is good. Two depths are studied, 200 and 450 metres. The first is today's maximum operational depth in the North Sea, and the second is the depth which most in the offshore industry believe has to be reached.

Technical Assumptions

Some technical assumptions have been postulated in order to make it possible to work at 450 metres depth and to operate a lock-out submersible in an efficient way.

Breathing system

A good breathing apparatus of rebreathing type is available. It gives an acceptable gas consumption and an acceptable performance concerning breathing resistance. The bail-out system gives an endurance of 15 minutes of 450 metres depth including the gas heater.

Flying bell

It is assumed that a flying bell of the Bruker concept is to be used and that its performances correspond to those stated.

Diver lock-out submersible (DLO)

The energy supply on board the submersible has a duration which allows for a minimum of 16 hours continuous work, supplying all energy-consuming systems on the vessel and four divers in saturation. The handling system is designed to allow for launch and recovery in a sea state between 5 and 6.

Autonomous submarine

It is assumed that an autonomous submarine of the Saga concept is to be used and that its performances correspond to those stated.

Support vessel

It is assumed that each transport system which requires a support vessel is supported by a modern monohull DSV.

POSSIBLE INFLUENCE FROM DIFFERENT DIVER TRANSPORTATION SYSTEMS ON THE PERFORMANCE OF DIVING OPERATIONS

The views presented in this chapter are based on compilation and analysis of interviews with executives representing major oil and diving companies. The material has been compiled and construed by the author.

The scenario presented showed little relevance when the interviews and discussions started. The idea to present the transportation system's influence on each task has to be abandoned since there was no influence of any significance.

	Safety				Reliability			Working Capacity				Adaptability				
	Diver During Transport.		Diver at Work	Install-ation	Weather Sensitivity		Techn. Level	Diver at Worksite		Hori-zontal	Verti-cal	Alter Ope-ration		Change Worksite		Distur-bance
	200	450	-	-	200	450	-	200	450	-	-	200	450	200	450	-
SDC	8	7	8	9	7	5	8	9	8	5	10	10	8	9	8	10
FB	6	6	9	7	8	8	7	8	7	7	10	10	8	10	10	9
DLO	4	4	8	7	3	3	5	6	4	8	1	3	3	8	8	5
AS	10	10	10	6	10	10	4	10	6	10	1	3	3	3	3	2

Fig. 4 Summary matrix. Grades from 1 to 10 are given, where 1 is poor and 10 is optimal.

The results are therefore presented in another manner than first anticipated. They have been divided into the following categories: safety, reliability, working capacity and adaptability. The relevance of the results has not been reduced by this.

A summary matrix (Fig. 4) based on the interviews and available literature has been compiled. Grades from 1 to 10 have been given in order to try to obtain a significance in the differences between the transportation systems. The different grades given are commented on.

Safety of Divers During Transportation

During the transportation from the support vessel to the worksite, the most dangerous situation is the passing of the surface and the docking manoeuvres. The surface zone should be passed as quickly as possible, as done with the SDC. The docking and attachment manoeuvres of the flying bell and the DLO are potentially dangerous operations, particularly in high sea states.

Since the divers of the autonomous submarine do not pass through the surface, these potentially dangerous situations do not exist regarding that system.

The SDC must pass through the surface zone which is potentially dangerous. If the SDC is used of 450 metres depth, there is an increased risk of entanglement with the guide wires, lift wire and umbilical due to their length and stretching.

The flying bell must accomplish two potentially dangerous phases: the passing of the surface zone and the docking manoeuvres. It is not more dangerous to deploy divers at large depths with a flying bell.

The surface operation of the DLO is highly dangerous and can only be performed in calm seas.

Safety of Divers at Worksite

When the divers are at their worksite, their safety is mainly dependent on the distance to the relative safety of the bell or sub and the time it takes for the diver to reach it or to be brought there. This makes the flying bell safer than the other systems. The distance between the worksite and the transportation system is not always significant.

Another risk is being pulled off the worksite by a transportation system in disarray. The SDC is run from a dynamically positioned support vessel. These vessels have been known to suffer from computer malfunctions which momentarily take the ship off position. Such malfunctions are very rare today since the position systems consist of at least three independent systems which individually can keep the vessel in position if the others fail.

Because the other systems are locked to a structure or standing on the bottom, they do not suffer from this risk. If, however, a diver locks out from a flying bell or DLO, it has a major influence on the buoyancy of the unit. If, during this phase, the buoyancy systems fail or the operator shows lack of attention, a very dangerous situation of rapid ascent can occur. Such a rapid ascent will inevitably be fatal to a saturation diver in the water and to the divers in the pressure compartment if the hatch cannot be closed.

The risk of being pulled away by a transportation system in disarray must be regarded as small for the SDC, flying bell and DLO, and as practically non-existent for the autonomous submarine.

With regard to the diver transportation system, the safety of the divers at the worksite is depth independent.

Safety of Installations

In general, the offshore operators are not worried about the diving contractor damaging platforms or bottom installations. Concerning the flying bell and the DLO, the associated masses are relatively low and the impact forces moderate. However, damage can be caused to exposed and vulnerable equipment such as a satellite well-head.

Concerning the SDC, there is a small risk of damaging an installation when lowering the clump weight or when repositioning the support vessel with the clump weight not on the bottom.

The other three systems may collide with, for example, a well-head, although modern obstacle sonars are fairly accurate. The flying bell may also cause damage to an installation when locking on to it. Hoses, piping and other components of the installation may be squeezed by the attachment claw.

With regard to the diver transportation system the safety of installations is depth independent.

Weather Sensitivity

Regarding its own operation and divers the autonomous submarine is totally surface independent and therefore weather independent. During an operation, where it depends on surface support, it might suffer from weather downtime owing to the support vessel's inability to work.

The launch and recovery of the SDC are well-known operations. The limiting factor is the behaviour of the support vessel and its bell handling system. During operation, the SDC is influenced by the movements of the support vessel if the heave compensation system is insufficient or erroneously operated.

The flying bell is not influenced by the movements of the support vessel during operation but is weather sensitive during the launch and recovery operations. It has still to prove its ability to perform the docking manoeuvre in high sea states.

If the operation can be performed successfully in rough weather, it will be possible to use the flying bell in higher sea states than with the SDC. The Perry MDU claims to be operational from MSV Tharos in sea state 7. If the flying bell has the same ability, it is superior to the SDC.

The DLO will be inferior to the other systems if a major breakthrough is not made in the launch and recovery operation. Such a breakthrough must be based on a docking manoeuvre similar to that of the flying bell.

To get a good correspondence between answers on this often disputed subject is close to impossible. There tend to be as many different answers as there are companies answering.

Technology Level

The risk of downtime due to an increased level of sophistication in the diver transportation systems is the same as with any other system development. All forms of diver transportation systems are continuously becoming more sophisticated owing to improved knowledge and changes in requirements and regulations. However, the risk of downtime increases with the degree of technical complexity of the hardware. Since 'Murphy's Law' is well established offshore and since things get worse under pressure, the technical sophistication of the flying bell is something of a drawback, even though most of the additional equipment is rather straightforward and should not be expected to cause much downtime.

It is evident that the DLO must improve all sides of its system. The transfer of divers under pressure will demand a dedicated support vessel with the capacity to handle such a large device. The servicing and maintenance between dives must be more efficient and less time consuming.

There is no substance for a discussion concerning the autonomous submarine in the same way as above. There is no experience whatsoever in the operation of such a system. It is undoubtedly so, that the SAGA project will improve diving technology further, even if it will cause problems for the operator during its initial missions.

The influence from the technological level is depth independent concerning the different diver transportation systems.

Working Capacity

The working capacity of the divers when transported in one of the four different ways was discussed from both qualitative and quantitative points of view. However, the quality of the work performed by the diver does not seem to depend on how he reached his worksite.

The amount of work performed by the single diver and by the system as a whole can be affected by the transportation system. The ability to place the diver close to his worksite will reduce time- and energy-consuming swimming. If the horizontal distance is significant, problems can occur with the transportation of additional equipment.

Keeping a Diver at the Worksite

The avoidance of crew changes gives the autonomous submarine an advantage when performing long-duration tasks without any need for repositioning.

Each change of crew means loss of time and efficiency. The turnaround time differs between the systems, particularly concerning the DLO. There is also a difference between the SDC and the flying bell, but that time is not possible to determine today. The flying bell loses time to the SDC when it flies and locks on to its worksite, but it delivers the diver closer.

The autonomous submarine is able to keep a diver at the worksite around the clock at 200 metres depth if one diver locks out at the time. At 450 metres depth there will be a six-hour gap before diver number one can continue after diver number six due to the necessary resting time.

The SDC and the flying bell are not quite equal in this regard. The flying bell loses time during the flying mode and the docking manoeuvre. The amount of this time is hard to determine. Occidental claims 30–45 minutes to the lock-on from the launch depending on the location of the worksite, but gives no record of the reverse operation.

Regarding the DLO, it is hard to envisage the authorities approving the substantial 'bell-runs' of more than 16 and 8 hours respectively at 200 and 450 metres depth.

Horizontal Capacity

The horizontal capacities of the autonomous submarine and the DLO are obvious. The submarine has a higher endurance compared with the DLO.

During a horizontal operation, for example, flowline inspection, the SDC and the flying bell are connected to their support vessels. This is a drawback for both systems, but if compared, the flying bell is more adapted to such an operation.

The autonomous submarine is the prime system here. The DLO does not have the endurance of the submarine or its payload capacity. The flying bell is still connected to a support vessel which must follow it, though not as much as the SDC.

The horizontal working capacity of the transportation systems is depth independent.

Vertical Capacity

The autonomous submarine and the DLO are both strictly bottom oriented, and their vertical capacity is confined to the upper limit of the diver's current saturation depth. At 200 metres depth, this is 40 metres.

The SDC and the flying bell have no vertical restrictions when placing a diver at his worksite. If the support vessel of the flying bell is at a significant distance from the installation and if the worksite is at a minor depth, problems can occur in running tools and equipment to the diver.

The vertical working capacity of the transportation system is depth independent.

Ability to Alter Work Operation

Assuming that all tools and other equipment are surface supplied, there is no essential difference between the bell concepts, other than the horizontal running of tools and equipment.

Concerning the DLO and the autonomous submarine, it must be assumed that a change in the working operation is a time- and strength-consuming alteration.

The demobilization and mobilization of different sets of equipment in the underwater environment is hard work.

Ability to Change Worksite

The question of the ability to change worksite arose some disagreement. One side maintains that the SDC can be relocated as fast as the flying bell or with an insignificant time difference. This operation is performed by simply moving the support vessel while the SDC is submerged. The technique is well proven within moderate distances, such as along a jacket structure. The other side claims the superiority of the flying bell and to some extent that of the DLO.

The autonomous submarine is restricted to bottom orientated work and is not supposed to change its position continuously.

The changing of the worksite is supposed to call for a new position for the transportation device, meaning also the repositioning of the SDC support vessel.

Sensitivity Regarding Outside Disturbances

The system's sensitivity regarding outside disturbances is depth independent.

The SDC is a robust and insensitive system. Errors in matters of tools and construction parts can easily be corrected with assistance from the surface.

The flying bell can be somewhat isolated if working close to the surface with its support vessel at a substantially horizontal distance. This might imply difficulties in sending tools and equipment to and from the worksite.

The DLO must return to the surface if something is wrong. This operation might take a few hours, but it will not disqualify the system.

The operation of an autonomous submarine demands extensive management and careful planning. It leaves no room for mistakes such as bringing the wrong tool or the wrong nuts and bolts. An erroneous specification, however minor, can send the submarine back to its base, which may be days away.

Transportation System Related Auxiliary Systems

The auxiliary system that has the largest influence on the operation is the handling system. This is evident for the three surface-supplied systems. If the transportation system cannot be launched or operated owing to high sea states or insufficient handling systems, the obvious result is downtime.

The ability to place the diver as close as possible to his worksite is considered to be of high significance. This is highly dependent on the function of the handling system.

A well-functioning and accurate positioning system is essential if the right position is to be reached and maintained. Most offshore installations do not have a system of marking and identifying the different parts of the structure such as nodes and bracings. The difficulties in navigating in and close to an offshore structure are evident.

The flying bell is dependent on its propulsion system to reach the worksite. When there, it has to lock itself to the structure with its attachment claw or, if it is a bottom oriented task, position itself on the bottom.

Yet another system necessary for successful operations in the flying mode is the buoyancy system. The accuracy of that system is crucial for docking, attachment and lock-out operations.

The performance of the DLO is influenced by the same systems as the flying bell. Since it is surface independent in operation, there are further systems which have an effect on the operation. The two most important are the storage of energy and the gas supply system. They affect the duration of the operation and the accessible payload.

Vital System Components Not in Operation Today

There are components which are not in operation today and components which for some reason or another are disqualified.

Mutual to all the systems is the lack of a diver bail-out system and emergency heating for the breathing gas, both with a duration of 15 minutes.

The DLO and the autonomous submarine lack a low gas consumption breathing apparatus to keep the necessary gas storage at a minimum. The DLO is in need of an improved handling system, a through-water transmission system for all data signals, including video, and a heating system for the divers adjusted to the energy storage system.

System components that in some ways are inferior to the standards include breathing apparatus, heating system, and diver communication system. Other systems that are not fully proven but which do not have standards to meet include SDC handling system, flying bell docking system and DLO energy storage.

Concerning the diver's personal equipment, it is possible to develop sufficient systems with today's knowledge and technology. The problem is the financing of this small volume of very expensive equipment.

The development of the handling system of the SDC is harder to predict. Most of the work seems to be the upgrading and refinement of existing equipment and procedures. If this is sufficient, only time will tell, as with the performance of the docking system of the flying bell, whose abilities will be shown in the coming extended operation of that system.

There is nothing that indicates a further development of the presently insufficient system components of the DLO.

The transportation systems are affected by these components. If the components had been fully accounted for in the analysis, the DLO would have been the one most affected. However, the general differences between the systems would have been about the same.

DISCUSSION AND CONCLUSIONS

In the cases discussed, divers who are locked out to perform work on underwater installations on different depths are used. Their work capacity and the quality of their work will be essentially the same in all transportation schemes. Differences come about with regard to the change-out of divers and to the possibility to place the diver close to his worksite.

It is necessary to return to the surface for each change-out of divers when using a SDC or a flying bell. Since this and the return trip take time, efficiency is lost. Depending on the conditions for the alternate team in a DLO and its questionable capacity concerning operation duration, a more rapid exchange of divers can be performed. However, such an operation is not possible with today's level of technology regarding the DLO concept.

The autonomous submarine has the advantage of not requiring a support vessel and the capacity to operate continuously regardless of sea conditions. Offsetting this is the high capital investment and the lack of local support. If an unplanned situation occurs, the submarine will have to return to its base.

The submarine is a part of the operation when large horizontal distances are to be covered and various spots on the way need diver attention, eg a series of satellite well-heads and the connecting flowlines.

The submarine is also strictly bottom oriented. Since work performed at an offshore installation is bottom oriented during installation and more midwater oriented during operation, a specialized installation and possibly dismantling of offshore installations are recommended to the operator.

The flying bell will be able to place the diver closer to his worksite compared to the SDC, but this is not to say that the SDC will not be able to get close enough. It is difficult to envisage how the minor improvements in operational capacity, provided by a mobile system can be justified from a technical point of view in the tight margin, competitive world in which the diving industry operates today.

REFERENCES

Bursaux, A. G. and Fleury, G. J. M. 'Assistance to 3300 Feet Depths.' OTC 3721, Proceedings of OTC, Houston, 1980.

Cook, R. W. 'Technological Advancements of Lock-Out Submersibles.' Proceedings of Underwater Operations and Techniques, conference held in Paris, December 6–8, 1982.

Haas, J. 'Flying Bell Mobilizes Diving.' *Marine Technology* 1985 **16:** 87–92.

Jacquet, E. and Rosenberger, R. 'Practical Experience with the Mobile Diving Unit.' OTC 4210, Proceedings of OTC, Houston, 1982.

Lynnes, J. 'A Season's Diving Operations from an SSV Working in DP Mode.' Proceedings of Offshore Göteborg '85, conference held in Göteborg, February 25 to March 1, 1985.

Nilsson, S. Å. 'Diver Delivery Vehicles: a Way of Assisting the Deep Diver.' Proceedings of Offshore Göteborg '85, conference held in Göteborg, February 25 to March 1, 1985.

SAGA 1: Principal Characteristics and Performances. Published by Comex Industries, April 13, 1984.

11

Standards for Underwater Breathing Apparatus

C. Tarmey, Gas Services
Offshore Limited, UK

Before retiring from the Royal Navy this year to join Gas Services Offshore Ltd, I spent my last two years as a Project Officer at the Navy Experimental Diving Unit (NEDU) in Panama City, Florida. As part of my duties there, I was responsible for the test and evaluation of diving equipment for potential US Navy use. During the latter part of 1986, I evaluated a wide range of military and commercial diving equipment, looking for a new saturation diving rig. As a result of that evaluation, we saw that there had been some very significant advances in breathing apparatus technology, particularly with the arrival of the Gas Services Ultraflow 500 Demand Regulator. This led us at NEDU to take a fresh look at our published unmanned acceptance criteria (NEDU Report 3-81).

At the same time Dr Clarke, at the Naval Medical Research Institute at Bethesda, Maryland, was conducting a review to see what acceptance criteria they would like to recommend for future procurements or designs of underwater breathing apparatus. Shortly, we shall see how NEDU and NAMRI arrived at similar recommendations.

But first let us discuss how unmanned testing is done, how we measure the results, and how to interpret those results. Then we shall look at some test results and consider NEDU's recommended new performance goals. Then briefly we shall see what has made the Ultraflow 500 the super performer that it is.

In conclusion I would like to introduce the secondary life support system (SLS) that Gas Services has developed to provide up to 15 minutes' bail-out duration at 500 MSW, with gas supplied to the diver at a safe minimum inspired temperature—even with total umbilical severance.

UNMANNED TESTING

Figure 1 shows a basic breathing simulator. A variable speed motor drives a piston with a variable stroke length to 'inhale' and 'exhale' from the breathing apparatus fitted to a manikin head. In this way we can simulate various breathing rates and tidal volumes and so RMV.

Figure 2 shows the standardized RMVs and how they relate to diver work rate. I have highlighted 62.5 RMV and 75 RMV because they will be significant in our later discussion of US Navy acceptance criteria.

Advances in Underwater Technology, Ocean Science and Offshore Engineering, Volume 14: Submersible Technology

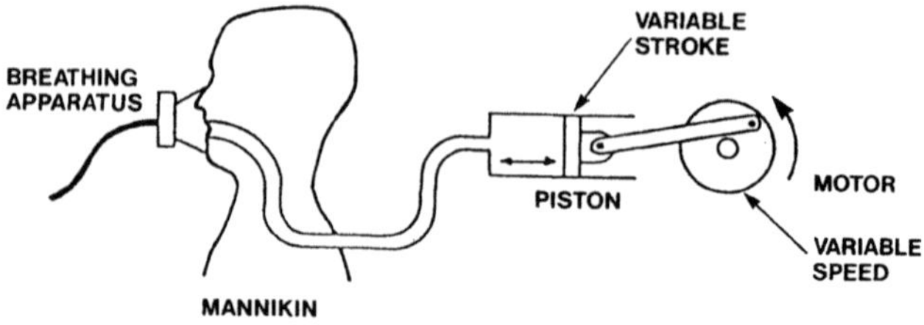

Fig. 1 Basic breathing simulator.

Tidal volume (litres)	×	Breaths per minute	=	RMV	Equivalent Diver Work Rate
1.5		15		22.5	Light work
2.0		20		40.0	Moderate work
2.5		**25**		**62.5**	**Moderately heavy work**
3.0		**25**		**75.0**	**Heavy work**
3.0		30		90.0	Extreme work

Fig. 2 Standardized RMVs and their relation to diver work rate.

Figure 3 shows a schematic of a typical unmanned test facility. This facility would automatically record differential (or changes) in pressure at the oral nasal or mouthpiece and plot a loop against tidal volume as the breathing simulator inhales and exhales. The result is the loop shown in Fig. 4.

From the loops, which are taken at various depths and at each of the RMVs (or diver work rates) that were shown earlier, we can measure the peaks of *inhalation resistance* and *exhalation resistance*. The total differential between these peaks is known as *the breathing resistance* of the UBA. The total areas of the loop represent the total additional work that the diver would have to perform to breathe the UBA. This is known as the rig's *work of breathing*.

CURRENT ACCEPTANCE CRITERIA

US Navy

The US Navy performance goals for every type of breathing apparatus are published in NEDU Report 3-81. For the purpose of this discussion, let us look at the acceptance criteria for Category 2 UBA (Fig. 5). These are umbilical supplied demand UBA. From this table, extracted from the report, it can be seen that acceptance is based upon 62.5 RMV—moderately heavy work.

The reason for this being the chosen RMV is explained in the note below the table, but could be summarized as: although man can do 75 RMV, umbilical supplied demand regulators can do no better than 62.5 RMV at depth, so we must accept a 62.5 RMV performance goal.

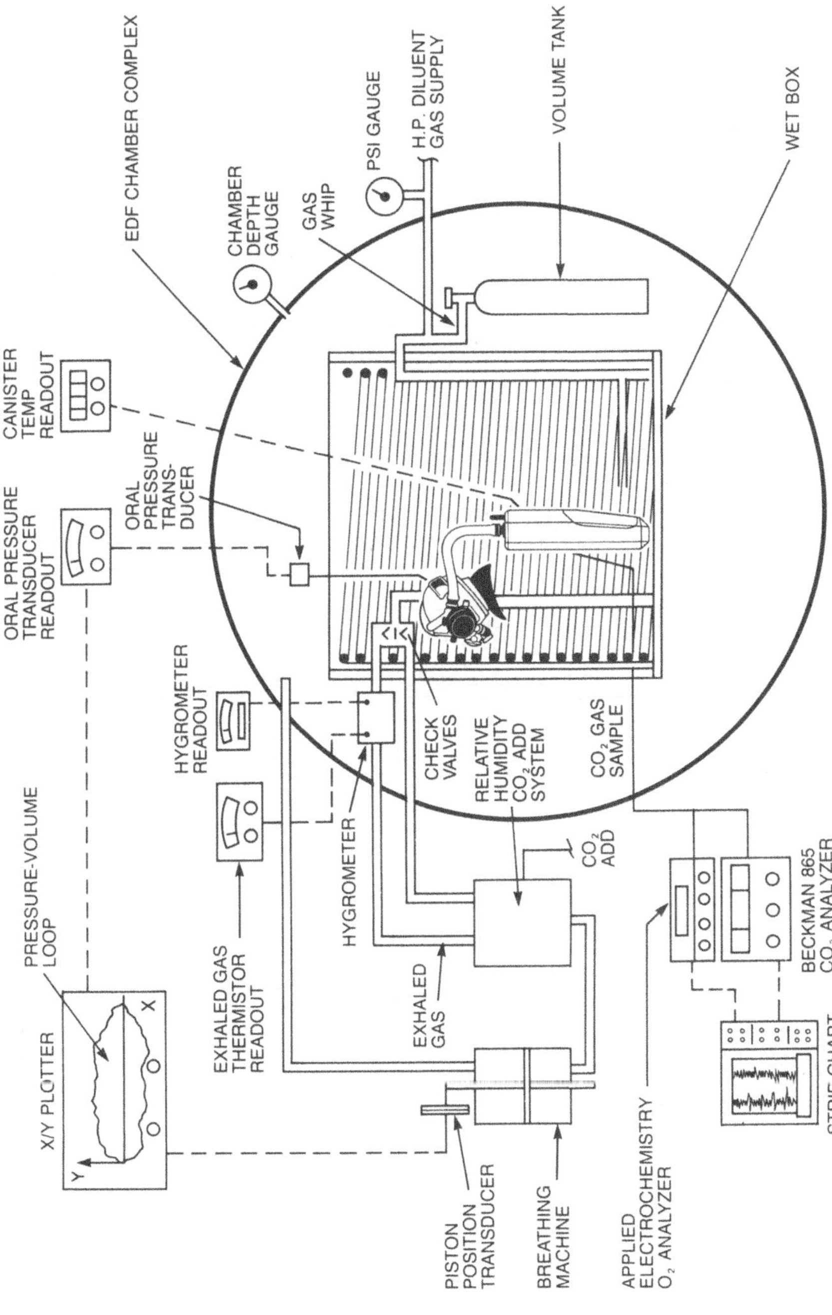

Fig. 3 Test set-up.

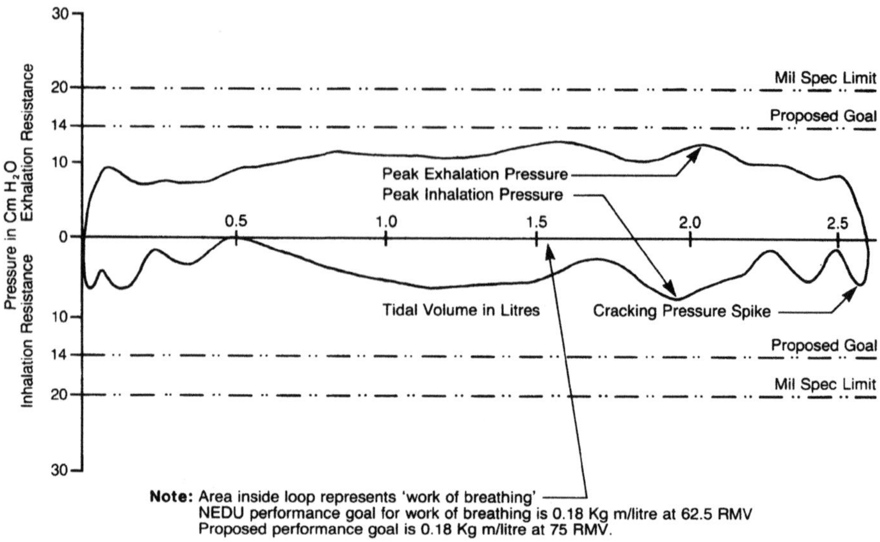

Fig. 4 Typical pressure: tidal volume loop (work of breathing, 0.15 kg m/l; breathing resistance, 24 cmH_2O).

					CATEGORY 2 1000 FSW HeO_2 (132 FSW–AIR)	CATEGORIES 3&5 1500 FSW HeO_2 (200 FSW–AIR)		CATEGORY 4 1500 FSW HeO_2 (200 FSW–AIR)	
LPM	RMV LPM	TV ℓ	FREQ	PEAK FLOW RATE LPS	WORK/ kg-m/ℓ	ΔP** CmH_2O	WORK/ kg-m/ℓ	ΔP** CmH_2O	WORK/ kg-m/ℓ
0.90	22.5	1.5	15	1.18		1.5	0.02	1.5	0.02
1.60	40.0	2.0	20	2.09		4.0	0.06	4.0	0.06
2.50	62.5	2.5	25	3.27	0.18*	10.0	0.15	10.0	0.15
3.00	75.0	2.5	30	3.93		14.0	0.22	14.0	0.22
3.60	90.0	3.0	30	4.71		20***	0.32	20***	0.32

Fig. 5 1981 NEDU performance goals.

D of E/DNV

Figure 6 shows D of E and DNV with two levels of acceptance. A 'recommended' line which is close to NEDU's criteria and an 'acceptable' limit which allows an extremely high work of breathing at the higher RMVs. Frankly, these upper acceptance limits are simply unachievable by divers.

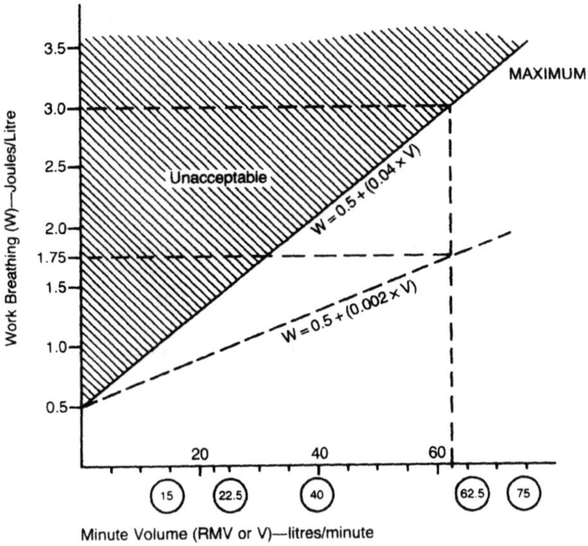

Fig. 6 D of E and DNV with two levels of acceptance.

Results of NEDU Testing from NEDU Report 11-86

Figure 7 shows the results of testing of a DSI Superlite 17B using HeO_2 mix to 1000 FSW (305 MSW). From an inspection of the graphs, we immediately see that deeper than 250 FSW (75 MSW), inhalation resistance is increasing significantly and this is reflected in work of breathing results exceeding NEDU's 62.5 RMV goal. In D of E or DNV terms, this would appear in the acceptable (not the recommended) areas until about 850 FSW, when it would be unacceptable.

Looking at the 1000 FSW loop for 62.5 RMV (Fig. 8), it can be seen how inhalation resistance is far beyond a diver's capability.

The Ultraflow 500 Demand Regulator shows another picture (Fig. 9). Inhalation is superb all the way to 1000 FSW (350 MSW), even at 75 and 90 RMV. Again this reflects in excellent work of breathing results to 1000 FSW.

Looking at the breathing loops (Figs. 10 and 11), again excellent results for breathing work and breathing resistance can be seen.

It is therefore not surprising that the US Navy is now procuring Ultraflow 500 Helmets and Bandmasks for their saturation diving programme. Indeed, their evaluation has just been completed of the Helmets and Bandmasks for Open Circuit Air and HeO_2 surface supplied diving.

Just as significant, however, is that for the first time an umbilical supplied demand regulator has been shown to be capable of matching the performance of free flow helmets and new 75 RMV performance goals have been recommended in the US Navy. Technology has caught up with what is physiologically desirable. To quote NEDU Report 11-86, 'a diver's ability to perform useful work need no longer be limited by the performance of his equipment'.

Figure 12 shows NEDU's proposed new performance goals for umbilical supplied demand UBA.

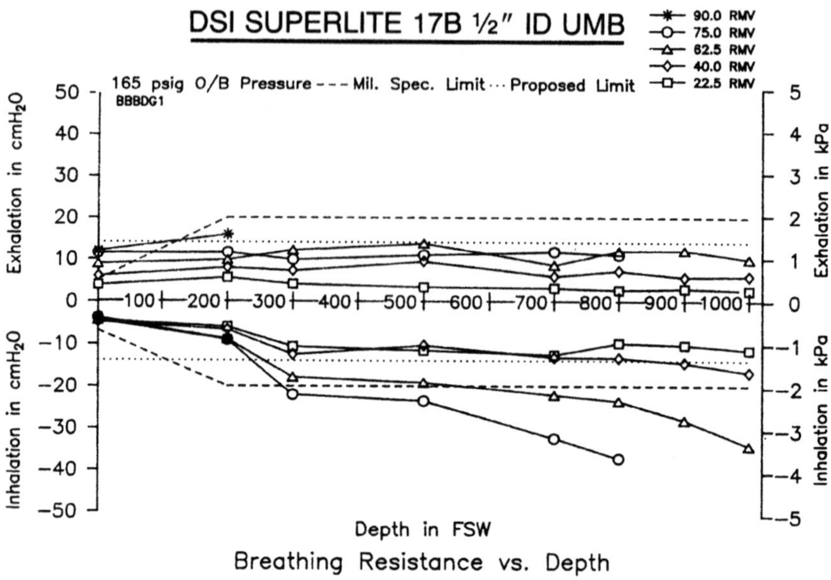

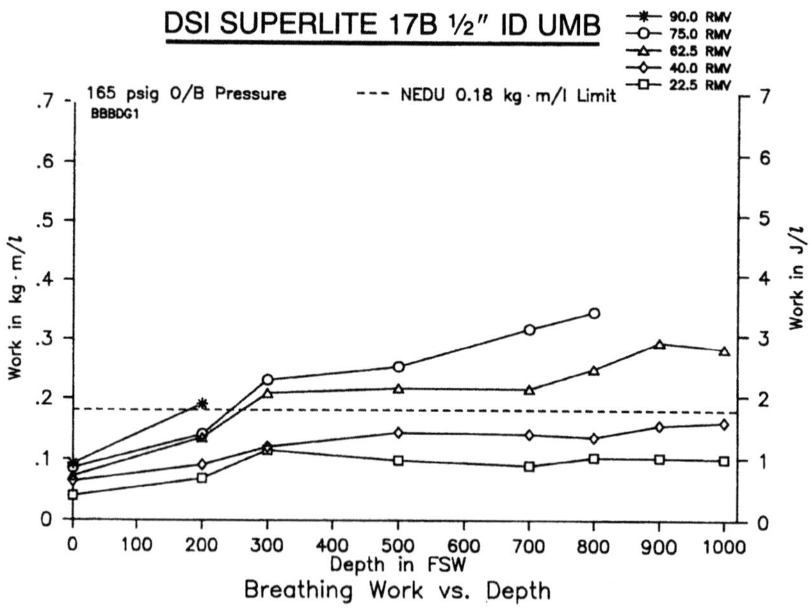

Fig. 7 Results of testing a DSI Superlite 17B using HeO_2 to 1000 FSW.

While this was going on at NEDU, Dr Clarke at NAMRI was preparing a paper to recommend new UBA performance criteria. With increasing depth, divers find breathing harder, even in the chamber complex, due to the increased gas density, reduced gas viscosity and the effect of pressure on them. So with increasing depth they have less capacity for external work of breathing. From a

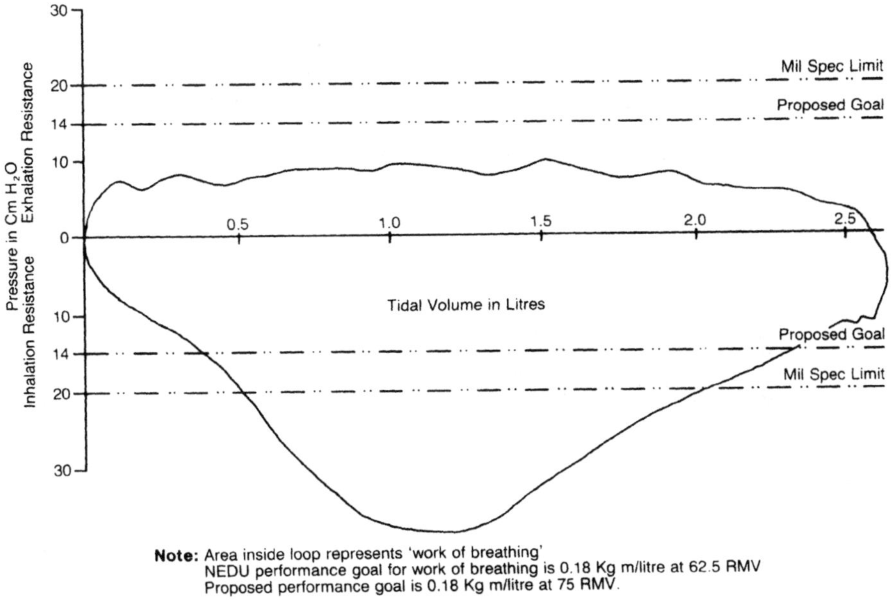

Fig. 8 DSI Superflow: $\frac{1}{2}$-inch I/D umbilical, 165 psi O/B, 1000 FSW, 62.5 RMV (work of breathing, 0.30 kg m/l; breathing resistance, 49 cmH_2O).

review of manned data available—and there is not a lot of it—it seems that a goal of 14 cmH_2O resistance would ensure that divers would avoid any CO_2 retention. Up to 20 cmH_2O resistance would result in some CO_2 retention with some possible dyspnoea in some subjects. More than 20 cmH_2O and up to 30 cmH_2O was feasible but unacceptable. Accordingly NAMRI recommended a desirable 14 cmH_2O limit with 20 cmH_2O resistance as an acceptable maximum.

This agrees nicely with NEDU's recommendations.

WHY IS THE ULTRAFLOW 500 SO GOOD?

Conventional demand regulators (Fig. 13) are balanced during pre-dive set up, so that the combined spring forces P_2 and P_3 equal the overbottom gas supply force P_1. During inhalation the diver's suction on the diaphragm opens the inlet valves against P_2 and P_3. As gas flows to the diver, the pressure in the diver's umbilical falls so that the force P_1 is reduced. The greater the demand for breathing gas, the greater the umbilical pressure drop will be and so P_1 will be even more reduced.

As P_1 reduces and P_2 and P_3 are relatively increased, the diver has to perform increased inhalation work to keep the inlet valve open.

Typically, conventional regulators fail at RMVs over 62.5 (moderately heavy work) at depth. Hence the current NEDU performance goal being limited to 62.5 RMV—however undesirable this is.

Looking now at the Ultraflow 500 (Fig. 14) we see a different arrangement. The difference is in the inlet valve assembly. P_3 is a very light spring opposing

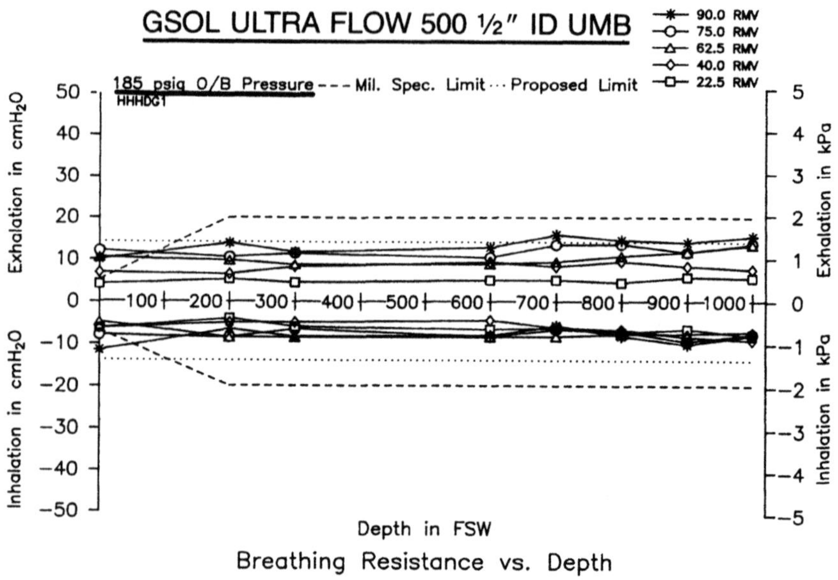

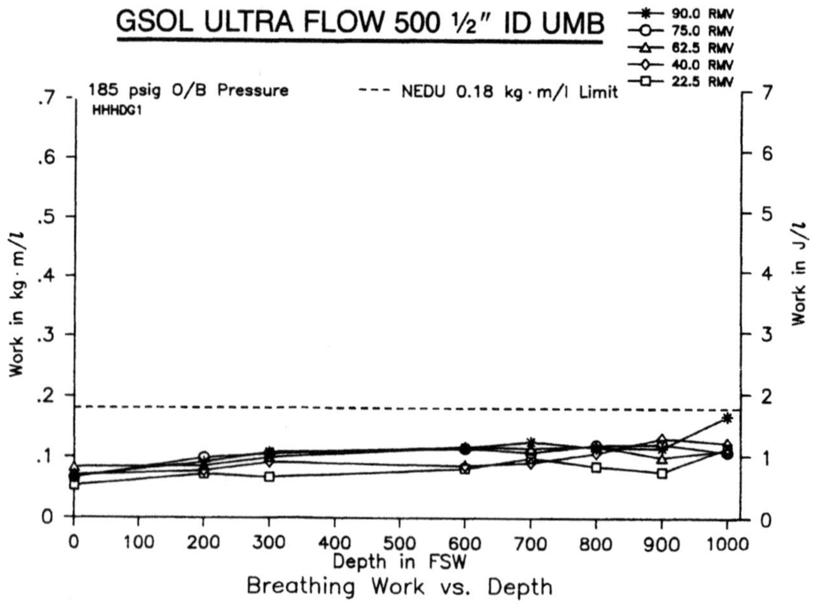

Fig. 9 Ultraflow 500 Demand Regulator.

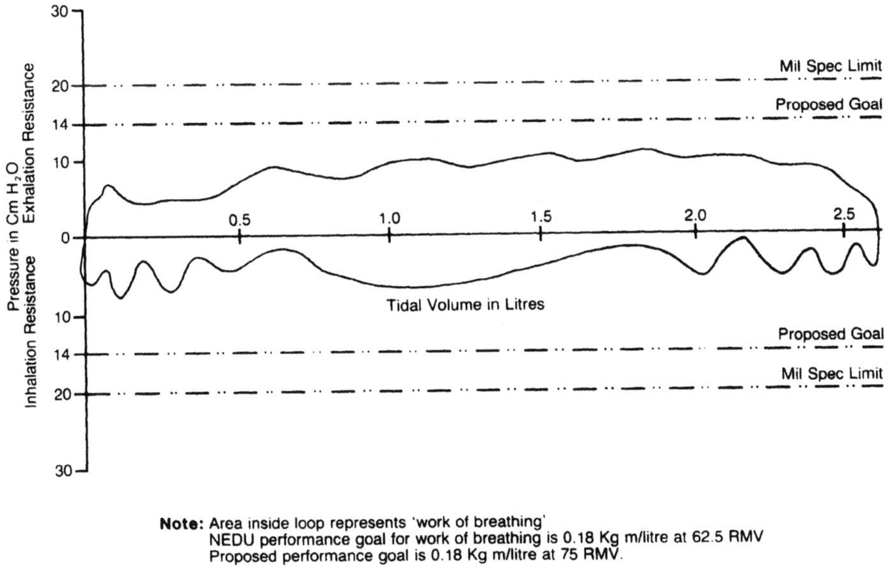

Fig. 10 GSOL Ultraflow 500: $\frac{1}{2}$ I/D umbilical, 185 psi O/B, 1000 FSW, 62.5 RMV (work of breathing, 0.12 kg m/l; breathing resistance, 19 cmH_2O).

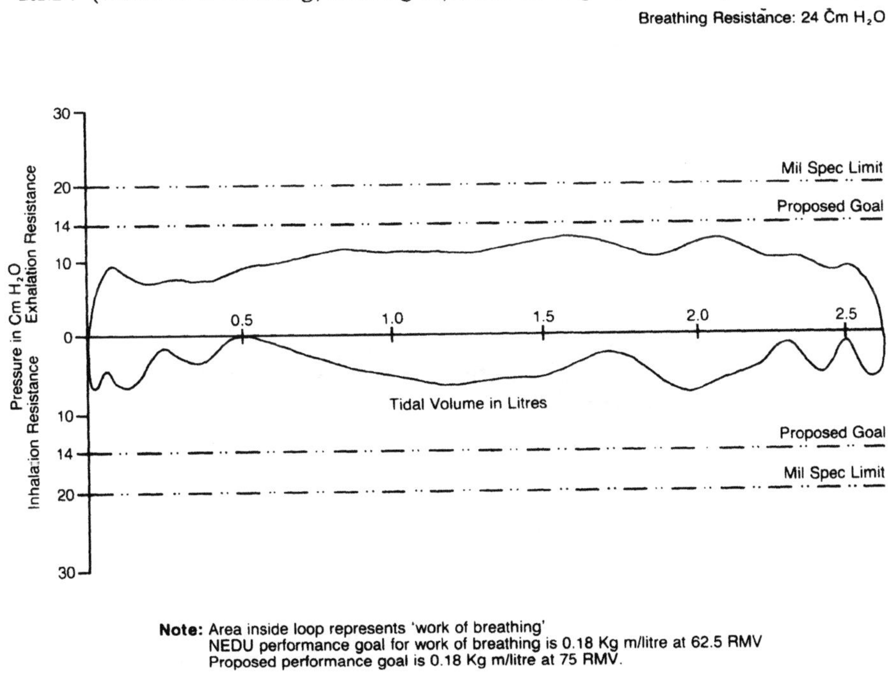

Fig. 11 GSOL Ultraflow 500: $\frac{1}{2}$-inch I/D umbilical, 185 psi O/B, 1000 FSW, 75 RMV (work of breathing, 0.15 kg m/l; breathing resistance, 24 cmH_2O).

					CATEGORY 2 190 FSW (TO MAXIMUM OPERATING DEPTH)	
VO_2 LPM	RMV LPM	TV ℓ	FREQ	PEAK FLOW RATE LPS	ΔP** CmH_2O	WORK/ kg-m/ℓ
0.90	22.5	1.5	15	1.18	1.5	+
1.60	40.0	2.0	20	2.09	4	+
2.50	62.5	2.5	25	3.27	10	0.18
3.00	75.0	2.5	30	3.93	14	0.18
3.60	90.0	3.0	30	4.71	20	+

Fig. 12 Proposed performance goals.

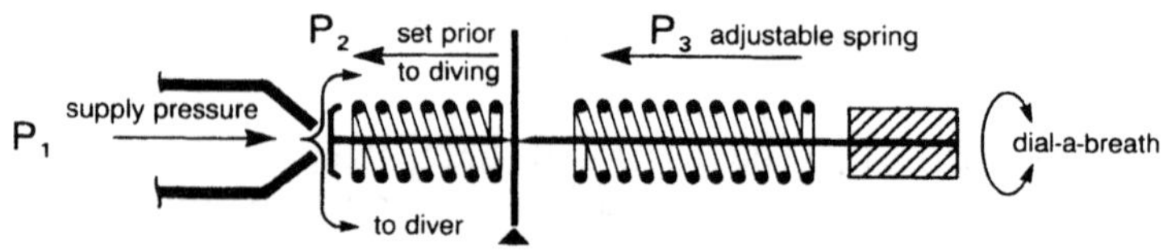

Fig. 13 Conventional demand regulator (e.g. KMB TO, Superflow, USN Mk 1 MOD O/S, Ultraflow 350).

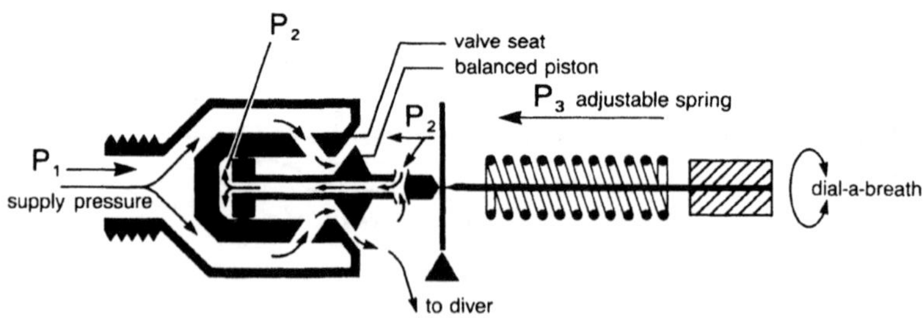

Fig. 14 Ultraflow 500 'balanced' demand regulator.

the inlet supply force P_1, which is allowed to equalize through and around the valve stem and seat so that P_2 and P_1 are equal. Theoretically, the valve seat needs only a light P_3 force to close the valve at the end of inhalation. In practice the 'balanced' valve is designed to be slightly out of balance so that in the event of regulator failure it would fail to open.

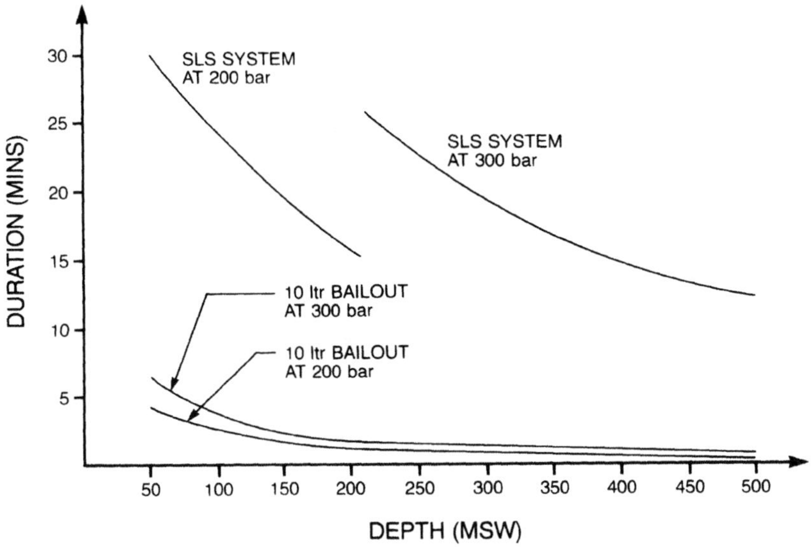

Fig. 15 Duration comparison graph for open circuit bail-out versus SLS system (RMV = 75 LPM).

During inhalation the umbilical supply pressure always drops as gas flows to the diver. But now as this happens P_2 also drops as the piston balances to the new supply force. This means that the Ultraflow 500 is very tolerant of changes in supply pressure and this means that greater RMVs become possible, and this is the justification for NEDU's new performance goals.

SECONDARY LIFE SUPPORT SYSTEM (SLS)

A major problem has always existed for divers working deeper than, say, 50 MSW. This has been the provision of a bail out with a suitable duration to allow them to return safely to the bell, should their normal supply be interrupted for any reason.

Until now, the normal practice has been to use an open circuit Scuba cylinder connected to the helmet regulator via the side block assembly.

Figure 15 shows that even if a 10 litre cylinder were charged to 300 bar, the bail-out duration would only be about $\frac{1}{2}$ minute at 500 MSW, or $1\frac{1}{2}$ minutes at 250 MSW. At 50 MSW less than $6\frac{1}{2}$ minutes is achieved.

The second problem that arises with an open-circuit bail-out cylinder is providing gas at a safe minimum inspired temperature when hot water is lost at the same time that the main gas supply is interrupted. Gas heating is essential at deeper depths if thermal shock to the diver is to be avoided.

Figure 16 shows the Gas Service's SLS, which provides a minimum 15 minutes' duration at 500 MSW or 30 minutes at 50 MSW. Gas is maintained at a safe minimum inspired temperature, even with total umbilical severance. No longer is there any excuse for failing to provide an adequate bail-out facility for divers' work in an already hostile environment.

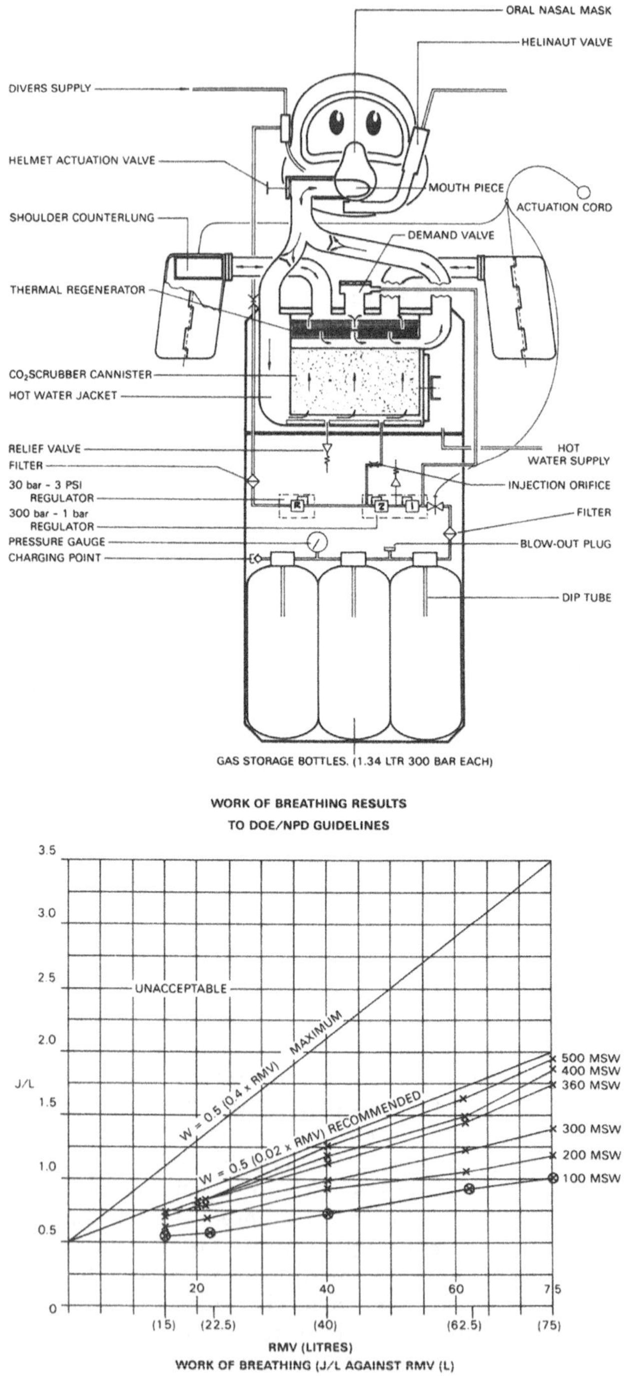

Fig. 16 Schematic of Gas Service's SLS.

12

Monitoring of Bacterial Status and Antibacterial Cleaning of Saturation Diving Systems

S. T. Brooke, P. R. Blewett and C. M. Childs,
Wharton Williams, UK

Infection of the outer ear (otitis externa) is a common problem during saturation diving. The moist, warm environment of high-pressure heliox facilitates the growth of the gram-negative bacillus *Pseudomonas aeruginosa*. Symptoms of itching, burning, pain and discharge can become intense and result in the diver being unable to work (Farmer, 1982). If, as a result, divers have to leave saturation, this causes inconvenience and loss of earnings to them and disruption of the work schedule they were undertaking.

Normal methods of prevention of infections include cleaning of the interior surfaces and furnishings of the chambers regularly, careful attention to hygiene by divers, including cleaning of personal diving equipment, and avoidance, if possible, of the introduction into the chamber of divers with existing infections. At present, some screening of divers is carried out by culture of ear swabs, but divers are not routinely monitored for asymptomatic carriage of likely pathogenic organisms.

To investigate whether the current techniques of chamber and equipment cleaning were adequate, a study was conducted during a routine commercial diving operation. This study examined the effectiveness of three different disinfectants (Hycolin, Tego and Panacide) in reducing detectable contamination of five areas inside the saturation chambers, and tested for the appearance in two body sites of likely pathogenic bacteria on the divers. External ear bacteriological swabs and swabs from predetermined sites within the chambers were taken and cultured every 48 hours. Faecal samples were cultured pre-dive, and rectal swabs were taken and cultured once a week throughout the diving operation. Facilities were available on the diving-support vessel for culture and screening of culture plates for likely pathogens, and positive cultures were put into transport medium for later laboratory analysis in Aberdeen.

MATERIALS AND METHODS

During the operation by Wharton Williams (diving contractor) to recover the remaining gold bullion from H.M.S. *Edinburgh* in the Barents Sea between

Advances in Underwater Technology, Ocean Science and Offshore Engineering, Volume 14: Submersible Technology

August 30 and September 24 1986, bacteriological monitoring of the dive was carried out. The operation was chosen as suitable for the study because the site of diving was remote from normal laboratory resources in the event that otitis externa became troublesome, and the depth and possible duration of the operation made this quite likely. The work was carried out from the diving-support vessel *Deepwater II*.

SATURATION SYSTEM

Three interconnected chambers on the diving-support vessel served as the living quarters for three teams of three divers during the operation (Fig. 1). The divers spent 25 days under pressure in the chambers in a heliox atmosphere. This was composed of 98.2% helium with 1.8% oxygen, a partial pressure of 0.41 bar oxygen being maintained throughout the operation. A pressure of 0.8 bar nitrogen remained in the chamber gas from the initial compression of the chamber system. Carbon dioxide was constantly removed from the chamber atmosphere by passage over soda lime. The saturation system on the diving-support vessel was maintained after initial compression at a pressure equiva-

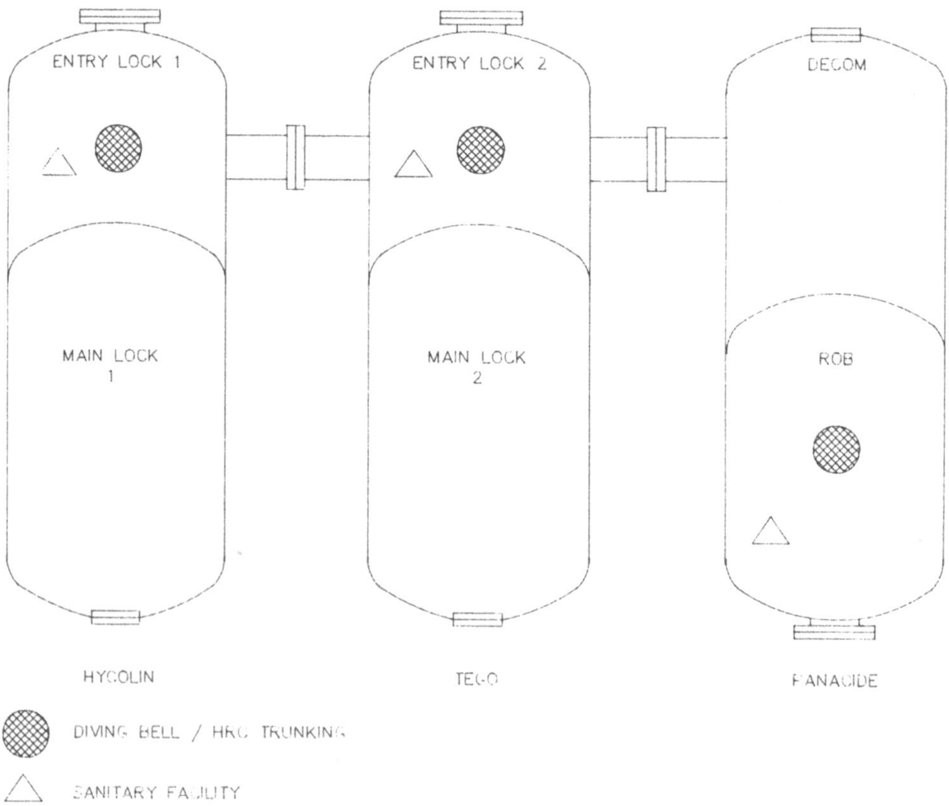

Fig. 1 Diving saturation system.

lent to 230 MSW, and the depth to which the divers descended to work varied between 230 and 245 MSW.

A diving bell, for transport of the divers from the surface chambers to their worksite at depth, could be attached to two of the three surface chambers to allow entry of the divers, resulting in more diver movement and the production of a more humid environment because of water brought in by the divers and their equipment.

Prior to blowdown, swabs were taken from all chamber sites and from the divers, who were informed of the importance of the cleaning regimes and maintenance of the sampling schedule. Chamber sites included the showerhead, sink, bilge, and bulkhead. Each site in the chamber was marked to ensure the swabs would be taken from the same area. Ear samples were collected every 48 hours with sterile cotton-tipped swabs, and rectal swabs were taken once a week. Chamber-site samples were also taken every 48 hours with Panatest Dipslides and any growth rating more than 10^5 on the infection scale was sloped for on-shore analysis.

Rectal and ear samples were plated out onto Blood Agar, McKonkey's, Sabaraud and Pseudomonas plates. The inoculated media were then incubated at 34 °C for 24 hours before further inspection. Gram stains were carried out on any pathological growth and the bacteria were put into a suitable transport medium of ThioBroth or an Agar slope. These were then analysed in more detail by the City Hospital, Aberdeen.

RESULTS

All chamber sites that were swabbed had a pre-dive infection level of zero. The pre-dive stool samples and ear swabs revealed the normal range of flora. None of the divers was carrying *Pseudomonas aeruginosa*. (Two divers failed to return stool samples.)

Ear, rectal, and chamber swabs throughout the dive contained the following range of flora: *Klebsiella* sp., *Candida albicans, Staphylococcus albus, Enterobacter aerogenes, Escherichia coli, Proteus* sp., *Enterobacter cloacae, Staphylococcus pyogenes, Streptococcus faecalis* and *Pseudomonas aeruginosa*. There were no major infections that prevented diving during the operation and no other medical problems other than two divers with minor scalds to the hands and legs caused by hot-water suits.

On day 1, one diver in Chamber 3 had *Pseudomonas aeruginosa* in his left ear. This was treated with polymyxin–gentamicin drops, and the normal cleaning regime with Panacide continued. There was no further incidence of *Pseudomonas aeruginosa* either on the divers or any of the chamber sites in Chamber 1 for the remainder of the saturation.

Pseudomonas aeruginosa occurred in Chambers 1 and 2 during the first half of the dive. In Chamber 2 it appeared only on chamber sites showerhead, sink and bilge but not on the divers. In Chamber 1, one diver had *Pseudomonas aeruginosa* in the left ear for six days and it also appeared in the showerhead, sink and bilge. On day 2 of the dive the showerheads and pipes were renewed in Chambers 1 and 2, as these appeared to be a possible source of infection. The pipe that ran from the main water supply to the showerhead and toilet was also replaced. This was a pipe common to both appliances and it was thought that it may have been possible for bacteria to have passed from the toilet up to the

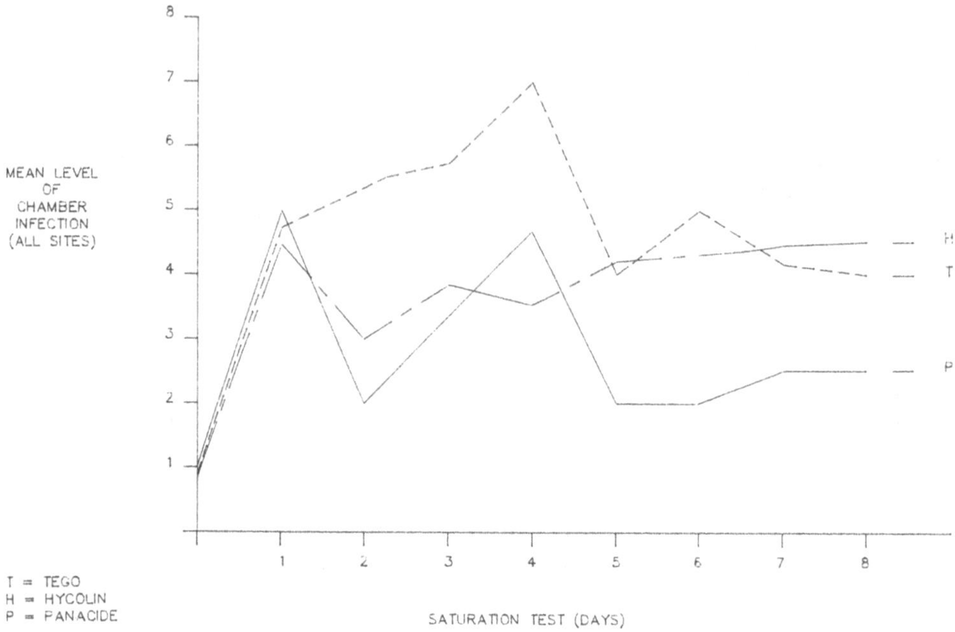

Fig. 2 Level of chamber infection.

showerhead. It was also possible that bacteria had remained in the moist environment of the pipes between operations and had infected the incoming divers.

The means for the level of chamber infection across all sites are presented in Fig. 2. The overall level of infection between chambers was not significant ($F > 1$). Although there is no significant difference, the means appear to show that Panacide maintains the lowest level of infection throughout the dive.

There was a significant effect for days on the level of infection ($F(2,7) = 5.37$, $p < 0.0001$). The level of infection therefore differed significantly across days, and the means in Fig. 2 show that a peak occurred at the midpoint of the dive. There was a significant interaction between days and disinfectant ($F(7,14) = 3.52$, $p < 0.0001$). This shows that the change of level of infection over days is different for each disinfectant. It was decided not to carry out univariate follow-up analyses as the adjusted Huynh–Feldt value ($p < 0.5911$) was relatively low and there was a risk of inflating the Type 1 error rate.

LIMITATIONS

The ecological validity of the data collected is strong due to its applied setting. It should, however, be noted, that in a field study such as this, the reliability of the data is limited by the loss of control concomitant with field research. The LSTs monitored the cleaning regimes and swabbing schedules as rigorously as possible. Diver fatigue, however, can result in the reduction of complete thoroughness in such a study. Diver movement between chambers, the common use of diving bell and equipment, and the common use of the showers in Chambers 1 and 2 by all three teams of divers, are possible sources of cross-infection.

The data should therefore be examined with these limitations in mind. It should also be remembered that the data have strong validity and are more likely to be representative of what actually happens in operational diving than are the results of an in-vitro study.

DISCUSSION

The relatively low incidence of infection throughout the dive perhaps reflects the benefits of maintaining strict cleaning regimes and having an on-board laboratory to monitor the level of infection and prescribe treatment before symptoms become troublesome. The results show that it might have been possible that bacteria were incubating in the water pipes serving the shower and toilet in Chambers 1 and 2 and that *Pseudomonas aeruginosa*, which was not present on any of the divers pre-dive, originated there rather than having been brought into the chamber from outside.

The significant interaction of days and disinfectant shows that at certain times Panacide was the most effective cleaning agent in terms of overall chamber cleanliness. The level of infection peaked mid-way through the dive and the patterns from Fig. 2 reveal that levels continued to rise for Tego while those for Hycolin and Panacide dropped off.

After an initial incidence of *Pseudomonas aeruginosa* in Chamber 3 there were no further occurrences of the pathogenic bacteria, thus lending further evidence to the efficacy of Panacide. It should be recalled that there was no shower or bell trunking in Chamber 3 and there was therefore a lower level of dampness and a reduced risk of infection in this chamber.

CONCLUSION

The absence of any debilitating infections resulting in loss of diver time supports the existence of an on-board laboratory constantly monitoring bacteriological growth. It appears that Panacide was the most effective disinfectant in the containment of both personal and chamber infection levels. A controlled laboratory study replicating this field research would increase the reliability of the results reported here.

ACKNOWLEDGEMENT

The authors acknowledge the assistance given by Dr T. M. S. Reid and the staff of the Regional Laboratory, City Hospital, Aberdeen.

REFERENCE

Farmer, J. C. Otologic and paranasal sinus problems in diving. *In* P. B. Bennett and D. H. Elliott (Eds), *The Physiology and Medicine of Diving*, Baillière Tindall, London, 1982.

13

Hydra 8: Pre-commercial Hydrogen Diving Project

J. P. Imbert, C. Gortan, X. Fructus, T. Ciesielski and B. Gardette,
Comex Services, France

Deep divers are exposed to three types of environmental stresses:

- the high-pressure nervous syndrome (HPNS), which is related to the effects of the hydrostatic pressure on the central nervous system. These appear at around 200 m and increase with depth, causing motor disturbances that may impair divers' efficiency.
- stress caused by the density of the breathing mixture, which increases with depth. Breathing dense gas requires increased ventilatory efforts that may reduce divers' work capacity.
- the fatigue of long saturation periods (confinement, thermal stress, bad sleep, lack of appetite, etc.) that is generally related to body-weight losses.

To overcome these environmental stresses, scientists have used the properties of the different constituents of the breathing mixtures.

The first deep dives were conducted using helium as a diluent. Helium has a low molecular weight but no specific anti-HPNS properties. The only way of controlling HPNS was to slow down the compression. In 1972, helium permitted 610 m to be reached during the Comex *Physalie VI* dive.

In 1975, Duke University introduced the concept of the 'pressure-reversal effect' (Simon *et al.*, 1975). A significant reduction in HPNS was obtained by adding to the helium/oxygen mixture a given amount of a narcotic gas, in that case nitrogen. The new mixture was called 'trimix', but we would rather call it 'nitrogen trimix' to be more specific.

The possibilities of nitrogen trimix were further investigated by different research centres. The two milestones of these experiments were the Comex *Janus IV* open-sea dive to 500 m, in 1977, and the Duke University *Atlantis III* simulated dive to 685 m, in 1983.

However, the benefit of using nitrogen to reduce HPNS is counterbalanced by a significant increase in gas density (nitrogen is seven times heavier than helium). During the *Janus IV* dive, the breathing gas, which contained 5% nitrogen, had a density of 10.5 g/litre. It would have been 8.2 g/litre with heliox (22% less). For this reason, a large amount of nitrogen cannot be used without affecting the diver's ventilatory function.

Advances in Underwater Technology, Ocean Science and Offshore Engineering, Volume 14: Submersible Technology

COMEX *HYDRA* PROGRAMME

The idea underlying the *Hydra* programme was to investigate a new diving gas that would be both light and narcotic. The only possibility left in the Mendeleev table was hydrogen!

Since 1983, four onshore dives using hydrogen have been conducted at Marseille Hyperbaric Centre in collaboration with the US Navy (NMRI), the Swedish Navy (NMD), the French Navy (CERB, CEPISMER, GISMER, CERTSM) and the Centre National de la Recherche Scientifique (CNRS).

The final aim of this programme is to demonstrate the operational feasibility of hydrogen diving with a deep, open-sea dive. The code name of this dive will be *Hydra 8*.

This paper is intended to present the studies and the developments that have permitted Comex to step from the onshore experimental phase to the offshore operational phase of deep hydrogen diving.

ANIMAL STUDIES

Although hydrogen is a very active chemical element, previous experiments with monkeys had shown that hydrogen could be considered as an inert gas from a biological point of view (Brauer *et al.*, 1966; Rostain, 1973).

Comex carried out complementary studies to confirm that hydrogen is a non-toxic molecular gas. In a first series of experiments, 40 mice were exposed for 48 hours at 600 m, breathing hydrogen/oxygen. Comparison with a reference group exposed to helium under the same conditions showed neither behavioural change nor any histological sign of toxicity.

Further experiments were then performed at extreme depths to determine any possible limit to hydrogen use. A group of 120 mice was exposed to depths ranging from 1300 to 2000 m, either with helium or hydrogen. The survival rate at 1800 and 2000 m was much higher with hydrogen than with helium, which tends to prove that hydrogen does not develop any toxic effect up to these depths (Gardette, 1986).

Depth (m)	Helium	Hydrogen
1800	60%	0%
2000	100%	60%

Fig. 1 Fatality rate among mice exposed to extreme depths, breathing either helium or hydrogen mixtures.

MANNED HYDROGEN DIVES

Following the animal studies, hydrogen diving appeared safe and practical in the depth-range that covers human intervention. In June '83, 18 persons, including Mr H. G. Delauze, chairman of the company, performed short bounce-dives to 75–90 m offshore Marseille, with hydrogen as bottom mix (*Hydra 3*). As no incident was recorded (Fructus, 1983), the programme proceeded with deep saturation diving.

Dive	Depth (m)	Method	Studies
Hydra 3	75 90	Hydrogen bounce dives	First observations on man
Hydra 4	120 300	Heliox saturation with short exposures to hydrogen (24.5 bar max.)	quantification of the narcotic potency of hydrogen in man; divers' ergonomy in dry and wet conditions
Hydra 5	450	Hydrogen trimix saturation (25 bar)	pressure-reversal effect; divers' ergonomy in dry and wet conditions; switch from hydrogen to helium; hydrogen saturation decompression
Hydra 6	520	Hydrogen trimix saturation (25 bar)	validation of procedures for *Hydra 8* offshore dive; divers' work capacity in water; selection of divers
Hydra 7	180 260	Hydrogen saturation (25 bar max.)	selection of divers; narcotic potency of hydrogen

Fig. 2 Summary of the Comex *Hydra* programme

Hydra 4

The primary objective was to quantify the narcotic potency of hydrogen in man. The dive was a heliox saturation during which 6 divers were exposed to short periods (1 to 6 hours) of hydrogen breathing in a specially designed dome. The

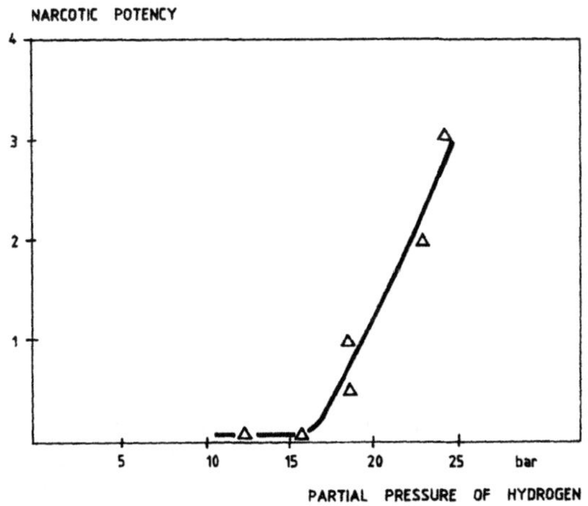

Fig. 3 Quantification of the narcotic potency of hydrogen in man according to the scale developed by Fructus (1983), during the *Hydra 4* dive.

depth of exposures progressively increased from 120 to 300 m. The narcosis appeared at 180 m in some divers and definitely affected all divers at 240 m, i.e. 24.5 bar of hydrogen (Carlioz *et al.*, 1984; Fructus *et al.*, 1984; Gardette *et al.*, 1984).

It thus became apparent that hydrogen was too narcotic to be used as a dilutant and that helium would have to be added to the hydrogen/oxygen mixture for further deeper experiments. Hydrogen diving thus became a new version of trimix diving, which we would like to call 'hydrogen trimix'.

This experiment also showed a lot of variation in individuals' susceptibility to hydrogen narcosis and the need was recognized to select divers with a large saturation experience, who tend to be less sensitive.

Hydra 5

The main objective was to reach a significant depth, 450 m, in order to verify the anti-HPNS effect of hydrogen. The dive was a full-scale hydrogen trimix saturation and 25 bar of hydrogen were injected into the chamber during the compression. Effectively, none of the symptoms of HPNS (tremor, dysmetria), or articular pain were observed. Divers did not present the characteristic muscle contractions and restlessness, and felt more relaxed and comfortable. Because of the low gas density, divers could breathe through the nose and did not complain of the 'stuffed nose' feeling as in other deep experiments. For the same reason perhaps, they had better appetite and minimized their weight losses after saturation (Gortan and Delauze, 1986).

The second objective was to study a procedure that would permit divers to switch from hydrogen to helium at the end of the bottom period, thus allowing heliox decompression. Although the inert-gas switch was progressive, using an intermediate gas mixture for 8 hours, divers entering the helium atmosphere experienced a sudden rise in circulating bubbles. The contingency procedures were immediately applied and the divers were recompressed 20 m with hyperbaric oxygen on mask, which stopped the bubble formation.

Similar inert-gas switches had been performed during the *Hydra 4* dive and had caused no problem of isobaric counter-diffusion. It was thus concluded that switching from hydrogen to helium could be dangerous and the maximum permitted hydrogen pressure drop was set to 13 bar in helium. For deep dives involving more than 13 bar of hydrogen, final decompressions would have to be performed, with a progressive replacement of hydrogen by helium inside the chambers.

Hydra 6

This dive was meant to be the onshore validation of the planned open-sea dive to 520 m. Its objective was to demonstrate divers' work capacity during simulated dives in cold water. Divers wearing full equipment were asked to perform increasing work loads on an ergonometer. The respiratory-function tests indicated that divers never experienced ventilatory limitation for graduated work up to 150 W. Divers also carried out specific tasks such as pipe-puzzle assembly, and their performances appeared to be significantly better with hydrogen trimix than with nitrogen trimix (Fructus *et al.*, 1987).

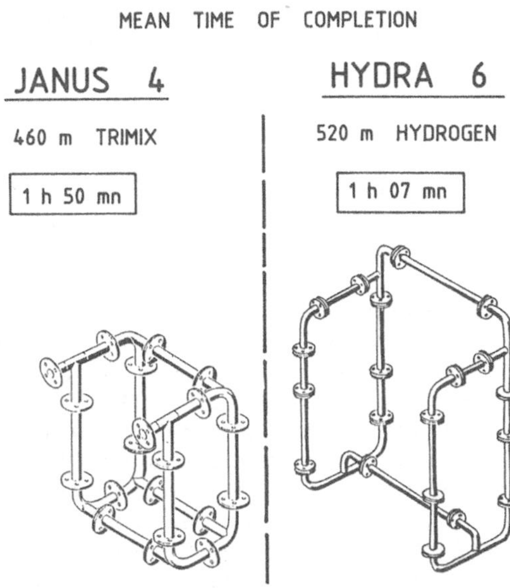

Fig. 4 Comparison of the time required to carry out the pipe-puzzle tests during *Janus IV* experiment on nitrogen trimix and during *Hydra 6* experiment on hydrogen trimix at 450–500 m.

Hydra 7

This next experiment was designed to define a method of selection of divers. During *Hydra 6*, one of the divers experienced psychological problems of a psychotic type. Medical investigation of his case concluded that the problems could be explained by a lack of saturation-diving experience. Four highly experienced saturation divers were thus exposed to the same experimental protocol as during *Hydra 4*, i.e. increasing depths from 180 to 260 m, but were this time fully saturated with hydrogen. No sign of narcosis could be measured in the divers, which tends to indicate that the narcotic potency of hydrogen depends on whether the diver is abruptly switched from helium to hydrogen or progressively exposed to increasing hydrogen pressures.

Hydra 8 Offshore Dive

This dive has been designed to demonstrate the feasibility of hydrogen diving during a full-scale deep offshore dive.

This operation is planned for the 1987/88 winter in the Mediterranean. The underwater site will be located in a depth range of 500–520 m. Two teams of divers will carry out various tasks, such as handling, mechanical connection, cutting, etc. on a base platform lowered from the vessel. Each team will conduct one dive a day, during which two divers will perform a minimum of three hours of work in the water. The aim is to achieve at least 60 hours' work at the bottom in 5 days.

The vessel selected is the D.S.V. *Orelia*. The ship, which is under the British flag, is presently being upgraded for hydrogen diving. The modification of the

Orelia will be done under the supervision of British Department of Transport which has already called on the Department of Energy for assistance with the diving part of the project. The vessel and the diving equipment will be certified by the Lloyd's RS.

Equipment and procedures for Hydra 8

Hydrogen diving could only be considered after an extensive study of the risks relevant to the handling of hydrogen and its use in chamber systems initially designed for heliox diving.

Fire risk outside the chamber system

Fire risk is the obvious concern of hydrogen diving. The lower explosion limit (LEL) in air is 4% hydrogen at atmospheric pressure, and the maximum permitted hydrogen level has been set to 2% (50% of the LEL) for hydrogen-diving operations. The margin is thus extremely reduced.

Moreover, hydrogen is a fast-diffusing gas and tends to leak through cylinders, piping and chambers. The saturation system can be designed to minimize the leaks but cannot eliminate them all. Thus the risk must be prevented of any local formation of hydrogen pockets in the chamber hall.

The safety of hydrogen diving first relies on a system of hydrogen sensors fitted all over the hyperbaric complex to detect minute leaks of hydrogen. Then, a powerful complex to detect minute leaks of hydrogen. Then, a powerful system of ventilation must be installed to collect and evacuate the hydrogen leaks. This approach is sufficient for a well-controlled system in an onshore installation and has been retained for the Marseille Hyperbaric Centre.

For the planned offshore dive on the *Orelia*, the problem is more complex. Most of the diving equipment (gas storage, dive control, umbilical storage, life-support units, etc.) will be installed on the deck, in the open air. However, for the chamber system, which is located below the deck, no ventilation system could be defined that could cope with a system defect causing a massive leak or an auto-ignition of hydrogen. It was thus decided to inert the chamber room, and

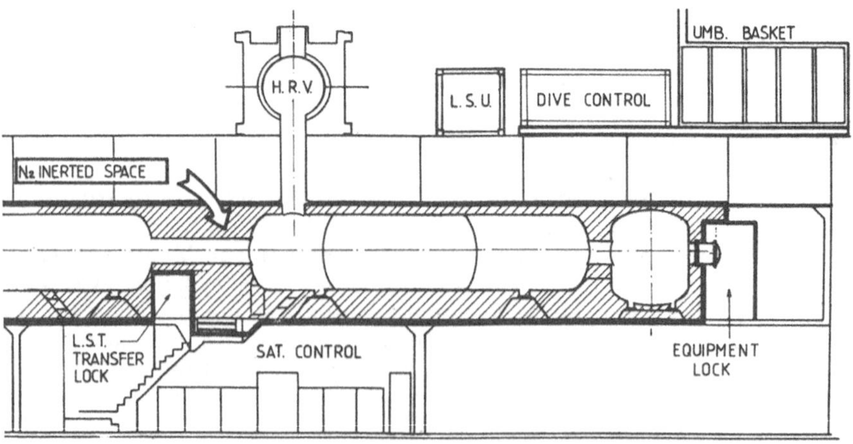

Fig. 5 *Orelia's* saturation system as modified for deep hydrogen diving.

liquid nitrogen will be stored on board for this purpose. The inerted system is inherently safer than a vented system in that both oxygen and hydrogen have to reach critical levels before ignition can take place. Detectors continuously monitor the gas levels and in the event of the flammable limit being approached, steps can be taken in good time to prevent such ignition. The method has also the advantage of being well documented in the literature of transportation of natural gas by ship.

Inerting will require a lot of modifications in the saturation system. Part of the chamber valves will have to be remotely controlled and a special cabinet will be installed around the main lock to allow its use from outside the inert area. For spot operations, or in case of an emergency, the LSTs will enter the chamber room through a special lock and 'dive' the inert area with smoke-diving apparatus.

Fire risk inside the chamber system

Under pressure, and for different gas concentrations, very little information was available in the literature on the flammability of Hydrogen. Comex had to carry out specific tests to determine what would be the safe limits of use of hydrogen/helium/oxygen mixtures at different depths.

Experiments were conducted in a small explosion chamber where different concentrations of the gases were ignited by high-energy welding electrode arcs. The lower explosion limit of hydrogen in heliox was found to range from 5.5% at 15 bar to 4.35% at 75 bar, for the concentrations relevant to hydrogen trimix diving (Gortan, 1983).

The procedures and equipment were thus designed to insure that at any time, at any location, there is no risk of forming a flammable gas mixture. The maximum permitted oxygen concentration was set to 2.5% (50% of LEL) inside the saturation system.

Using this value and considering a 500 mbar p_{O_2} as commonly used in saturation operations, it can be seen from Fig. 6 that hydrogen diving is safe and practical for depths in excess of 200 m (for instance, a typical bottom mix at 300

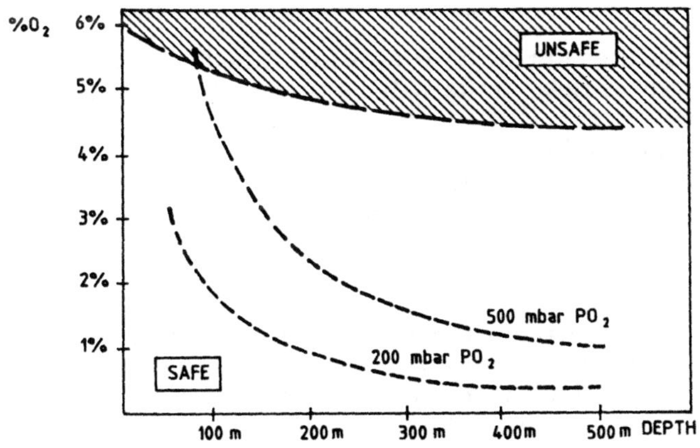

Fig. 6 Flammability threshold of hydrogen/helium/oxygen mixtures versus depth.

m contains only 2% oxygen) and the deeper the saturation depth, the lower the oxygen concentration and the safer the operation.

Piping, locks and trunking

The basic rule in adapting the *Orelia* saturation system to hydrogen diving was to limit to a minimum the number and the length of lines carrying hydrogen mixtures. Each of these circuits will be flushed with either pure nitrogen or heliox, before and after its use.

To prevent hydrogen/air contact, locks and bell trunking will be flushed with nitrogen prior to being opened and all the dumped gas will be collected and evacuated outside through a special venting circuit. The same precaution will be taken on pressurization of the locks or the bell trunking to avoid air being introduced in the chamber's atmosphere. The procedure will consist of a cycle of pressurization/decompression on heliox before pressure equalization to cut down the oxygen concentration. Similarly, during a dive, the hub of the bell will have to be flushed or flooded before opening of the lower door to prevent contact of a trapped air pocket with the bell atmosphere.

A special semi-integrated umbilical will be used to prevent hydrogen build-up in the inter-hose space. To avoid the use of pneumatic/electrical slippering ring, no umbilical winch will be used, and the bell umbilical will be handled with a block and stored in a basket on deck.

Life support units

The main difficulty in atmosphere control comes from the need to inject pure oxygen to compensate for leaks and divers' metabolic consumption.

For the chambers, Comex designed and tested a special life-support unit adapted to hydrogen diving. The unit has a high-power control of temperature and humidity, as required for deep diving operations, but its key part is the oxygen addition system. Pure oxygen is effectively injected into the gas stream, but a series of safety devices are installed for the following reasons:

- to ensure good oxygen mixing;
- to prevent any direct connection between oxygen supply and chamber atmosphere;
- to shut off the system immediately in case of an increase in temperature.

The system was used during *Hydra 5, 6* and 7, and proved to be very safe and efficient, and two of these units will be installed on the *Orelia* saturation-chamber system.

For the bell atmosphere, the oxygen level will be simply controlled by ventilation with bottom mix. The size of the bell is such that with an initial 900 mbar p_{O_2}, there will be sufficient oxygen inside the bell for three men for 30 hours.

Hydrogen removal unit

The design of this unit is related to the problem of decompression. On one hand, abrupt switching from hydrogen to helium cannot be considered because of isobaric counter-diffusion effects, and divers need to start their decompression on hydrogen. On the other hand, hydrogen must be removed before the chamber reaches 200 m for safety reasons.

During the *Hydra 5* dive, one team of divers was decompressed with progressive ventilation of the chamber with helium. This procedure proved to be efficient but was rather costly.

Comex thus developed a special unit to remove selectively the chamber hydrogen during the first phase of the decompression. A stream of chamber gas is directed over a catalyser that combines hydrogen and oxygen into water. By controlling the gas flow rate, one can control the amount of hydrogen removed and thus the chamber ascent rate. The unit was used during *Hydra 6* and thus became the first chemical decompression ever performed. It will be redesigned to be adapted to operational use and installed on the *Orelia*.

Divers' equipment

Divers' equipment was tested during the *Hydra 6* dive to 520 m. Divers' thermal protection did not differ from conventional deep-diving equipment, except that a new Comex Pro silicone hot-water suit was tested during the dives.

However, for the underwater breathing apparatus (UBA), special developments were required. The UBA must function in a gas-reclaim mode to prevent hydrogen bubbles coming up to the surface and accumulating in the moon-pool of the vessel. Moreover, the UBA should not work in closed circuit, to avoid the problem of pure oxygen injection.

Two types of equipment were tested—the Comex Pro X-lite helmet associated with the BOS bail-out system, and the Lara semi-closed circuit system, which were evaluated during normal and emergency diving situations. Both systems performed well and received good comments from the divers, but the Comex Pro system is likely to be selected because of its compatibility with the gas-reclaim system already installed on the vessel.

Divers' monitoring and communications

The divers working in the water will be equipped with a special monitoring system to control the safety of their operation. Individual monitoring will include relative depth to the bell, hot-water temperature at suit inlet, inspired gas temperature, rectal temperature and breathing frequency rate.

No special problem of communication has been identified with hydrogen. Divers' voices are distorted, but conventional helium unscramblers seem to work satisfactorily, even though the setting is slightly different from that for helium.

CONCLUSION

The operation comes at a time when deep commercial diving is becoming a reality. Operations are carried out on a routine basis at 300 m in the Campos field in Brazil, for instance. New projects in Norway and in Brazil will extend this activity to the 300–450 m depth range.

Comex expects that the *Hydra 8* dive will demonstrate the advantage of hydrogen diving, i.e. increased work capacity and reduced saturation fatigue, and hopes that this technique will open up new markets to the diving industry.

REFERENCES

Brauer, R. W., Jonhsen, D. O., Pessoti, and Redding, R. W. 1966. Effect of hydrogen and helium at pressure to 67 atmospheres on mice and monkeys, *FEDN Proc.* **25,** 202.

Carlioz, M., Gardette-Chauffour, M. C., Rostain, J. C. and Gardette, B. 1984. La narcose à l'hydrogène: étude psychométrique et neurophysiologique, Xth Congrès E.U.B.S., Marseille, 4–6 Oct. 1986.

Fructus, X. 1983. Hydra 3: 16 dives in open sea with hydrogen/oxygen mixture, *Pressure* **12**(4), 14.

Fructus, X., Gardette, B. and Carlioz, M. 1984. Narcose à l'hydrogène, Xth Congrès E.U.B.S., Marseille, 4–6 Oct. 1986.

Fructus, X. and Gardette, B. 1987. *Hydra 6*, Comex internal report.

Gardette, B., Gortan, C., Carlioz, M. and Fructus, X. 1984. L'hydrogène: un gas nouveau pour la plongée profonde humaine. Caractéristiques et résultats généraux de l'expérience HYDRA 4. Xème Congrès E.U.B.S., Marseille, 4–6 Oct. 1986.

Gardette, B. 1986. Plongées réalisées chez la souris à 1300, 1400, 1500, 1800 et 2000 m aux mélanges héliox et hydréliox, Comex internal report, July 1986.

Gortan, C. 1983. System design and safety considerations for Hydra experiments, Comex internal report.

Gortan, C. and Delauze, H. G. 1986. Hydra 5: hydrogen experimental dive to 450 metres. Offshore Technology Conference, Houston, May 5–8, 1986.

Rostain, J. C. 1973. L'effet des hautes pressions avec divers mélanges gazeux chez le singe *Papio papio*. Thèse de doctorat de spécialité en physiologie animale. Université de Provence.

Simon, S., Katz, Y. and Bennett, P. B. 1975. Calculation of the percentage of a narcotic gas to permit abolition of the High Pressure Nervous Symptom, *Undersea Biomed. Res.* **2**, 229–303.

14

Diver Selection for Deep Dives (below 300 msw)

S. T. Brooke, University of Wales
Institute of Science and Technology,
Wharton Williams, UK

INTRODUCTION

Diver selection up to current working depths in the North Sea (200 msw) has been predominantly determined by a process of self-selection. Successful completion of training requirements, strong motivation, establishment with a diving contractor and efficient underwater performance are perhaps the essential 'selection' criteria used for contemporary saturation diving. The safety record is at present good in the North Sea and this in itself vindicates the 'selection' process which has evolved through some natural form of diver attrition and ascendancy of the most competent.

The future of deep (below 300 msw) diving ultimately rests in the ability of the divers to perform the tasks required by the client. The move to explore and develop deeper oil and gas reserves is inevitably going to require the involvement of a freely moving independent diver in tandem with the use of RCVs, ROVs and possibly one-atmosphere divers. Depths beyond 300 msw are going to place increased hazards on the diver and therefore increase the possibility of dangerous incidents occurring.

The relationship between the diver and his environment becomes more hazardous as a function of the increase in depth. Amplification of diver risks and diver-related problems at greater depths highlights the need for a more stringent and methodical diver-selection process. How then, for example, does a diving contractor select a team of eight divers from a pool of 50 experienced and competent saturation divers for an operational dive at 400 msw?

A range of experimental dives has been carried out at various depths up to 686 m over the last decade—*Deep Ex. 1 and 2, Physalie, Atlantis, Janus*—and these have demonstrated that manned intervention can be achieved beyond the current working depths. However, apart from the *Janus IV* series, which carried out an open-sea dive to 460 msw, and a recent Norwegian dive to 300

Advances in Underwater Technology, Ocean Science and Offshore Engineering, Volume 14: Submersible Technology

msw, all the dives were experimental and completed in dry, tightly controlled and relatively safe conditions. There is regular diving being carried out now in Brazil to depths of 300 msw, but information regarding these dives is sparse.

Regular, operational diving with tight schedules to be met is a very different prospect to laboratory dives, and something of a quantum jump (Bentley, 1984). The effects of cold, isolation, water and anxiety are likely to amplify any problems experienced in experimental work and, in all probability, introduce more stressors than exist in artificial dives (Baddeley, 1966; Idzikowski and Baddeley, 1983).

One of the conclusions to the *Deep Ex.* series of dives in Norway was that further research and development was necessary for operational diving at depths up to 500 msw (Tonjum, 1984). In a workshop on the long-term neurological consequences of deep diving (Shields *et al.*, 1983) the ethical point was raised as to the safety of submitting man to ever-increasing pressures and stressors while not fully understanding the physiological and psychological effects in both the short and the long term. Smith-Sivertsen (1983) was convinced that diving beyond 300 msw should be halted until diving physiology had caught up with technology and it had been proved that working beyond 300 msw was reasonably safe. A structured approach to diver selection may preclude the ethical issues raised.

There is a scarcity of literature on diving selection for several reasons. First, the self-selection process to current depths has apparently worked well; secondly, selection has proved in the past to be a notoriously difficult subject and therefore not an attractive proposition; and thirdly, historically selection procedures have rarely had better than chance probability of success.

The selection research that has been carried out has centred largely on selecting recruits for initial diver training or Scuba (Biersner and Ryman, 1974; Hogan, 1985; Kragh, 1962; Vaernes, 1982). Vaernes (1982) identified specific defence mechanisms which predicted poor performance in initial diver training and suggested that this may also apply to certain situations with experienced saturation divers. There has been little work on developing selection criteria to determine what is the make-up of a successful deep diver (> 300 msw). Rostain *et al.* (1983) attempted to predict susceptibility to high-pressure neurological syndrome (HPNS) from rapid compression to 180 m and then comparing performance at 450 m, and concluded that predicting behaviour at 450 msw by looking at variations at 180 m was not possible.

Much of the experimental work has used different subject samples, for example: scientists, SCUBA divers, commercial divers; different compression profiles and gas mixes; and different research protocols. While this provides a good range of information, it makes cross-comparison and the identification of patterns of behaviour and cognitive dysfunction, as a function of depth, gas and chamber environment, difficult. Physiological functions are more constant across humans than psychological variables and it is therefore unwise to generalize results, particularly psychological ones, from one diver to another. However, certain patterns have emerged and these will be covered later in the paper. One message that is common to much of the research is that diver selection for deep work is of considerable importance (Biersner, 1984; Logie and Baddeley, 1985; Logue *et al.*, 1986).

The aims of this chapter are threefold:

1. to review the research covering deep diving over the last decade;

2. to establish selection criteria that are the most valid for operational deep diving;
3. to outline a research programme designed to validate a selection battery which can be easily and inexpensively administered for deep-diver selection.

The credibility of the diving industry and the safety of the divers are dependent on the safe and efficient performance of dive teams, both at the surface and underwater, and the successful completion of tasks. Competition to deep diving from RCVs and ROVs presents another threat to manned intervention. Selection procedures can help to minimize the risks inherent in deep diving and thus promote the efficiency, reliability and safety of saturation diving already demonstrated at depths of up to 200 msw in the North Sea.

DEEP-DIVING RESEARCH

Research over the last decade examining the effects of deep diving has provided a wealth of information on physiological, behavioural and psychological variables (Carter, 1979; Curley *et al.*, 1979; Hugon *et al.*, 1983; Lewis and Baddeley, 1981; Logie and Baddeley, 1983; Logie and Baddeley, 1985; Logue *et al.*, 1986; Vaernes *et al.*, 1987). All these studies report on experimental dives to a minimum depth of 300 m, with the deepest going to 686 m. These dives and the work currently taking place off Brazil demonstrate that it is physiologically possible to compress and decompress man to such depths. However, performance impairments do occur and the long-term neurological consequences have yet to be fully established (Shields *et al.*, 1983).

Behnke (1945) carried out one of the earliest studies examining the psychological and psychiatric effects of diving. The data revealed an impairment in memory and concentration, an increase in anxiety, depression and euphoria, and a range of covert and overt somatic disorders such as dyspnoea, tremors, speech disturbance, nausea and fatigue. Extra effort was needed to perform sensorimotor tasks but the actual level of performance did not decrease. These are not dissimilar to many of the findings of more recent research, which points to some degree of predictability in response to a pressurized and isolated environment. Selection procedures involving physiological and psychomotor tests frequently failed to produce data of value, according to Behnke (1945). He continued by arguing that those in daily contact with the divers, i.e. the Chief Petty Officers, were so effective in screening out the unfit that their judgement was as useful as the test batteries.

Forty years of research has significantly increased the available information on the physiological and psychological response to an hyperbaric environment, and has consequently improved the sophistication of the tools we use to measure them. It is, however, still vital to obtain the information available from the dive supervisors and life-support technicians who are in daily contact with the divers.

Responses to the Hyperbaric Environment

Logue *et al.* (1986) concluded their study of the record-breaking dive to 686 m with the following: 'Healthy, bright, motivated and well-trained divers can continue to function at extreme depths. However, in the absence of HPNS, changes become apparent on tasks requiring fluid, rapid problem-solving abilities or extended long-term memory performance. Clearly, unusual or emergency situations would require skills that might have been significantly attenuated by depth' (p. 233). Two major issues arise from this research: first, significant performance impairments do occur at depth and secondly, emergency situations call for special training and preparation at such depths.

Table 1 presents on overview of the major areas of impairment experienced at depth. The intensity and duration of the impairment varies not only as a function of individual differences but also due to the differing dive profiles and test protocols employed. The inter-individual physiological symptoms tend to remain more constant than the psychological variables which display a large degree of variability not only at an inter- but also at an intra-individual level.

The effects of HPNS now appear to be able to be controlled by slow compression rates and specific holding periods. Vaernes *et al.* (1987), in a study at NUTEC, evaluated a compression profile over three different dives with six different divers to 360 msw, breathing heliox. Mild impairments of peripheral motor function, memory, perceptual speed and reasoning occurred but in only one of the 18 divers was there a marked impairment on EEG, cognitive performance and tremor simultaneously. These dives demonstrate that HPNS can be controlled and only produces mild symptoms in fit and healthy divers.

Although a general pattern of responses occurs, large inter- and intra-individual differences do exist as a reaction to the environment of pressure, mixed gas, isolation and confinement (Logue *et al.*, 1986; Rostain *et al.*, 1983). As mentioned earlier, the ability to screen for HPNS susceptibility and other reactions to pressure, by procedures such as rapid compression to shallower-than-working depths has not proved successful (Rostain *et al.*, 1983).

Vaernes *et al.* (1987) reported that nine divers completed an open-sea dive to 300 msw after completing on-shore trials satisfactorily to 350 m. There were no significant problems or symptoms reported. The divers experienced similar symptoms to the dry dive but reported that all tasks carried out required a larger degree of physical and mental effort.

Imbert (1984) has shown that divers can operate at depths of between 250 and 307 msw by reporting that Comex has successfully completed 43 dives up to 1984 at these depths with 107 bell runs. However, he added that deeper work would be harder and more demanding, and highlighted HPNS, thermal stress, long saturation times and increased gas density as the main stressors. Patterns of cognitive and motor performance emerge from the research dives and present the following picture:

1. Performance deteriorates during compression, tends to stabilize after 3–4 days at storage and then returns to normal or near-normal (i.e. pre-dive baseline) during and following decompression. Task completion requires more effort at depth than at either 1 atm or dry conditions under pressure, to achieve the same standards as at shallower depths.
2. Physiological responses are more constant across individuals than are either psychological or behavioural responses.
3. Reaction to emergency at depth is a serious problem and requires special training and preparation.
4. Large individual differences and the extra demands of deep diving necessitate effective screening and selection procedures.

TABLE 1 *Experimental dives below 300 msw*

			Psychological and physiological impairments[a]						
Study and max. depth	*Days at max. depth and gas mix*	*Subject number*	Memory	Reasoning	Reaction-time	Visuo-spatial ability	Sleep quality	Handgrip strength	Manual-dexterity
Biersner, R. J. and Cameron, B. J. (1970) 300 m	3 $He–O_2$	5	+			+			
Townsend, R. E. and Hall, D. A. (1978) 300 m	5 $He–O_2$	24			+		+		
Lewis, V. J. and Baddeley, A. D. (1981) 540 m	? $He–O_2$	2	+	+			+		
Nutec Deep Ex 81 (1981) 500 m	7 $He–O_2$	3	+		+	+			+
Logie, R. H. and Baddeley, A. D. (1983) 660 m	1 $He–N–O_2$	2		+			+		
Huson, M. *et al.* (1983) 300 m	15 $He–O_2$	?		+					
Logie, R. H. and Baddeley, A. D. (1985) 540 m	6 $He–O_2$	2		+	+				
Logue, R. *et al.* (1986) 686 m	1 $He–N–O_2$	3	+	+					
Vaernes, R. *et al.* (1987) 350 m	11 $He–O_2$	6	+	+		+		+	

[a] The intensity and duration of impairment vary between studies

SELECTION

The literature on diver selection is relatively small and very varied. Attempts to select trainees by standard psychological screening have been only marginally successful and physical tests attempting to predict trainee ability have also met with little success (Biersner *et al.*, 1978; Biersner and Ryman, 1974; Gunderson *et al.*, 1972; Hogan, 1985; Rostain *et al.*, 1981; Vaernes, 1982).

Biersner (1984) in reviewing the research sees two distinct areas emerging in the field of diver selection:

1. Performance, physiological and psychological attributes; physical and medical fitness, cognitive and psychomotor tests linked to specific underwater tasks and actual underwater task performance.
2. Social attributes: group cohesion, motivation, risk taking, demographics (education, experience, birth order) and personality.

Field independence, that is the ability to extract stimuli from an embedded context, mechanical and arithmetic aptitude, social adjustment, attitudes and personality have all been linked to successful diver performance, especially in the diver trainee. The majority of standard personality inventories have proved of little value except the Edwards Personality Preference Schedule which has shown divers to be lower on scales of affiliation, intraception and nurturance, and higher on scales of change and aggression than a control group (Biersner and Cameron, 1970). These differences do not necessarily reflect attributes of a good diver but they do add to the relatively small knowledge we have of a diver's personality profile. Further research using these scales as independent variables would validate their use as predictors of performance and descriptors of the diver profile.

Biersner and Larocco (1987) extended the study of diver personality by examining 30 USN divers on four personality and two demographic variables, exposing them to a stressor (pressurization in a small chamber for several hours) and measured their reaction on three physiological and four subjective scales just prior to and immediately after the stressor. Diving experience, level of education, control and sensation-seeking were shown to be effective measures for depicting individual differences in acute physiological and emotional response to a diving stressor. Measurements collected during the actual 'dive' situation would perhaps have increased the validity of the findings.

Bachrach (1984) regards the level of control of a diver as a key factor in diving safety and behaviour underwater. Locus of control is important in two main areas: (1) during the long periods of boredom in decompression and during long saturations when control over the environment is very limited; (2) in an emergency situation where, by definition in deep work, bale-outs are short and the margin of error is very small. An individual's locus of control, that is whether somebody perceives events as being under his own control (internal), or under the control of fate, luck or powerful people (external), is therefore important in determining behaviour, not only in day-to-day existence in the chamber but also in an acute-stress environment.

Control also has an important role in 'near-miss' accidents which can have debilitating effects on the central nervous system when compounded by decompression sickness (Vaernes and Eidsvik, 1982). Peters *et al.* (1977) also report that frightening experiences in a diver's history can lead to CNS

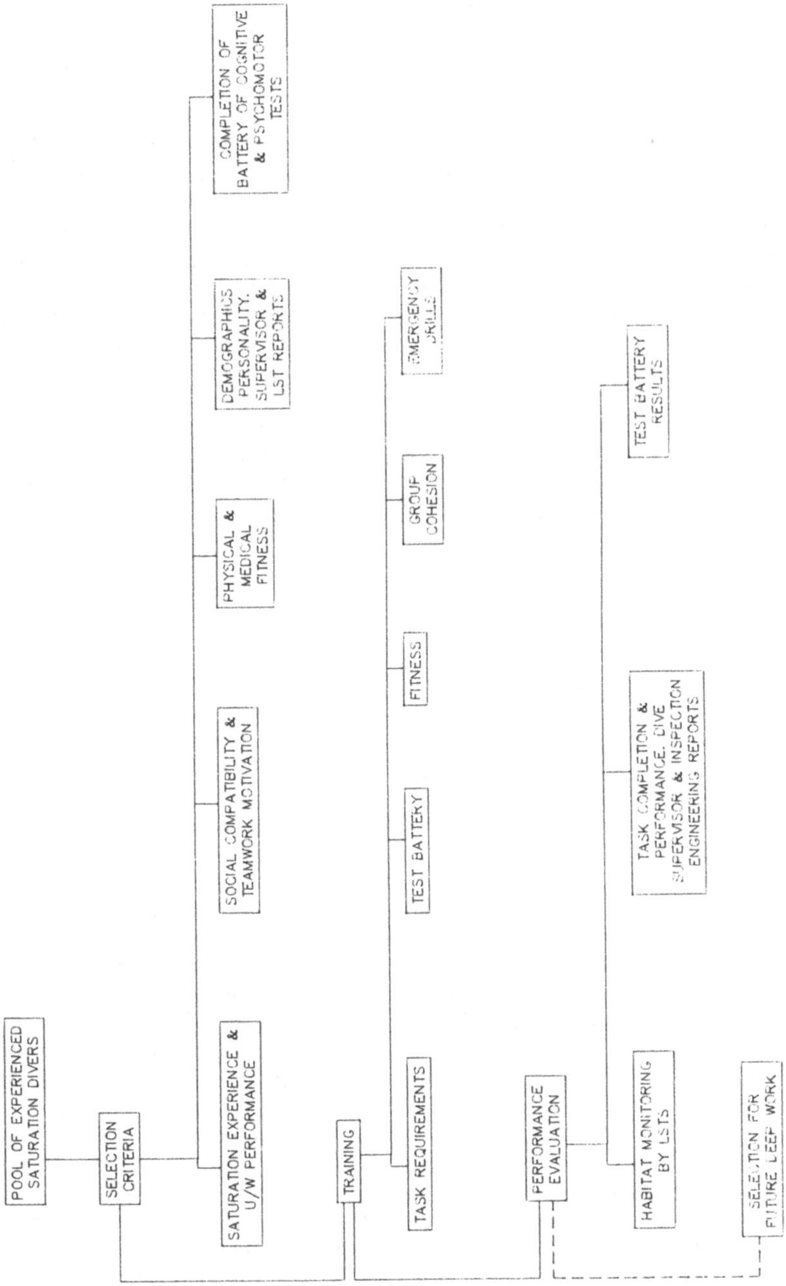

Fig. 1 Diver selection for deep dives

disorders. It is therefore important to screen for any incident that took place off the records. A potential scheme to enable the reporting of 'near-miss' accidents in diving is in the process of being set up.

The Defence Mechanism Test (DMT), developed by Kragh (1960) was used by Vaernes *et al.* (1982) to predict performance under stressful conditions. The stressors of nitrogen narcosis and a novel and frightening environment are significant threats to novice divers. High defence scores on the DMT predicted poor performance in novice divers and showed that they were a high-risk potential in diver training. The DMT may also predict poor performance, especially in emergency situations, in experienced divers.

Hogan (1985) examined seven dimensions of physical fitness to determine if they would predict success in diver training. The following tests were predictive of performance on the course as rated by three instructors on nine performance scales: 1 mile run; sit and reach; and arm endurance. Major fitness-test batteries are generally poor at predicting diver performance due to their lack of specificity. Physical fitness is obviously an important asset in commercial diving where efficient oxygen metabolism and good muscular endurance are vital for high levels of performance during long saturation dives. Fitness also has other indirect effects, such as enabling individuals to cope more efficiently with psychological stress (Brooke and Long, 1987); and generally improving self-confidence, well-being and mood state (Folkins and Sime, 1981).

Figure 1 presents a schema for diver selection.

Experience, Information Processing and Anxiety

The inter-relationship of experience, information processing and anxiety is important when considering diver selection. Superior performance by experienced divers can be explained by their superior information-processing capability. There is less demand on an experienced diver's information-processing mechanism, and the redundancy in the system can be used to cope with an increase in environmental-information load or in emergency situations where the increased load is of an emotional nature (Jorna, 1984). Sanders (1983) developed a model of stress and human performance (see Fig. 2). Stress results whenever the effort mechanism is either seriously overloaded over time or falls altogether short in accomplishing the necessary energetical adjustments. The multidimensional model is based on additive-factors logic (Sternberg, 1969) which states that if any one of the stages in information-processing is affected, then, as a serial process, the overall response is affected. Therefore the cognitive and motor performance requirements can be disrupted at various stages. In an example of a diver carrying out a water-jetting job at 180 msw, the information being processed for the job is termed 'environmental' and includes details such as direction and force of the jet, depth of marine growth, location and time. At the same time the diver has to monitor his own vital support such as gas and hot-water flow. This can be described as a secondary task which requires less processing for the experienced diver. A further load is the 'emotional' load which covers factors such as cold, fatigue and anxiety.

The information-processing capacity of the diver is therefore linked to his experience and current emotional state. Performance breakdown occurs when any of the stages in the model is overloaded, and therefore a certain capacity of reserve is vital to cope with unexpected situations. If a problem arises, for

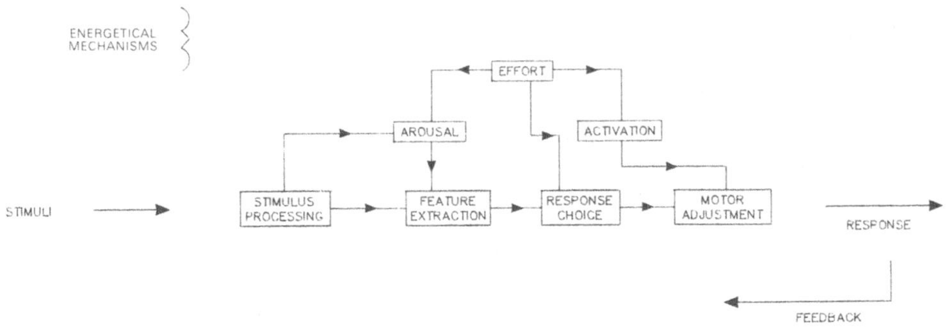

Fig. 2 Information processing model

example the umbilical back to the bell is snagged, the diver has to cope with an increased environmental load, a snagged umbilical, and the emotional load of anxiety about being able to free the umbilical. In this situation an imbalance of the energetical system could arise resulting in stress and a breakdown of rational behaviour. The deeper incidents such as these occur the greater the informational load and therefore the greater the likelihood of an accident happening. The stages in the model can be examined in a variety of tests and when correlated to a range of other predispositional characteristics of the diver it might be possible to make a judgement as to his performance in an overload situation.

Jorna (1984) carried out a study examining the effects of experience and anxiety on diver performance. The performance of novice and experienced divers was compared in cold water (4–8°C), and very poor to zero visibility in depths of 3–5 m. The depth of the dive obviously poses limitations on the generalizability to deep work, but the comparison of novice and experienced divers provides some interesting information. A continuous memory task, reaction time and heart rate were used to compare the two groups. The experienced divers had stable underwater performance which demonstrates that they could easily cope with the extra load of the task. Inexperienced divers started with low levels of performance and slowly improved over the three trials. A control group showed no such improvement over time, thus supporting the hypothesis of an interaction of experience and information processing. An increase of informational load interacts with experience, i.e. ability to process information, and training improves processing ability and thus improves performance. This is not, perhaps, entirely surprising. It is important, though, to understand the theory behind the stress and performance degradation. The performance of well-learned skills either at 1 atm or shallow depths will be significantly affected by a large increase in depth. Skills could be affected by the extra load from novel stimuli and possibly anxiety. Other research has also shown how experience can help divers to cope, not only with normal tasks but also with emergency situations (Biersner *et al.*, 1978; Radloff and Helmreich, 1968; Weltman *et al.*, 1970).

The level of anxiety can also have negative effects on diver performance (Baddeley and Idzikowski, 1985; Mears and Cleary, 1980; Vaernes, 1982; Weltman *et al.*, 1971). Whilst an experienced commercial diver will generally overcome the effects of anxiety in a non-threatening situation, anxiety may well affect his performance in emergency or in the novel environment of extreme depth.

SUMMARY OF RESEARCH PROGRAMME

The research programme can be summarized as follows:

Phase 1: to review the deep-diving research literature and to identify the most sensitive indicators of successful performance at depth; to collect information from diving supervisors on their views on selection criteria for deep diving.

Phase 2: to develop a test battery using information gained from Phase 1. The test battery is to include tests of cognitive and psychomotor performance. The battery will be administered in a series of dry dives to 100 m and 350 m. In one of the 350-m dives 12 divers will have to be selected from a pool of 50. The information gathered from this dive will be used to validate the selection criteria.

Phase 3: to evaluate the final selection package on an operational dive to a minimum depth of 300 m. Diver performance will be monitored by supervisors and LSTs and also by successful task completion. On the basis of this information a user-friendly computer package which will be easy and cheap to administer will be developed.

CONCLUSION

Selection of personnel is a complex task. Selection of highly specialized saturation divers for deep work involves a very large range of factors. The establishment of a selection process may be used to screen out those who are not capable of performing at depths beyond 30 msw, and thus ensure the continued record of safety and professionalism in the North Sea.

REFERENCES

Baddeley, A. D. 1966. Influence of depth on the manual dexterity of free divers: a comparison between open sea and pressure chamber testing, *Journal of Applied Psychology* **50,** 81–85.

Baddeley, A. D. and Idzikowski, C. 1985. Anxiety, manual dexterity and diver performance, *Ergonomics* **28,** 1475–1482.

Behnke, A.R. 1945. Psychological and psychiatric reactions in diving and submarine warfare, *American Psychiatry* 720–725.

Bentley, K. T. 1984. The future role of the diver, *Advances in Underwater Technology and Offshore Engineering* **1,** 27–32.

Biersner, R. J. 1984. Psychological standards for diving, in C. W. Shilling, C. B. Carlston and R. A. Mathias (Eds), *The Physician's Guide to Diving Medicine.* Plenum Press, New York, pp. 520–530.

Biersner, R. J. and Cameron, B. J. 1970. Cognitive performance during a 1000-foot helium dive, *Aerospace Medicine* **41,** 918–920.

Biersner, R. J. and Ryman, D. H. 1974. Prediction of SCUBA training performance, *Journal of Applied Psychology* **59,** 519–521.

Biersner, R. J. and Cameron, B. J. 1970. Betting preferences and personality characteristics of Navy divers, *Aerospace Medicine* **41,** 1289–1291.

Biersner, R. J. and Larocco, J. M. 1987. Personality and demographic variables related to individual responsiveness to diving stress, *Undersea Biomedical Research* **14,** 67–73.

Biersner, R. J., Hall, D. A. and Linaweaver, P. G. 1976. Occupational differences between conventional and saturation divers, *Aviation, Space and Environmental Medicine* **47,** 29–32.

Biersner, R. J., Hall, D. A., Linaweaver, P. G. and Neuman, T. S. 1978. Diving experience and emotional factors related to the psychomotor effects of nitrogen narcosis, *Aviation, Space and Environmental Medicine* **49,** 959–962.

Brooke, S. T. and Long, B. C. 1987. Efficiency of coping with a real-life stressor: a multimodal comparison of aerobic fitness, *Psychophysiology* **24,** 173–180.

Carter, R. C. 1979. Mental abilities during a simulated dive to 427 metres underwater, *Journal of Applied Psychology* **64,** 449–454.

Curley, M. D., Berghage, T. E., Raymond, L. W., Sode, J. and Leach, C. 1979. Emotional stability during a chamber saturation dive to 49.5 atmospheres absolute, *Journal of Applied Psychology* **64,** 548–557.

Folkins, C. H. and Sime, W. E. 1981. Physical fitness training and mental health, *American Psychologist* **36,** 373–389.

Gunderson, E. K. E., Rahe, R. H. and Arthur, R. J. 1972. Prediction of performance in stressful underwater demolition training, *Journal of Applied Psychology* **56,** 430–432.

Hogan, J. 1985. Tests for success in diver training. *Journal of Applied Psychology* **70,** 219–224.

Hugon, M., Fagni, L. and Seki, K. 1983. Deep-sea diving: human performance and motor control under hyperbaric conditions with inert gas, *Advances in Neurology* **39,** 829–849.

Idzikowski, C. and Baddeley, A. D. 1983. Fear and dangerous environments, in R. Hockey (Ed.), *Stress and Fatigue in Human Performance.* Wiley, Chichester, pp. 123–144.

Imbert, J. P. 1984. Programme for working dives to 400 msw. II. Technical aspects, *Advances in Underwater and Offshore Technology* **1,** 38–61.

Jorna, P. G. A. M. 1981. Stress, information processing, and diving. Report from Institute for Perception, National Defence Research Organization TNO. 1—27.

Kragh, U. 1962. Predictions of success of Danish attack divers by the defense mechanism test (DMT), *Perceptual and Motor Skills* **15,** 103–106.

Lewis, V. J. and Baddeley, A. D. 1981. Cognitive performance, sleep quality and mood during deep oxyhelium diving, *Ergonomics* **24,** 773–793.

Logie, R. H. and Baddeley, A. D. 1983. A trimix saturation dive to 660 m. Studies of cognitive performance, mood and sleep quality, *Ergonomics* **26,** 359–374.

Logie, R. H. and Baddeley, A. D. 1985. Cognitive performance during simulated deep-sea diving, *Ergonomics* **28,** 731–746.

Logue, P. E., Schmitt, F. A., Rogers, H. E. and Strong, G. B. 1986. Cognitive and emotional changes during a simulated 686-m deep dive, *Undersea Biomedical Research* **13,** 225–235.

Mears, J. D. and Cleary, P. J. 1980. Anxiety as a factor in underwater performance, *Ergonomics* **23,** 549–557.

Peters, B. H., Levin, H. S. and Kelly, P. J. 1977. Neurologic and psychologic manifestations of decompression illness in divers, *Neurology* **27,** 125–127.

Radloff, R. and Helmreich, R. 1968. *Groups Under Stress: Psychological Research in Sealab II,* Appleton Century Crofts, New York.

Rostain, J. C., Lemaire, C., Gardette, M. C., Doucet, J. and Naquet, R. 1981. Criteria analysis of selection for deep diving: EEG and performance, in A. J. Bachrach and M. M. Martzen (Eds), *Underwater Physiology VII,* pp. 435–443.

Rostain, J. C., Lemaire, C., Gardette-Chauffour, M. C., Doucet, J. and Naquet, R. 1983. Estimation of human susceptibility to the high-pressure nervous syndrome, *Journal of Applied Physiology* **54,** 1063–1070.

Sanders, A. F. 1983. Towards a model of stress and human performance. *Acta Psychologica* **53,** 61–97.

Shields, T. G., Minsaas, B., Elliott, D. H. and McCallum, R. I. (Eds) 1983. *Long-term Neurological Consequences of Deep Diving,* E.U.B.S. Workshop, Verbum, Stavanger.

Smith-Sivertsen, J. 1983. Deep diving: a controversial medical problem, in T. G. Shields *et al.* (Eds), *Long-term Neurological Consequences of Deep Diving,* Verbum, Stavanger, pp. 13–17.

Sternberg, S. 1969. On the discovery of processing stages: some extension of Donders' method, *Acta Psychologica* **30,** 276–315.

Tonjum, S. 1984. Norwegian deep-diving trials, *Philosophical Transactions of the Royal Society of London* **304,** 143–149.

Townsend, R. E. and Hall, D. A. 1978. Sleep, mood, and fatigue during a 14-day He–O_2 open-sea saturation dive to 850 fsw with excursions to 950 fsw, *Undersea Biomedical Research* **5,** 109–117.

Vaernes, R., Bennet, P. B., Hammerborg, D., Ellertsen, B., Peterson, R. E. and Tonjum, S. 1982. Central nervous system reactions during heliox and trimix dives to 31 ATA, *Undersea Biomedical Research* **9,** 1–14.

Vaernes, R. J. 1982. The defense mechanism test predicts inadequate performance under stress, *Scandinavian Journal of Psychology* **23,** 37–43.

Vaernes, R. J. and Eidsvik, S. 1982. Central nervous dysfunctions after near-miss accidents in diving, *Aviation, Space and Environmental Medicine* **53,** 803–807.

Vaernes, R. J., Aarli, J. A., Klove, H. and Tonjum, S. 1987. Differential neuropsychological effects of diving to 350 meters, *Aviation, Space and Environmental Medicine* **58,** 155–165.

Weltman, G., Christianson, R. A. and Egstrom, G. H. 1970. Effects of environment and experience on underwater work performance, *Human Factors* **12,** 587–598.

Weltman, G., Smith, J. E. and Egstrom, G. H. 1971. Perceptual narrowing during simulated pressure-chamber exposure, *Human Factors* **13,** 99–107.

15

Saturation Diving: Diving Deeper Than 300 m

P. Rosengren, Head of Section,
The Norwegian Petroleum Directorate,
Norway

The drilling for oil and gas on the Norwegian Continental Shelf started in the summer of 1966. The first operation was carried out in 94 m of water.

As the drilling started offshore, the requirement for diving services developed. The diving in Norway had up to then been limited to inshore diving using mainly standard diving gear. Divers and diving systems had to be brought in from the Mexican Gulf and the Mediterranean.

The equipment and procedures used in the first years were quite simple and were meant to be used in less harsh environments than the North Sea. The diving on the Norwegian Continental Shelf entered a period with very rapid development of both equipment and procedures. The main bulk of diving changed from bounce diving to saturation diving.

DEVELOPMENT

During the first ten years the drilling was concentrated in rather shallow water depths, but the average water depths have increased considerably in the last years.

The Ekofisk Field, at about 70 m, was discovered in 1969 and the Troll Field, at approximately 350 m, was discovered in 1979. Up to now fields that have been built out have been at depths shallower than 200 m.

In 1981 it was decided to lay the Statpipe pipeline over the Norwegian Trench. This meant that pipeline-repair capability was needed for repairs at the maximum depth of 300 m. Statoil started the development of repair procedures, diving equipment and procedures.

This initiative was followed by Shell who wanted to have divers as back-up in case the need should arise at the Troll Field at 350 m. In recent years Norsk Hydro have continued the development because of the planned pipeline over the Norwegian Trench at a maximum water depth of 360 m.

Advances in Underwater Technology, Ocean Science and Offshore Engineering, Volume 14: Submersible Technology

FUTURE FIELDS

In the future, the fields will be built out at greater and greater water depths. In some of the new areas where exploration drilling is taking place, the water depths are now between 400 and 500 m.

In the southern part of the North Sea the Snorre Field, at water depths ranging from 300 to 370 m, will probably be built out in the next five years. A tension leg platform and subsea completed wells with templates will probably be installed.

In the same area is located the enormous Troll Field. The maximum depth in the area where the first platform may be placed is about 350 m. It is estimated that the Troll Field contains approximately 1288×10^9 Sm^3 gas and 41×10^6 Sm^3 oil. The field may be in production around 1995.

Further north at Haltenbanken many discoveries of both oil and gas have been made. The water depth in the area ranges from 250 to 350 m. Some of the discoveries may be built out and in production before 1995.

In the Troms area the water depth goes all the way down towards 500 m, and discoveries have been made of both oil and gas. It is difficult to say when these fields may be built out.

FUTURE DEVELOPMENT OF INTERVENTION METHODS

The cost of diving services goes up tremendously as the water depths increase. The oil company will always choose a method which in total will give the least expensive solution. This means that fields deeper than 250 m will be planned to be built out, maintained and inspected without divers.

The engineers will put much more effort into the development of systems which can be operated by remotely operated vehicles and tools. This has always been a challenge to engineers and I believe it will also be one in the future.

We have seen a considerable development in remotely operated tools over the last years. The ROV market has boomed, but is now suffering from the same slump in activity as the rest of the industry.

The ROVs should be able to take over many of the routine tasks the divers are asked to carry out today. If a subsea installation at great depth has been constructed with the clear aim that a specialized ROV should be able to carry out most inspection and maintenance tasks, the need for diver intervention should be very limited.

EQUIPMENT

In spite of the development of equipment over the last years there are still limitations in connection with deep diving:

- A lot of effort has been spent in order to improve the breathing systems. The external breathing work is now getting down towards zero. Most of the systems in use today are based on the demand principle, and in order to make further improvements, new principles will probably have to be used.
- The communications between divers under pressure and surface have been a disaster for years. The lack of proper communication puts serious limits

on the effectiveness and it also seriously affects the safety of the divers. The communication system must be improved in years to come so that diving to greater depths can be carried out safely and effectively.

- The thermal problems increase with depth. The different systems designed for keeping the diver's temperature within safe limits will have to be improved in years to come.
- Lots of new developments have been made in equipment over the last few years, but all the new ideas should now be put together in one package. In order to make the diver as effective as possible the ergonomy of the individual equipment will have to be improved, e.g. it should not be necessary for the diver to use more than one movement to go from normal to bail out.
- The work performance of a diver at greater depths will decrease. This is due to the fact that the energy used to breathe at those depths will take a considerable amount of the total energy. The time in water will have to be reduced. The bell runs should be limited to a maximum of six hours.
- The effectiveness of the divers may be improved by developing more specialized tools for the divers to work with. Tools should be designed for the work to be carried out by a diver in water.
- The increased water depth and the requirement of the positioning of the bell will increase the requirements of the DP-system.
- During most deep dives divers lose weight. It is important that the divers are in top condition when they are asked to carry out work on the sea bed. This aspect needs to be studied further.
- Questions have been raised about the long-term effects of diving. The investigation into these matters must go on in order to ensure us that diving, and especially deep diving, does not give the divers any long-term effects.
- Other areas, for example blood changes, changes in lung function, thermal physiology, pharmacology, changes in the circulatory system, bacteriology, etc., are areas where we do not fully understand what happens to divers under pressure. During deep dives in the years to come we shall have an opportunity to study the problems and get answers to some of the questions.

Preparation before a Deep Dive

In NPD we have followed the philosophy that certain preparations have to be carried out before dives to a depth greater than 180 m. Dives to those depths are still not to be considered as routine operations in the North Sea. The following sequence should be followed:

- A diving contractor and a ship should be specially selected according to the work one wants to carry out.
- Divers with the experience needed and, if possible, with experience from other dives in the same depth range, should be picked out and they should go through special medical examinations.
- The equipment and procedures one chooses to use should be re-evaluated in the light of the planned operation.
- In order to increase the work performance of a diving system, four-man bells may be used. A diving vessel operating with two four-man bells may have four divers out in the water working together.

- All the divers in the water should at all times be watched by remotely operated television cameras. The safety margins at great depths are small and if anything is wrong, or goes wrong, the surface crew should know as early as possible.
- If a job of long duration is to be carried out at great water depths, the need for personnel under pressure will be much higher than today. The number of divers at working depth and decompressing may be between 50 and 100. A practical solution may probably be to compress and decompress the divers at land-based chamber complexes.

MEDICAL CONSIDERATIONS

The medical aspects of deep diving have been investigated during all the dives which have taken place in recent years. According to the doctors, we have no firm evidence that deep diving will have any permanent ill effects on divers. However, it is generally agreed that many questions are still unanswered and we need to do more research in order to understand what is happening and thus ensure that diving is safe.

Certain areas of concern have been highlighted:

- High-pressure nervous syndrome is not considered to be a major problem. By using conservative compression profiles and allowing the divers time to adapt to the depth, most of the divers are not noticeably affected by HPNS. When doing deep dives a test battery may be developed so that one can ensure that the divers who go to work do not have any HPNS problems.
- Equipment should be tested unmanned to the maximum working depth plus a safety margin to ensure that all equipment functions satisfactorily and within the performance criteria laid down.
- Manned verification dives should be carried out where ship, procedures, equipment and some of the divers prove that the total system really functions as expected at the depth where the operation will take place.
- If everything functions satisfactorily the offshore operation may start.
- After the deep-diving operation has finished the divers should go through special post-dive medicals in order to ensure that they have no long-term effects from the dive.

CONCLUSION

In the future the oil companies will try to build out fields in deep water without the use of divers. I believe most of the oil companies will prefer to have the divers as a back-up in case something goes wrong. We must be prepared for the tasks the diver will be asked to carry out. When the remotely operated systems fail and it is impossible to pull the problem to the surface, the divers will be called in to solve the problems for many years to come.

Part IV

Diving Shallower Than 50 m

16

Mathematical Modelling of Tissue Bubble Dynamics During Decompression

M. L. Gernhardt, Ocean Systems Engineering Limited, UK

It is well accepted that bubbles, through some mechanism, are the cause of marginal symptoms of decompression sickness (DCS). For this reason a theoretical model of extravascular gas bubble dynamics has been derived to aid in developing and evaluating decompression procedures. The model has been developed and refined over a number of years and has been aimed at both diving and aerospace decompressions.

The model assumes that gas nuclei exist, or through one or more processes form, in extravascular tissues during decompression. The model then describes the subsequent growth and resolution dynamics of a gas bubble during the course of decompression.

The model accounts for the diffusion of inert gas across the tissue/bubble interface, gas solubility and diffusivity, surface tension, tissue elasticity and a perfusion-limited inert gas exchange with the tissue. The model can account for multiple inert gas breathing mixtures, and variable ascent and descent rates. A graphic illustration of the model and a list of the assumptions and features is shown in Fig. 1.

Basic Assumptions

- Bubbles are the cause of marginal symptoms of DCS
- Gas nuclei exist or form in extravascular tissues during decompression
- The gas exchange between the extravascular bubble and the surrounding tissue is limited by diffusion
- The gas exchange between the blood and tissues is limited by the blood perfusion to the tissues

Advances in Underwater Technology, Ocean Science and Offshore Engineering, Volume 14: Submersible Technology

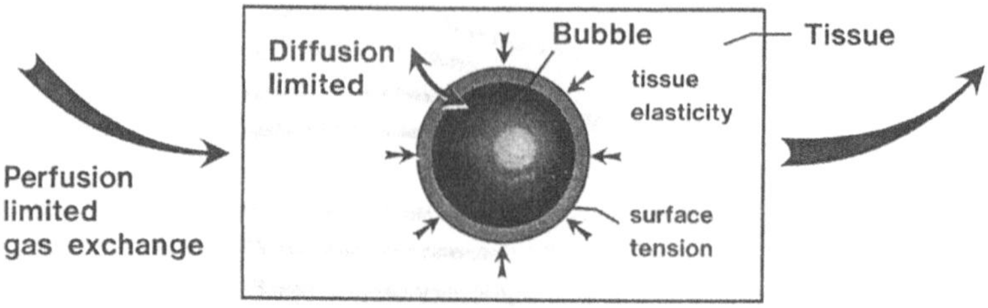

Fig. 1 Bubble dynamics model.

The model accounts for:

- Gas solubility and diffusivity
- The surface tension acting on the tissue/bubble interface
- The tissue elastic forces that would oppose bubble growth
- Multiple and sequential inert gas breathing mixtures
- Oxygen breathing
- Variable ascent and descent rates

The model does not account for the statistical nature of decompression sickness, or individual differences in the ability to withstand gas emboli. For this reason it is not thought that any critical theoretical bubble size would be predicted by the model that would always cause, or conversely always prevent, decompression sickness. Instead it is thought that the model could be used as one tool, in conjunction with empirical data and operational experience, to evaluate the relative severity of a variety of complex diving or aerospace decompressions.

The most basic assumption of the model is that gas nuclei exist or form in extravascular tissues during decompression. With this assumption, instead of trying to model the complex and poorly understood processes by which bubble nuclei form in tissues, their existence during decompression is assumed and used as a starting point for modelling the bubble growth and resolution phase.

This is a conservative and reasonable assumption supported by a wealth of data which suggest that bubble growth occurs at pressure differentials much less than the tensile strength of body fluids, thus implying the existence or formation of these bubble nuclei.

STANDARD DECOMPRESSION MODELS

Most standard diving and aerospace decompression models are based on modified Haldanian decompression concepts and incorporate a perfusion-limited inert gas exchange between the blood and the tissues. This assumption is common with the bubble dynamics model and allows the bubble model to be correlated with the vast database of information available from procedures developed with the Haldanian models.

The basic Haldanian model has been modified extensively over the years through a largely empirical process and has been adapted to both diving and aerospace applications. The primary difference between standard diving and aerospace models is not in the inert gas exchange, but in the criteria for a permissible decompression.

The commonly used diving models typically use a critical pressure difference, or supersaturation, between the calculated tissue tension and the ambient pressure.

Aerospace decompression models utilize a pressure reduction ratio, between the calculated tissue tension and the space-suit pressure, referred to as an R-value.

There is generally little direct correlation between the diving and aerospace models, because the diving models typically allow supersaturations of greater than one atmosphere for certain tissue compartments. This is clearly inconsistent with aerospace decompressions where supersaturations of much less than one atmosphere in all tissue compartments are necessary to protect against DCS. This is the reason the aerospace models rely on the pressure reduction ratios.

By modelling a gas bubble within a perfusion-limited tissue, the bubble dynamics model describes the bubble volume response to any change in pressure. For this reason, the model should be more continuous between diving and aerospace decompressions, allowing empirical data from both high pressure and low pressure environments to be analysed and correlated.

DERIVATION OF THE MODEL

For an extravascular stationary bubble, the flux of gas into the bubble is driven by the pressure gradient between the tissue and the bubble. The bubble then readjusts its pressure and volume to come to equilibrium. Neglecting convective effects, the flux of gas is given by Fick's law:

$$J = \frac{-\alpha D\,(P_{\mathrm{T}} - P_{\mathrm{B}})}{RT\ h} \tag{1}$$

Where

J = Molar flux of gas
D = Diffusivity
h = Effective thickness of the diffusion barrier between the bubble and the tissue
P_{B} = Pressure of gas in the bubble
P_{T} = Tension of gas in the tissue
R = Ideal gas constant
T = Temperature
α = Solubility of gas in the tissues

Multiplying equation (1) by the area of the interface will give the number of moles of gas per unit time diffusing into the bubble:

$$J \cdot A = \frac{\alpha AD}{RTh}(P_{\mathrm{B}} - P_{\mathrm{T}}) = \frac{\mathrm{d}n}{\mathrm{d}t} \tag{2}$$

If the separated gas is assumed to be an ideal gas, then the equation of state for an ideal gas equation can be solved for $\mathrm{d}n/\mathrm{d}t$, the number of moles of gas diffusing into the bubble with time. This expression can be equated to the equation (2) to yield an equation which describes the flux of gas into the bubble and the bubbles pressure/volume response (assuming constant temperature):

$$\frac{-\alpha AD}{h}(P_B-P_T)=P_B\frac{dV}{dt}+V\frac{dP_B}{dt} \tag{3}$$

Where P_T = the tension of inert gas in the tissues and P_B = the pressure of inert gas in the bubble.

The pressure of gas in the bubble will be equal to the sum of the pressure forces acting on the bubble. The three dominant forces would be:

- **Hydrostatic pressure** The pressure of surrounding environment.
- **Surface tension** The pressure due to the surface tension acting on the tissue/bubble interface would be equal to $2\delta/R$ where δ is the surface tension in dynes/cm and R is the bubble radius.
- **Tissue deformation** The pressure due to tissue deformation which will be equal to $\frac{4}{3}\pi R^3 H$, where H is the bulk modulus of the tissue expressed as $\Delta P/\Delta V$ (3).

Bubble expansion in tissues of high bulk modulus will result in greater increases in intra-bubble gas pressure than in tissues of lower bulk modulus. The increase in gas pressure within the bubble will tend to drive the gas out of the bubble, resulting in a smaller bubble radius.

The pressure of gas in the tissue P_T, can be calculated using equations derived from applying a mass balance to a perfusion-limited tissue. The result of this derivation is a form of the familiar equation for a perfusion-limited gas exchange (4)

$$P_{T_A}=P_T(0)+\{x(P_0-vT)-P_T(0)\}\times (1-e-\left(\dot{Q}\frac{\alpha_b}{\alpha_t}\right)t \tag{4}$$

Where,

P_T = Inert gas tension in the tissue
P_0 = Initial hydrostatic pressure
x = Mole fraction of inspired inert gas
$\dot{Q}$ = Blood perfusion
α_b = Gas solubility in blood
α_t = Gas solubility in tissue
t = Time
v = Ascent rate

The exponent in equation (4) can be related to the tissue half-time by:

$$\dot{Q}\frac{\alpha_b}{\alpha_t}=\frac{0.693}{\text{half-time}}$$

Where the half-time is the time for a tissue to reach half saturation with a given pressure.

The expressions for P_B and P_T are substituted into equation (3), and the resulting equation is solved for bubble radius as a function of time, using computer numerical solution techniques (second order polynomial extrapolation).

The equations can be modified to account for multiple gases by defining the total tissue gas pressure to be equal to the sum of the component gases. An effective gas diffusivity is also defined based on the component gases' diffusivity and concentration.

By varying the ambient pressure, breathing mixtures and ascent rates, it is possible to simulate the bubble dynamics for any complex decompression protocol. The computer solution is designed so that the final bubble radius from one exposure becomes the initial value for the next exposure. In this manner the results of decompressions, recompressions and gas switches can be simulated.

THEORETICAL SIMULATIONS AND DISCUSSION

The solution to the bubble radius equation is a plot of the bubble radius against time for any specified decompression protocol. The simulations discussed in this section are based on the theoretical model and *should not be construed as conclusive recommendations regarding the validity of any specific decompression protocol or collectively about any group of decompression tables*.

The simulations shown in Fig. 2 illustrate the theoretical bubble growth characteristics associated with a pressure reduction of one atmosphere. (From two to one atmosphere and from one to zero atmosphere.) The pressure reduction from one to zero atmosphere shows much greater bubble growth than the same pressure differential from two to one atmospheres.

Although this behaviour would be expected for an ideal gas undergoing an isothermal polytropic process, it is significant that the bubble model is able to distinguish between these two exposures. Standard diving models, based strictly on supersaturation concepts, would not predict any differences.

In the context of the bubble model, the theoretical tissue bubble would continue to grow until the tissues were completely desaturated with inert gas,

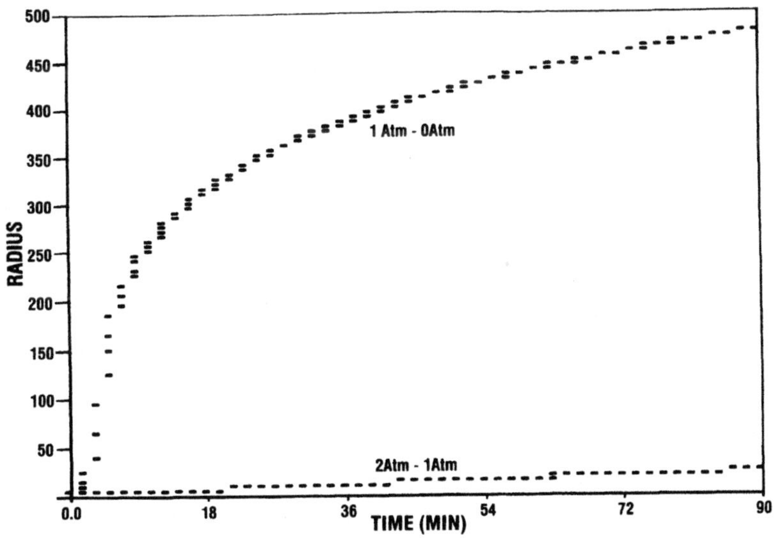

Fig. 2 Theoretical bubble growth associated with pressure reduction of one atmosphere.

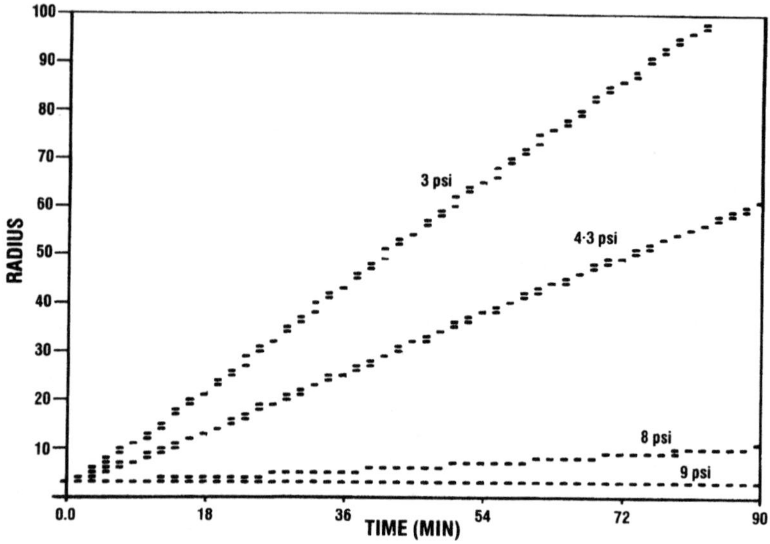

Fig. 3 Theoretical bubble growth for EVA excursions from one atmosphere to 3, 4.3, 8 and 9 psi with no O_2 prebreathing.

at which point the bubble would begin to decay under the resolving forces of tissue elasticity and surface tension, as illustrated in Fig. 2.

The next series of simulations (Fig. 3) were aimed at identifying the space-suit pressure (normoxic breathing mixture), which would protect against DCS, without the astronauts having to pre-breathe oxygen in order to lower tissue nitrogen tensions.

The case of an astronaut locking out of a one atmosphere space craft to work in a lower pressure space-suit would be analogous to a diver decompressing from a saturation chamber to work at sea level. The astronaut would be saturated with nitrogen at sea level and, depending on the suit pressure, a staged decompression and/or oxygen pre-breathing may be required to safely make a given pressure transition.

These simulations show significant bubble growth at 3 and 4.3 psi, some bubble growth at 8 psi, and little or no bubble growth at 9 psi, depending on the depressurization rate.

These predictions are confirmed by empirical data indicating that, without oxygen pre-breathing to lower tissue nitrogen tensions, a suit pressure of between 8 and 9.5 psi should protect against DCS when decompressing from sea level.

The next simulation (Fig. 4) shows the bubble growth characteristics associated with in-water decompression from a dive in the normal exposure and extreme exposure range. The bubble radius curves show much more bubble growth associated with the extreme exposure dive. It is well known that there are more risks associated with extreme exposure dives and they are almost universally not recommended for routine operations. However, it is significant that the bubble model is able to distinguish the relative severity of these dives, whereas an extrapolation of standard computerized diving models would not.

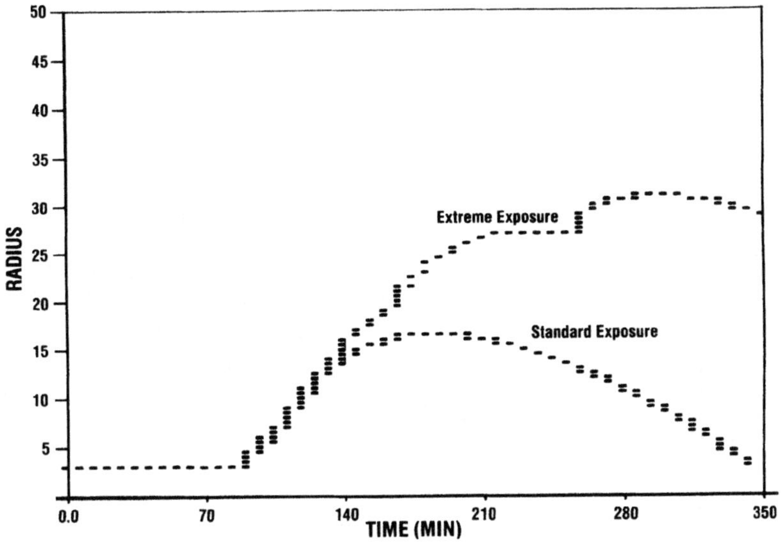

Fig. 4 Comparison of standard versus extreme exposure (in water decompression).

The next simulation (Fig. 5) shows the bubble growth characteristics associated with an air dive followed by surface decompression on oxygen. The two curves correspond to a modified and an unmodified procedure. Both curves show a period of bubble growth during the surface interval, followed by a period of bubble resolution during oxygen breathing under pressure in the chamber.

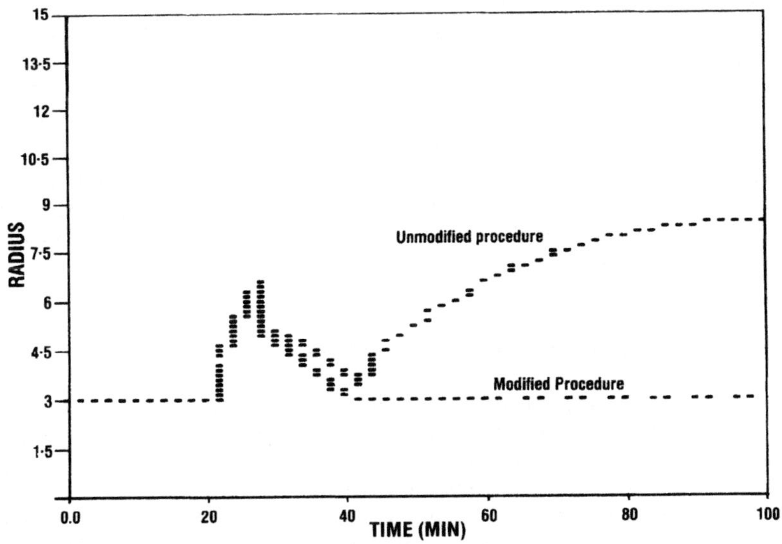

Fig. 5 Surface decompression on oxygen.

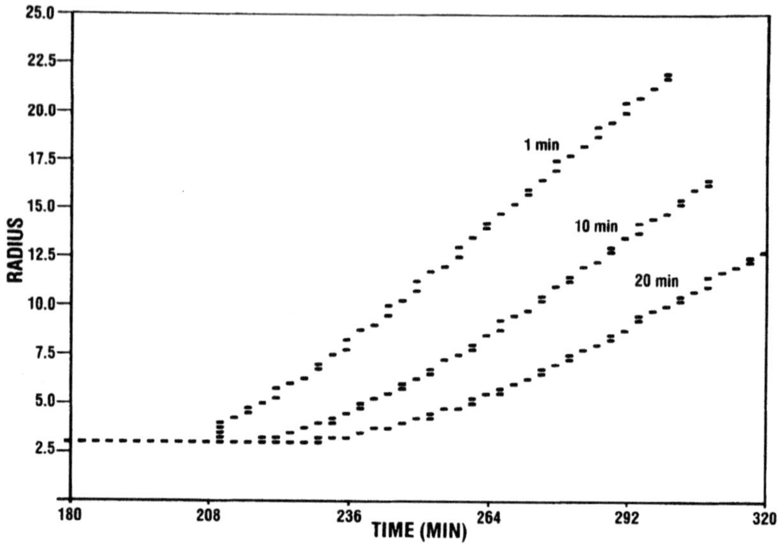

Fig. 6 Effect of depressurization rate for 1, 10, and 20 min ascents (14.7–4.3 psi) following 3.5 h of O_2 breathing.

For the unmodified procedure, the bubble is not completely resolved in the chamber and continues to grow after the ascent to the surface, reaching a maximum size in the time period generally associated with the onset of DCS symptoms, if they were to occur. For the modified procedure, which incorporates changes to the oxygen breathing and pressure protocol, the bubble is completely resolved during the chamber decompression period. Modifications of this type are generally felt to reduce the risk of DCS.

The simulation shown in Fig. 6 shows the effect of depressurization, or ascent rate, on the theoretical bubble growth associated with an aerospace decompression from 14.7 to 4.3 psi following 3.5 hours of oxygen pre-breathing. The curves shown are for ascent times of 1 minute, 10 minutes and 20 minutes. The 1 minute ascent time shows a short period of very rapid bubble growth which, with increasing time, results in a larger bubble size than does the slower ascent rates. This simulation suggests theoretically that a very fast ascent early in this type of decompression potentially results in larger bubbles much later in the decompression protocol.

Defining New Decompression Procedures

The simulations shown above have dealt with evaluation of the theoretical bubble growth characteristics associated with defined decompression protocols. The model could also potentially be used as one tool to define the basis for optimizing new decompression procedures.

One such example related to aerospace decompressions associated with extravehicular activity (EVA) is shown in Fig. 7. This simulation shows the effect of utilizing intermittent recompressions during the course of the EVA.

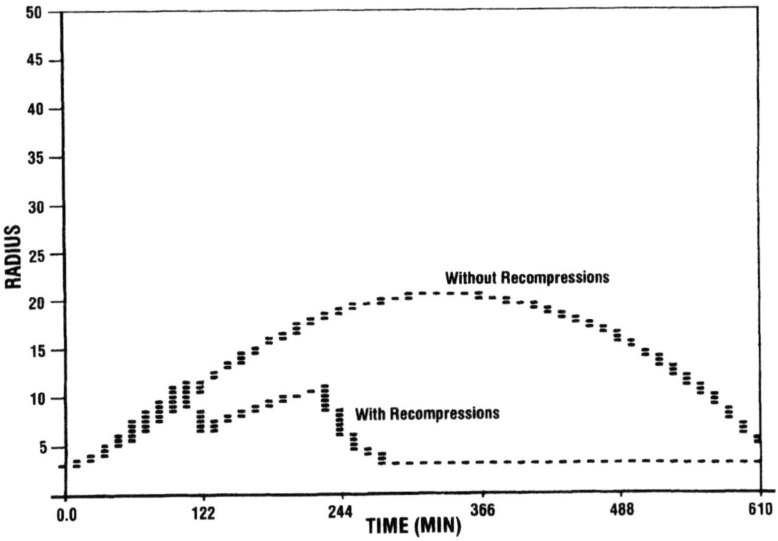

Fig. 7 EVA with and without intermittent recompressions.

The effect of the intermittent recompressions is to interrupt the gradual bubble growth that would normally be associated with this exposure. Since the EVA pressure is lower than the saturation pressure (sea level) and the simulated breathing gas is enriched with oxygen, continual off-gassing occurs during the course of the recompressions. The recompressions then help to resolve the bubble. The net effect is to keep the theoretical bubble smaller and resolve it sooner with only a minimal amount of appropriately chosen recompression periods.

This procedure is very different from that of repetitive diving where repeated exposures to higher pressures of inert gas can result in a cumulative effect, increasing tissue inert gas tensions and consequently the theoretical bubble size which would occur during subsequent decompressions.

This same theoretical concept illustrated for the aerospace case could potentially be applied to saturation diving.

Figure 8 shows the theoretical bubble growth characteristics associated with an older saturation decompression procedure which is no longer used operationally. The bubble dynamics illustrate that there is a period of slow but steady bubble growth early in the deeper stages of the decompression. The bubble growth is then well controlled in the longer and shallower phases; however, the bubble remains at a relatively large size. This is because there are no measures such as enriched oxygen breathing or recompressions to help resolve the bubble.

Figure 9 shows the bubble growth characteristics associated with a saturation decompression from the same depth, but this time incorporating minimal intermittent recompressions coupled with enriched oxygen breathing. The net effect is to control the theoretical bubble growth better.

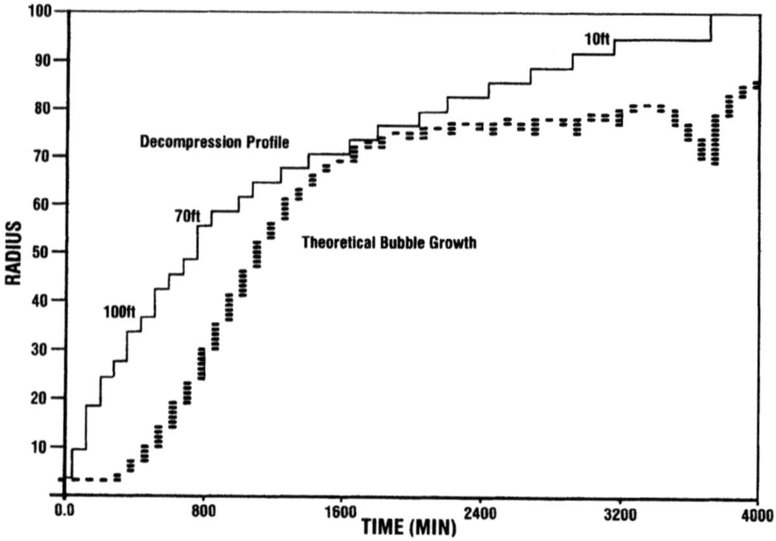

Fig. 8 Theoretical bubble growth: nitrox saturation decompression (schedule not approved for operations).

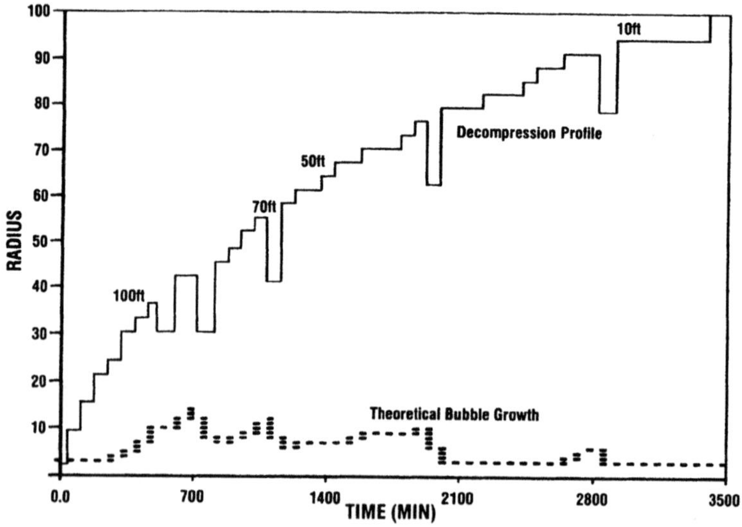

Fig. 9 Theoretical bubble growth: nitrox saturation decompressions (incorporated intermittent decompressions with enriched P_{O_2}).

CONCLUSIONS

The bubble dynamics model demonstrates qualitatively reasonable results which are generally consistent with a wide range of empirical diving and aerospace decompression data. The model integrates perfusion and diffusion-limited inert gas exchange and gas pressure/volume response into a single time-dependent criterion, which is potentially applicable over a continuous range of decompressions from hyperbaric to hypobaric environments.

Since the model does not account for the statistical nature of decompression sickness or individual differences in the population, it is not thought that a critical bubble size exists that would always cause or prevent decompression sickness.

Instead it is thought the model could be a useful tool to be used in conjunction with standard decompression models and empirical data. In this manner, patterns of bubble growth and resolution could be used as one index to evaluate and develop a wide range of decompression procedures.

17

An Evaluation of Automatic Recording of Commercial Air-Diving Data

J. Hendrick, Shell UK
Exploration and Production, UK

In 1983 A/S Norske Shell sponsored the development of an automatic dive recorder by A/S Sipro of Norway. At the time when the Sipro recorder was first introduced the diving industry was entering a period of financial constraint, and so it was not taken up by the diving contractors.

In 1986 the Shields/Lee report on a study into the incidence of decompression sickness in commercial air-diving operations was submitted to the Department of Energy. Close analysis of that report indicated that insufficient information was given in the dive records made available for the study to enable all potential causative factors of decompression sickness to be fully quantified and analysed. As a result of this, Shell UK Exploration and Production, operating in the UK sector of the North Sea on behalf of the Shell/Esso joint venture decided to evaluate the Sipro dive recorder so as to assess whether automatic recording of dive data would overcome this, and also to identify any limitations in the recorder and any improvements which needed to be made.

The purpose of this paper is to identify possible inadequacies in existing dive records and to report the findings from the evaluation trials. The opportunity will also be taken to present for discussion some ideas on how such a recorder might be further developed into a data-management system designed to assist both the diving supervisor and any analysis of diving activities.

EXISTING DIVE RECORDS

One of the main arguments in favour of recording dive data automatically is so as to provide consistent and accurate records, permitting simple checking and analysis as required. A study of dive records from various diving contractors shows a certain disparity between them, and that the majority of the record sheets are returned incomplete and give inadequate information. Among the points noted from the records studied were the following:

(1) Often only the name or title of the decompression table and schedule used is recorded in the diving operations log. The actual rates of ascent and times of departure from decompression stages are rarely recorded, so there may be no indication as to whether the proper decompression schedule was accurately

Advances in Underwater Technology, Ocean Science and Offshore Engineering, Volume 14: Submersible Technology

followed. The diver's ascent, surface interval and decompression profile actually followed is shown graphically on the automatic record. Where applicable, in-water stops would also be shown in the dive profile.

(2) In some cases, each phase of the dive is recorded on separate sheets, with the in-water phase appearing on one, the surface decompression on another, and any subsequent therapeutic decompression on a third. This makes any subsequent analysis of the dive unnecessarily difficult, and likely to be inaccurate.

The in-water and surface decompression phases are displayed together on the automatic record. The use of unique dive numbers would ensure that all information contained within the recorder's memory relating to a particular dive would be displayed as a complete package, regardless of any time lapse, or the recording of intervening dives, between completion of planned decompression and any subsequent therapeutic treatment.

(3) Entering all of the information required in the diving-operations log is time consuming, and therefore only those details which the supervisor considers to be important may be entered. This can also lead to the unauthorized but prevalent practice of keeping a 'rough' log, with the information being transferred to the fair log when the supervisor has the time. This not only provides inadequate information, but there is always the temptation to 'edit' the data when they are being transferred to the fair log.

(4) Existing diving-operations logs focus on the maximum depth and bottom time of a dive as the basis for selecting the particular decompression schedule to be applied. They rarely make provision for identifying or quantifying other factors which may affect that selection, such as the severity of work, ambient temperature and type of thermal protection.

THE SIPRO AUTOMATIC DIVE RECORDER

The Sipro recorder is intended to supplement the present forms of diving-operations log and is produced in two different versions, both of which record the diver's depth and ambient temperature against time.

On-line Version

On this, the sensor unit carried by the diver is connected via his umbilical to a computerized surface display and recording console. A separate sensor unit is required in the deck compression chamber to transmit data from the surface and therapeutic decompression phases of a dive, with the facility for the diving supervisor to select the relevant unit.

This version provides the diving supervisor with a real-time visual display of the dive profile as it progresses, and simultaneously records the information.

Off-line Version

The diver carries a battery-powered sensor and recording unit, which has a time range of 200 hours. On completion of a dive or series of dives within that timescale, the unit can be plugged into a jig, attached to the surface computer console, for the data to be decoded and recorded. Subsequently, the dive and ambient-temperature profile can be displayed on a VDU and printed in hard copy.

To avoid any error in relating a recording to the relevant dive in the diving-operations log, each unit should be allocated to one diver only per recording period. For surface or therapeutic decompression it is essential to ensure that sensor/recording unit is locked into the deck compression-chamber (DCC) with the diver.

EVALUATION TRIALS

The trials were conducted during 1987 at various Shell UK Expro locations by different diving contractors. The off-line version of the recorder was selected for the trial, primarily because this avoided the difficulty of having to provide divers' umbilicals with the extra electrical connectors which the on-line version would have required.

Technical Problems

Few technical problems manifested themselves during the trial, being generally limited to:

(a) loss of accuracy in recording ambient temperature and depth;
(b) failure of the recorder to 'initialize' properly;
(c) breakdown of three recorder units.

Procedural Problems

Proper use of the off-line recorder requires a certain amount of discipline in following correct procedures. A number of records were either lost, incomplete, or otherwise misleading due to the following:

(a) failure to attach the recorder to the diver before he entered the water;
(b) failure to transfer the recorder into the DCC with the diver for the surface decompression phase;
(c) leaving the recorder in the DCC on completion of surface decompression so that it was repressurized with the next diver and recorded an additional phase of surface decompression.

Improvements Required

(1) Real time should be displayed automatically.

(2) Simple functional checks are required at the start of a dive programme to ensure that the system is functioning correctly, and to check the accuracy of the sensors. Periodic calibration should also be possible.

(3) There should be the facility to overlay the recorded decompression profile with the planned profile. This would be particularly helpful to the diving supervisor when using the on-line version.

(4) The computer should be capable of sorting information so that it is grouped in accordance with discrete dive numbers for subsequent display and analysis.

POTENTIAL FOR FUTURE DEVELOPMENT

It would be possible for the automatic dive-recording system to be developed into a full dive data-management and recording system which would help the diving supervisor to keep track of all the variable factors affecting the safe and proper conduct of a dive. It would be essential for such systems to be inherently reliable, and the merits, demerits and practicability of providing such systems would need to be discussed among the various representative bodies associated with the diving industry. All that this paper sets out to do is to identify existing sensing units which could also be incorporated into the systems in order to provide the necessary information.

Anemometers

Measurement of wind speed is already available on sea-going vessels, and could readily be made available to the diving supervisor in dive control.

Tidal Current Meters

These are available commercially, and are already used on some air-diving support vessels to show the tidal current rate at the divers' work site.

Wave Radar

A fairly recent development, this is already in use on certain offshore platforms to monitor wave height and rate of surface current.

Each of these environmental monitoring devices could be incorporated into a diving data-management system. A display of both recent historical and current data from each of them would show developing environmental trends, and alert the diving supervisor as to when he should consider terminating a dive because of worsening conditions.

It is considered that, together with the existing provisions within the prototype, the incorporation of identified improvements, and the facility for typing in additional information, this would provide the diving supervisor with sufficient information to run a dive safely, and provide an accurate and complete record of all aspects of the dive.

SUMMARY

(1) There is a perceived need to improve existing methods of recording dive data in order to ensure that the records are both complete and accurate, and to assist the diving supervisor by ensuring that he has up-to-date information on the governing factors in a dive, presented in a form which he can readily interpret and understand.

(2) With some improvements automatic dive recorders of the type evaluated will provide a valuable supplementary record to existing diving-operations logs, but on-line versions are considered to be of greatest assistance to the diving supervisors.

(3) It would be possible to develop a full diving data-management system, but the practicability of doing so would have to be carefully assessed. If such a

system were to be accepted, the specifications would have to be carefully detailed in order to give confidence in its accuracy and reliability.

ACKNOWLEDGEMENT

The evaluation of the Sipro dive recorder could not have been carried out without the cooperation of the diving contractors concerned and the positive participation of their diving personnel. Their assistance is gratefully acknowledged.

18

Civil Engineering Working Practices and Standards

D. R. Shiers and P. Sellern,
Shiers Diving Contracts Limited, UK

(1) LEGISLATION AND AUTHORITY

The legislation applicable to all diving work in the UK and its territorial waters are the Diving Operations at Work Regulations 1981 No. SI 399. The legislation applies to all work inside docks, harbours, canals, inland waters, territorial waters, civil engineering sites, etc. and is enforced by the Health and Safety Executive with the Department of Energy Diving Inspectorate only advising as necessary. (Depths exceeding 50 m, DP vessels, etc.)

(2) DUTY TO ENSURE COMPLIANCE WITH THE REGULATIONS

In addition to specific duties placed on individuals by the various regulations, every person who to any extent is responsible for, has control over, or is engaged in the diving operation has a duty to ensure the regulations are complied with. In addition to the diving contractor, owners, and proposed owners of undersea cables and sewer pipelines, managers of civil engineering sites and inland waterways, harbour masters and all persons responsible for anything that effects the diving operations such as controlling water flow by means of sluices must also ensure compliance. These responsibilities continue through the contractual train from the ultimate client or owner to the diver and his attendant.

(3) QUALIFICATIONS AND TRAINING

All divers working in the civil engineering industry are required to have valid certificates of training and medical fitness. These would normally be either HSE

Advances in Underwater Technology, Ocean Science and Offshore Engineering, Volume 14: Submersible Technology

Part I or III, but due to the wide variety of work available they should also have the necessary competence to carry out safely the work which they are called upon to perform. Additional training and qualifications are also encouraged, and often required, for example Certificate for Weldment Inspection Personnel (CSWIP) 3.1U/3.2U certificates for underwater inspection.

(4) ENVIRONMENTAL AND WORKING CONDITIONS

Civil engineering by its very nature offers an extremely wide variety of working conditions together with their own particular difficulties and hazards. Poor access, poor visibility, strong river and tidal currents, debris, and heavy constructional plant and materials all present problems that require considerable experience and planning to ensure an efficient and safe operation. Particular care has to be taken in works associated with areas where differences in hydrostatic pressure could occur for example suction discharge/discharge culverts, penstocks and sluice valves.

(5) TEAM SIZE AND MEMBERSHIP

The number of persons in a diving team will vary with the type of underwater operation, the depth of water, the actual site and complexity of the operation. The minimum recommended team size would be a four-man team comprising diving supervisor, diver, standby diver, and tender. The use of a three-man team should only be used for short shallow diving operations in favourable water temperature, current and weather conditions, carried out primarily for search survey and inspection purposes. The diving supervisor should ensure that there is always a sufficient number of properly qualified divers and competent persons present to carry out the operation safely.

(6) PLANT AND EQUIPMENT

The selection of diving plant and equipment will also be dependent on the nature of the diving operation. Full LP air spreads complete with control cabins and decompression chambers may be applicable to long term civil engineering contracts, diving support vessels or for work in deep water, but for work of short duration in shallow water, mobile HP units would probably be more suitable.

Divers' personal equipment must also be chosen carefully, with free flow helmets, similar to the traditional Standard Hat such as the Swindell, which are selected for arduous construction work or for diving in polluted water and lighter demand helmets (KM17, KM10) for inspection and survey. Self-contained (Scuba) equipment, although not banned from commercial use is not generally used owing to the additional safety requirements needed.

Surface orientated dives require that air to the diver should be supplied from a suitable source through the necessary filter systems, via reservoirs which have sufficient capacity to supply air at the correct pressure and purity at an adequate rate to sustain prolonged physical exertion at the ambient pressure. A completely independent and separate air supply should be available for the standby diver. In addition to separate surface reserves, the diver also carries his own emergency reserve, which in most cases is a charged bale-out bottle.

The majority of civil engineering contracts are in water less than 10 m deep and decompression sickness is unlikely; however, the possibility of gas embolism must always be realized. For this reason, if a decompression chamber is not on site, the capability of providing suitable and rapid transport to a chamber capable of carrying out therapeutic treatment within two hours from the time when the need for recompression was realized must be planned into the diving operation. This capability is obligatory in the Diving Regulations for all work in depths exceeding 10 m. This arrangement requires considerable planning, especially when operations in coastal waters from small vessels or in remote areas inland. If a chamber is sited on the shore, it cannot always be assumed that access is possible because this may be prevented by heavy surf on the beach.

The access entry and exit to the water are often the most critical phase of a diving operation. Consideration should not only be made to the diver but also to the possibility of having to recover an unconscious diver and standby diver in the event of an accident occuring. Ladders, A-frames, winches and cranes are all to be considered, and in the latter cases must be man-riding certified. The anchorage of such systems can cause problems because quaysides and dock copings are frequently stone or concrete and the use of kentledge blocks or rock bolts may have to be considered.

SUMMARY OF TYPICAL DIVING OPERATIONS

Construction

Construction underwater demands considerable expertise and highly specialized tools which are being continually developed for the industry. Welding to steelwork, pipes, piles and offshore structures is now quite regularly achieved underwater, and with the aid of bolt guns, high-tensile bolts can be fired into steel plate up to 25 mm thick to provide fixings for anodes, structural steelwork or eyebolts. Divers fix reinforcement and help place concrete underwater using tremmie pipes or concrete pumps. They position piles and block for jetties, breakwaters and sea defences and can erect steelwork for slipways and other structures.

Excavation

Underwater excavation demands techniques much the same as more specialized land methods. Hard rock is drilled, and explosives are used to break it up. Softer shales and clays can be removed with ordinary pneumatic or hydraulic tools. One specialist tool used by the author's company is the high pressure water jet which uses a thin stream of water at pressures up to 10 000 psi and is capable of cutting through timber or soft rock with ease. For levelling formations or cleaning up canal and river beds, airlifts and jetting pumps are used in addition to the simple shovel and screed board.

Demolition

The removal of coffer dams and other temporary or permanent steel structures is usually carried out using oxy arc or thermic lance/rod cutting equipment. For the demolition of concrete structures and reinforced concrete piles it is generally necessary to use explosives, a method which is also used flattening dangerous wrecks and other underwater obstructions.

Inspection

Welds on pipelines, rig legs and all underwater steel structures need to be inspected on a regular basis, and Shiers Diving Contracts are fully experienced in the use of specialized non-destructive testing methods. Given the visibility, the condition of the concrete weight coating of pipelines and the amount of scour around structures and outfalls are periodically checked by divers using closed-circuit underwater television apparatus. This has the added benefit, that a complete videotape record can be provided for the client's files. Dams, bridge abutments, piers, locks, wharfsides, and weirs also have to be regularly inspected and maintained in order to forestall dereliction.

Surveying

As on land it is essential, before any construction is attempted, for a site survey to be carried out, whether the site is an old dock or the proposed route for an underwater pipeline. The surveys are designed to locate any obstacles or to highlight problems which might delay or endanger the projected scheme; these may include seabed samples obtained from boreholes. The line and level for dock walls and pipelines can be set out by the diver using similar techniques to those on land.

Maintenance

For offshore structures, berths and terminals, regular maintenance is essential. This work involves descaling and removal of marine growth, painting, installation and monitoring of cathodic protection systems, scour prevention, and any necessary repairs. Power station and pumping station inlets require regular attendance, and similarly lock gates, weirs, sluices and waterways must be kept free of obstruction and in good working order. Divers also assist in the placing of gabions or mattresses under water for the protection of canal and river beds. Damaged piles in jetties are repaired by welding, splicing or concrete encasing using grouting techniques in fabric membranes, and considerable savings in dry dock expenses are made by cleaning ships hulls while afloat.

CONCLUSIONS

As a result of ever-increasing quality control standards the role of the civil engineering diver has changed over the years from that of an underwater labourer to that of an engineer diver. Future projects that will require his basic skills as well as his technical expertise are:

- Tide generating barriers (the 'Severn' barrier is now close to becoming reality).
- Wind generating islands (a possible use for obsolete oil production platforms).
- Offshore airport islands.

With the continuing requirement for the construction, inspection and repair of dams, harbours, piers and offshore structures, the diver's work seems to be assured well into the twenty-first century.

19

The Future of Diver Training

A. D. Bax, Fort Bovisand
Underwater Centre, UK

Before considering the future of diver training, my intention is to take a good look at the past, consider the present, and then follow the trends that appear into the future.

THE PAST

The salvage of wreck must be the classic incentive which drives men and to a lesser extent women underwater. Throughout history this has been the case, and even the *Mary Rose* when it first sank in the mid sixteenth century was visited by 'Venetian' divers. Large bronze cannon which were readily seen or felt, plus coin and jewellery of any description were the lure, much as they still are today.

Training was minimal, if not non-existent. Equally by its very nature—hazardous, exciting, adventurous, crazy, foolhardy, isolated—diving attracted the type of person who would have made classroom discipline a definite challenge to any instructor.

The idea of teaching anyone this rather curious occupation probably first came to light with the exploitation of the first designs of the classic standard diving dress by the Royal Engineers in the 1830–40s. As the relative safety and efficiency of this dress became known, so its formal use developed with the added advantage of being nurtured in a disciplined service.

Each modification brought another wave of users, until it was common in military and civilian circles in the UK and many other parts of the world.

The trend tended to be similar in all countries: the military had the major need and so developed equipment, techniques and training standards. Industry then recruited the majority of their trained personnel from the military, and backed up their knowledge with in-house training when necessary.

Two world wars in Europe came and went with very little seeming change. However, change there was: between the world wars a young French naval diving officer had got together with an engineer and had developed a new piece of diving equipment which gave independence from an umbilical and hinted at a freedom of movement underwater of which all divers dreamed and which only the amahs of Japan and their counterparts all round the World tasted briefly.

Advances in Underwater Technology, Ocean Science and Offshore Engineering, Volume 14: Submersible Technology

Additionally new materials were becoming available which gave thermal protection without too much clumsiness, and lightness with strength. Change was definitely in the wind.

The aqualung as we now know it spread out of Europe, and in the two Gulfs (Arabian and Mexican) the standard dress was threatened for the first time in the 100 years of its history. The aqualung, with its limited endurance and tell-tale bubbles, did not appeal to the military, certainly in Europe. But its simplicity and cheapness made it available to the man in the street, and for the first time in history he/she was able to go underwater. Before he did, however, he had to be taught something. This requirement was in fact taken up by the world of the man in the street itself, and amateur diving training organizations were created by some competent and far-sighted pioneer divers. Meanwhile, in the Arabian and Mexican Gulfs technology was pushing the quest for oil slowly deeper and also encouraging new concepts in the design and mobility of surface supplied equipment.

Even after World War Two, during which many epic salvage operations were carried out, training did not come easily to the diving industry, for although the demands of new technology were heavy, the traditions that military training was a panacea and that if it was not quite all then in-house training could fill the gaps prevailed. Commercial—that is, non-military schools—grew up in both the UK and the USA, but not easily; frequently their competence was as suspect as their bank balance. In the United States the offshore diving industry also developed an apprenticeship scheme where the product of a civilian school spent several years as a tender learning the job before becoming a fully fledged diver.

Then came the revolution. The exploration which had begun in the early 1960s in the Southern North Sea bore fruit, and the great oil bonanza hit the UK as it had hit other parts of the world. Underwater construction work involved the extensive use of divers. Try as the design engineers sensibly did, the work could not be done without them. The supply of personnel from both the traditional Royal Navy and Army sources and the inshore diving industry were soon exhausted. Where would the new recruits come from? The sports organizations which had organized themselves so well? Early experience proved that leisure training helped, but certainly did not guarantee suitability for industrial work.

Such commercial schools as there were in the UK were totally inadequate to fill the gap. It was in this atmosphere that two very far-sighted men who are still active and extremely well known in the industry today decided that some sort of selection was essential from the only source of recruits that they had, that of the amateur. They sent several groups to our school, then in its infancy, with the instruction that as a form of selection they should be training to the highest standard possible in two weeks. The drop-out rate was well over 50%, but some of the men so selected remain in the industry today.

It was at about this time—1972—that the Society for Underwater Technology (SUT) produced a report which highlighted the need for training standards and adequate training.[1] This was followed in 1973 by a report by a government interdepartmental working party which studied the whole need for training offshore and found it necessary to include a recommendation for advanced diver training facilities.[2]

These studies took place against a background of increasingly serious accident statistics which were being given strident publicity. Then in 1974 a

Government Task Group was formed under the chairmanship of the Training Services Agency—an arm of the Manpower Services Commission which no longer exists.[3] Two points emerged from this meeting;

- The need for a *deep* diving school which would operate commercially but be encouraged initially by grant aid to students via the Training Opportunities Scheme (TOPS) and the European Social Fund.
- The need for training standards which could be made available internationally, as commercial training standards did not exist in the UK or in any other part of the world.

After much deliberation, and consultation these recommendations bore fruit, and in 1975/76 there appeared

- Comprehensive diving regulations—albeit in several statutes.[4–6] These regulations were sound but applied to the offshore industry *only;* the inshore industry was still governed by the regulations set out in 1960.[7]
- A government-funded underwater training centre in Western Scotland. This was contrary to the recommendation of the 1974 Task group,[3] which proposed commercial funding.
- Training standards.[8,9] As the new regulations only covered offshore activities, the standards which were established were limited to those considered applicable offshore, and were:

  Part 1 Air Diving: Scuba, Surface Supply, and Tools[8]
  Part 2 Mixed Gas: Oxy-helium and Closed Bell[9]

- Inshore activities continued under the 1960 regulations, under which experience only was required.

The training standards[8,9] also set out the facilities, personnel, etc. required for teaching the course. There were at the time—1975—about ten schools[3] of one sort or another in existence in the UK. Schools wishing to teach the standards were asked to apply, and after due consideration three schools were approved. Two of the ten existing schools, my own Fort Bovisand at Plymouth, and Prodive, now in Falmouth, and the newly created Centre at Fort William. The two southern schools were approved to teach Parts 1, and Fort William—because of its selected position with ready access to the necessary deep water—was approved to teach Parts 1 and 2.

All three approved schools benefited from the government decision to implement the recommendation of the 1974 Task Group[3] i.e. to 'pump prime' training for the diving industry. Would-be divers for the Basic Air Standard, and experienced divers for the Mixed Gas Standard, were selected by panels consisting of a representative from the Manpower Services Commission (MSC), Training Services Agency (TSA—now defunct), and the school. Those selected had their full residential fees for either the Part 1 or the Part 2 course paid.

Enshrined in the Mixed Gas Standard[9] from the beginning was the concept that it is necessary for divers to experience time at a depth of over 100 metres during training. The argument for this is that the trainee should experience a dive during which he undergoes the stress of knowing that his only safe haven is the bell, and not the surface. The argument against is that experienced divers are extremely unlikely to suffer any such 'psychological hang-ups' and indeed the impression gained is that few, if any, have ever stopped training for this reason. Equally, all the emergency, safety, and operating drills can be learnt and practised more than adequately at shallow depths. Whatever, it is undoubtedly clear that the present requirements of the Part 2 standard make it prohibitively expensive.

To continue with history. It was generally agreed that the piecemeal arrangement of the 1975 regulatons[4–6] and the discrepancy between them and the 1960 regulations[7] was unsatisfactory, and that combined regulations should be introduced. To this end in 1978 the Health and Safety Commission circulated a Consultative Document proposing unified regulations.[10] This document caused considerable discussion. Many and varied were the reactions from all sections of the diving community—except perhaps those offshore, for it effectively took the 1975 regulations and applied them inshore. By implication, among other things the Part 1 and Part 2 Training Standards were to be applied inshore. The realization that after a suitable 'grandfather' period *all* divers wishing to qualify, whatever their employment, would have to take a basic course lasting some 10–12 weeks brought particularly heavy comment. It had very serious financial implications for organizations not concerned with offshore budgets, and was obviously 'overkill'. Many meetings followed, and eventually in March 1981 the unified regulations were born[11] accompanied by a set of Guidance Notes.[12] The Regulations made provision for two new standards which were:

Part 3 Inshore. Scuba and Surface Supply *not* Tools
Part 4 Scuba only

The eventual implementation of the regulations was completed very rapidly in order to obtain the necessary parliamentary time, and the consideration of the four standards as integrated modules one leading to another was not possible. They remain therefore in their present illogical order—probably for all time.

One of the Regulations—No 14[11]—allows the granting of exemptions to the regulations, and associated with the new statute were four such exemptions. They remain in being and are for diving in connection with archaeology, journalism, scientific research, and the training of amateur divers.[15]

In June 1981 the Parts 3 and 4 Standards were published,[13,14] and subsequently in September a revised Part 1 Standard,[16] the Part 2 Standard having been revised in 1980.[17] An interesting omission from these revised standards was any description of the facilities, personnel, equipment etc. required by schools or other organizations wishing to train to them.

As in 1975, another point of training became enshrined in the standards, in that the Part 3 Standard, which is essentially the standard for inshore commercial divers, contains no requirement for tool training. Inshore commercial divers have an undoubted requirement for tools training, and at present the only way this can be achieved is by taking the complete Part 1 course. The diver or his inshore employer then has to pay for deep and offshore operational training, which the diver may well not use. It is difficult to see how this problem can be resolved without a change in Schedule 4 of the regulations themselves. When we look into the future later in this paper we will consider this.

In terms of diver certification and the accompanying regimentation imposed on an inherently independent strata of society, the 1980s will be unique.

Apart from the establishment of the Diving Training Standards, 1980 saw the implementation of the first specialist training scheme for divers. It was concerned with the levels of competence which the offshore operators felt strongly that divers should hold in the application of visual and NDT inspection techniques and procedures.

The CSWIP Scheme, as it is now known, was introduced. The major offshore operators had for some years been concerned for the quality of their inspection data. In cooperation with the diving contractors had utilized the framework of the existing scheme for the qualification of personnel operating in the 'dry' known as the Certification Scheme for Weldment Inspection Personnel (CSWIP), to create a scheme[18] to certify divers, at a Central National Examination Centre, a minimum level of practical and theoretical knowledge.

Nineteen eighty was also of considerable significance to the three approved schools in that for the first time the allocation of air students and, therefore funding, were limited. Subsequently in 1982 and 1983 the Manpower Services Commission put air diving training out to competitive tender between them. In 1984 government funding was withdrawn completely. However the numbers of trainees remained reasonably buoyant.

During this period the government also decided that it could no longer fund Fort William, and in mid 1982 funding was withdrawn and tenders invited to take over the centre as a private company. Training ceased for about one year, when it recommenced under its new private owners, who at that time ran the adjacent trials facility. However, students were now expected to pay their own fees, which they did with limited assistance from their industry.

To summarize the diving training situation; by 1984 all three government approved schools were run privately, and all government support for diving training had been withdrawn. With hindsight it might be said that the government pump-priming had gone on too long; it was certainly difficult to tell at the time. Whatever, the effect was to contribute to an oversupply of trained offshore divers. However, government funding was not alone in creating oversupply. Regulation 10(1) (b) of the 1981 regulations[11] allows divers to be qualified by virtue of their training or experience, or a combination of both. The intention was that UK divers who had been working abroad before the advent of the regulations (July 1981) would not be penalized, but would be able to qualify because of their experience. It was felt that in a year or two all these divers would have received their certificates and the majority issued would be 'by virtue of training'.

The actual effect of the inclusion of the word 'experience' in the regulations has resulted in a dramatically different situation. The HSE—presumably advised by their legal section—have felt that they are unable to interpret the regulation as it was intended and that experience gained at any time must be considered until such time as the regulations are amended. The practice therefore has been, and still is, that any diver who considers that his experience (which by the regulations[11] must have been obtained outside the UK and not from a UK-registered vessel) meets the requirements of any one of the Training Standards may apply to the HSE for the appropriate certificate. With few exceptions certificates have been issued after logbook inspection, but occasionally HSE have requested a theoretical examination, and it is assumed some have been refused. Perhaps the effect on training can best be illustrated by the available statistics:

1984 516 Part 1 Certificates were issued, of which 257 were recommended by schools after training, i.e. 50% of the certificates were issued to divers by virtue of their experience

1985 682 certificates were issued, of which 309 were recommended by schools, i.e. 57% of the certificates issued were 'virtue of experience'.

The industry had therefore no need to provide funding for diving training. A steady supply was provided by HSE from divers gaining experience and by the schools from training individuals who were prepared to find their own fees. At this time the offshore industry appeared to be in good health, and job prospects despite the oversupply seemed good enough to attract individuals willing to gamble on paying their own training fees and then earning them back in two or three years—some did. Training in the diving schools continued, but at a steadily decreasing rate, their market being reduced as the above figures show by some 50%. There was one insidious effect, perhaps not uncommon when competition is very heavy: course fees which had been kept down by the competitive tendering of the previous two years, remained down. As a result they are at the same level today. The future for schools could not be sound, undermined by the supply of divers qualified by experience and by fees which were too low.

Apart from diver training standards, 1984 saw the introduction by the Association of Offshore Diving Contractors (AODC), of a scheme for qualifying Life Support Technicians,[19] 1984 also saw the revision of the CSWIP Underwater Inspector scheme to take into account the 3 years' previous experience. In particular the scheme was split into two levels with the introduction of a second standard. There was now an Underwater Inspector Grade 1 Visual, and an Underwater Inspector Grade 2 NDT.

There followed three significant events.

(1) The disastrous collapse of oil prices in November 1985.

(2) HSE approval early in 1986 of an individual to run bell courses hiring everything necessary for the duration of courses only.

(3) The revision of all the Diving Training standards by the HSE (who had by now taken them over directly), to be implemented by November 1986.[20–23]

The fall in oil prices and the associated reduction in offshore activity severely curtailed entrants to the schools; individuals read in the press of cut after cut in manpower and realized that job possibilities were very few and far between.

The approval of an individual with no capital commitment led to a fierce price war, and the price of bell courses, far from remaining steady, dropped.

The revision of the HSE Training Standards particularly the Offshore Air or Part 1 Standard meant that Bovisand and Prodive would have to make, for them, massive capital investment in the face of a market decimated by the oil price if they wished to stay in the diving training business.

THE PRESENT

Since the advent of the first comprehensive regulations in 1975 the diving industry has changed from one with no qualifications and relatively little training to one which has a most sophisticated structure. It will help to lay out that structure at this point, including the AODC Offshore Diving Supervisor Grades introduced in mid 1986, which are planned to come into full operation in mid 1988.[24]

Diving Qualifications

Bell Diver	HSE Part 2
Offshore Air Diver	HSE Part 1
Inshore Air Diver	HSE Part 3
Scuba Diver	HSE Part 4

Offshore Diving Supervisor Qualifications

Diving Supervisor (Air)
Training Diving Supervisor (Bell)
Diving Supervisor (Bell)

Specialist Diver Qualifications

Inspection Diver

CSWIP Underwater Inspector Grade 1—Visual (3.1U)
CSWIP Underwater Inspector Grade 2—NDT (3.2U)

Life Support Technicians (LST)

Life Support Supervisor
Life Support Technician
Assistant Life Support Technician

There also exists a 'diver grading' scheme. This scheme should not be confused with the training schemes laid out above. It is concerned only with pay and has been drawn up between the two Unions concerned with diving—the Diving Section of the National Union of Seamen and the Professional Divers Association (PDA)—with the majority of the major offshore diving contractors. The contractors in this case worked individually and not as the Association of Offshore Diving Contractors (AODC).

The training structure exists and is being used. The question now is whether or not it can be sustained. Training costs money, and who will pay?—currently very few, only those individuals who can be reasonably sure that they have work guaranteed on completion of training, and a small number of foreign nationals—both self- and company-sponsored. The diving industry itself has no need to pay any more than in previous years. It is existing on the surplus of trained divers created in the past, funded predominantly by the government and the individual, as well as on the comfortable supply of divers certified by experience.

THE FUTURE

Nationalism, or something which might perhaps be called Europeanism, seems certain to play a dominant role in the future of diving training, certainly that which is concerned with Bell or Mixed Gas certification. The reason is that a philosophy which underlies all practical in-water training is that it should take place in as realistic environmental conditions as possible: it must be carried out in 'open water'. This philosophy was brought out in the wide-ranging discussions of 1973/74 and enshrined in all the UK standards and their revisions since they were first introduced in 1975. However 'open water' regrettably has not been defined, though the Association of British Diving Schools (ABDS) has long pressed that open water must be defined as 'tidal water'.

At present there is only one school in Northern Europe which clearly meets the requirement for mixed gas training if open water is taken as tidal—the Underwater Centre. Of the others, the excellent facilities at VROS, a covered mineshaft with a saturation system built over it in Melick-Herkenbosch, Netherlands, are restricted by the lack of open water. The Norwegian and French facilities have open water, but the Norwegians have simply hired a barge/vessel from industry when needed, and the French have not yet converted their barge to meet the the 1986 standard.[20]

The future of Bell Diver training will depend then on the definition of open water. It is currently being considered by the HSE in UK and by the Training Sub-Committee of the European Diving Technology Committee (EDTC). It is hoped that the deliberations of these two bodies will end with a similar decision, and there is a good chance of this because the UK Chief Diving Inspector is also the Chairman of both the EDTC and its Training Sub-Committee. The two schools in question were both established with the aid of very large government funds and are now in private hands.

If commercial decisions alone decide the future of Bell training and no school is provided with a State subsidy, then whichever way the decision on open water goes (it may have been made by the time this paper is published) the industry will probably be left with one Deep Diving School in Europe. If open water is tidal, VROS will find it very difficult to compete with the Underwater Centre as it will have to hire a barge/vessel for the open water work; if it is not tidal, the Underwater Centre will almost certainly not be able to compete with VROS, which has all the advantages of being a static, covered, inland facility.

The future of Air or Basic training, is most likely to be affected by the interpretation of the experience required for diver certification by HSE. An indication of the future already exists in the form of criteria set out by HSE in July 1987,[25] and a further proposal[26] later in that month to revise the whole question of qualifying by experience by amending the 1981 Diving Regulations.[11] The proposal recommends that the regulations are amended such that divers working in UK waters, and presumably from UK-registered vessels anywhere in the world, must hold a certificate issued by virtue of training only. It also goes on to propose that a board be set up, funded by industry, to establish a voluntary scheme whereby divers who are not intending to work in British waters and do not require an HSE certificate may receive a certificate of equivalence. An independent board assesses an applicant's experience and/or training and issues a certificate valid outside the UK indicating that the diver had reached an equivalent HSE standard. Such divers wishing to work in the UK would have to carry out an appropriate assessment or conversion course at an approved school. It would be hoped that:

- A representative of the Association of British Diving Schools would be included on the Board.
- That in assessing experience the Board, in addition to examining the logbook, arranges that practical ability is tested and theoretical knowledge examined. It would seem reasonable that the Board's assessments

should be as much in line with the assessments carried out by the approved schools during training as possible. In particular, practical ability should be checked, for although experience is valuable, all experience is not necessarily good and bad habits can become ingrained. There must not be one standard for those qualifying by training and another for those receiving equivalent certification by experience.

- Consistency is maintained by carrying out practical tests of ability at approved schools.

The HSE proposal[26] also proposes that schools in other countries should conform to the same standards as those in the UK and submit to similar inspections. This semi-formalization of an existing scheme may lead to recognized international standards of diving which can only be for the good of an industry which is essentially international.

Specialized training is also undergoing change, and the autumn of 1987 may see the introduction of an additional CSWIP standard 'Inspection Controller' (3.4U)[27] to be followed by a further standard 'ROV Pilot/Observer Inspector' (3.3U)[27] in due course. Both these standards are aimed at enhancing the quality of data collection and collation. Endorsements to the basic Visual and NDT standards—CSWIP Grades 3.1U and 3.2U—may also follow for concrete and pipeline inspection. Another future training need may also be generated over the next year or so when the proposals contained in an HSE Consultative Document[28] of 1984 for the introduction of Diving First Aid Regulations are eventually agreed. The document came under heavy fire when it was first circulated; however, a revised edition is now expected.

A final complication may arise which will affect all diving standards and therefore the future of all schools. If HSE repeat their policy (begun in 1986) of allowing an individual to hire all the necessary facilities—classroom, personnel, equipment—solely for the duration of a course, schools may become itinerant, as the lack of overheads makes such a concept apparently attractive. We shall have retreated to the situation which existed in the pre-regulation days of the early 1970s and the wheel will have gone the full circle.

CONCLUSION

In two or three years we may see the emergence of one diving training school in the UK maintaining its existence and a core of well qualified instructors all the year round by training in all specializations, probably hiring in a deep diving barge or vessel to run three or four Bell courses each winter. Having due regard for the number of divers required by the industry there may be room for one other similar establishment in Europe.

References

1. Report by the Committee on Diver Training and Performance Standards, Society for Underwater Technology, 1972.
2. Education and Training for Offshore Development. Report of a Government Inter-Departmental Working Party, Department of Employment, 1973.
3. Deep Diving Training in the United Kingdom. Report of Underwater Training Task Group, Training Services Agency, September 1974.

4. Statutory Instrument No. 1229. The Offshore Installations (Diving Operations) Regulations 1974.
5. Statutory Instruments Nos 116 and 2062. The Merchant Shipping (Diving Operations) Regulations 1975.
6. Statutory Instrument No. 923. The Submarine Pipelines (Diving Operations) Regulations 1976.
7. Statutory Instrument No 688. The Diving Operations Special Regulations, 1960.
8. TSA Training Standard Basic Air Diving & Underwater Working, TSA 1975.
9. TSA Training Standard Mixed Gas Diving. TSA 1976.
10. Health and Safety Commission Consultative Document. Health and Safety at Work (Diving Operations).
11. Statutory Instrument No. 399. Health and Safety, Diving Operations at Work Regulations 1981.
12. HS(R)8. A Guide to the Diving Operations at Work Regulations, 1981.
13. Manpower Services Commission Document. Air Diving Where No Surfaçe Decompression Chamber is Required on Site, June 1981.
14. Manpower Services Commission Document Air Diving with Self-Contained Equipment Where No Surface Decompression Chamber is Required on Site, June 1981.
15. Certificates of Exemption Nos DOW/1/81. Archaeology, DOW/2/81. Journalism, DOW/3/81. Scientific Research replaced by DOW 1/86, DOW/4/81.
16. Manpower Services Commission. Basic Air Training Standard, September 1981.
17. Manpower Services Commission. Bell Diving Training Standard, November 1980.
18. Document No. CSWIP-DIV-7-80, March 1980.
19. AODC document AODC 023, August 1984.
20. HSE Diver Training Standard Part II. Mixed Gas Diving.
21. HSE Diver Training Standard Part I. Basic Air Diving.
22. HSE Diver Training Standard Part III—Air Diving Where No Surface Compression Chamber is Required on Site.
23. HSE Diver Training Standard Part IV—Air Diving with Self-Contained Equipment Where No Surface Compression Chamber is Required on Site.
24. AODC document AODC 042, July 1986.
25. HSE letter and attachment of 3 July, 1987.
26. HSE letter and attachment of 21 July, 1987.
27. CSWIP Syllabus PH7 3.3U, 3.4U.
28. Health and Safety Commission Consultative Document. First Aid for Diving Operations Draft Diving Operations at work (Amendment) Regulations 1984.

20

Decompression Sickness in Commercial Air Divers

The Present Status of the Shields/ Lee Report

T. G. Shields, Hyperbaric Medicine Unit,
The National Hyperbaric Centre, UK

At the request of the UK Department of Energy, a survey was made of all commercial air dives carried out in the UK sector of the North Sea during 1982 and 1983, to determine whether or not this form of diving resulted in an unacceptable incidence of Type II decompression sickness (DCS). The report of the survey (Shields and Lee, 1986) was submitted to the Department of Energy in mid-1986, and resulted in a Diving Safety Memorandum (DSM 7/86) which placed restrictions on the permissible bottom time at given depths.

The survey was based on a retrospective analysis of dive logs submitted by the diving contractors. In all, 25 740 man-dive records were examined and analysed under 15 parameters, such as dive depth and time, table depth and time, type of thermal protection, etc. Surface decompression diving accounted for approximately 60% of the dives, no-stop diving for 30%, and in-water decompression for only 10%. The dives using in-water decompression had, in general, a much less severe hyperbaric exposure than those using surfacing decompression.

During the survey period, 79 cases of DCS were recorded; 44 with Type I manifestations, and 35 with Type II (neurological) manifestations. Although the nature of the data made conventional statistical manipulation impossible, an attempt was made to determine the significance of four contributing factors: the severity of the hyperbaric stress; the decompression procedure used; the extent of safety-factoring; and the type of thermal protection used. An index of dive severity (the Decompression Penalty Index) was proposed, based on a notional decompression time required for any given exposure; a dive with a DP Index of less than 30 was considered to be modest, one with a DP Index of between 30 and 60 to be of moderate severity, and one exceeding a DP Index of 60 to be severe.

Of these contributing factors, the most powerful influence was the severity of the hyperbaric stress (that is, increasing depth and/or time of the dive) (Table 1). As dive severity increased, there was a greater incidence of DCS. When this was

Advances in Underwater Technology, Ocean Science and Offshore Engineering, Volume 14: Submersible Technology

TABLE 1 *DCS related to hyperbaric stress and decompression procedure*

	Decompression penalty index					
	< 30		*30–60*		*> 60*	
	In-water decompression	*Surface decompression*	*In-water decompression*	*Surface decompression*	*In-water decompression*	*Surface decompression*
Number of dives	2 061	7 099	31	6 670	4	1 122
Number of cases of DCS	4	10	1	45	0	18
Incidence	0.19	0.14	3.22	0.67	0	1.60
Number of Type I cases	4	5	1	24	—	10
Number of Type II cases	0	5	0	21	—	1

TABLE 2 *Influence of the severity of hyperbaric stress and type of thermal protection on the incidence and type of DCS (surface decompression dives only)*

	Passive thermal protection			*Hot water thermal protection*		
	DP Index			*DP Index*		
	$\leqslant 30$	*31–60*	> 60	$\leqslant 30$	*31–60*	> 60
Number of dives	1 514	1 273	442	5 585	5 397	680
Number of cases of DCS	0	9	2	10	36	16
Incidence	0	0.7	0.5	0.2	0.7	2.4
Number of Type I cases	0	7	1	5	17	9
Number of Type II cases	0	2	1	5	19	7

controlled for, there was no obvious difference between the use of in-water stops and surface decompression, although this comparison was only possible on dives of modest severity. Thermal protection also had an effect on DCS, with a higher incidence resulting from the use of actively heated (hot-water) suits. This influence was only apparent, however, on dives of severe hyperbaric stress. Of particular concern was the increasing proportion of the Type II DCS on such dives, where hot-water suits were used (Table 2).

The conclusion from the report was that, although the overall incidence of DCS was commendably low, the total number of dives was such that an unacceptable number of individual cases of neurological DCS were accumulating. A 'no-Type-II' line was drawn on the depth/time plot of all the dives, to the left of which there would be an 'acceptable' incidence of Type II DCS (Fig. 1) and this line was used to determine a recommended bottom time at any depth. These times are the basis of DSM 7/86 (Table 3).

Although there were no cases of Type II DCS resulting from in-water decompression (possibly because 99% of such dives fell to the left of the DSM line in any case) it was decided to apply the restriction to this form of diving as well as to surface decompression diving. This was done because of a reluctance to encourage an increase in the commercial attractiveness of long-in-water-stop diving in the North Sea, with the potential dangers of long cold decompressions and the danger of pressure fluctuations on stops near the surface. It was recognized that these factors could be eliminated by the use of a closed diving bell, and that in these circumstances, the DSM 7 might not be appropriate. Indeed, evidence from other areas (Imbert and Bontoux, 1986) suggests that the use of conventional stage-decompression on long deep dives does not produce the same incidence of Type II DCS as occurred following such dives in the North Sea where surface decompression was used.

The imposition of the DSM was criticized by the diving industry on the grounds that it was based on information which was three years out of date, and

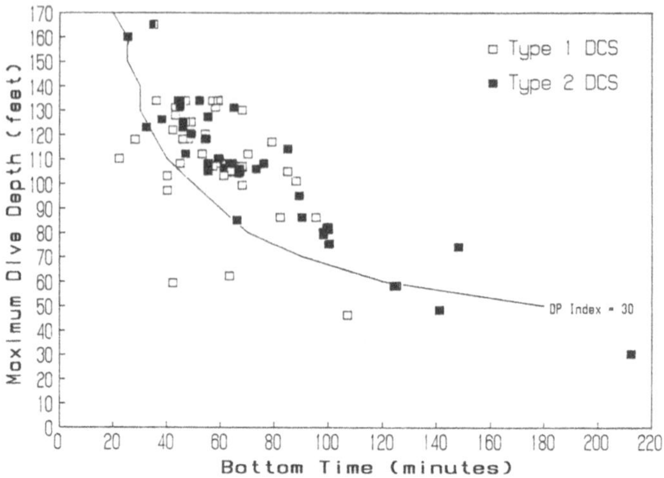

Fig. 1 Depth–time plot of all dives.

TABLE 3 *Maximum bottom times for dives within DP limits stated*

Depth (feet)	*Bottom time (minutes)*	
	DP limit of 30	*DP limit of 20*
40	360[a]	300[a]
50	180[a]	150[a]
60	120	100
70	90	80
80	70	60
90	60	50
100	50	40
110	40	35
120	35	30
130	30	25
140	30	25
150	25	20
160	25	20
170	20	15

[a]Note that these dives might exceed a maximum bottom-time limit.

that diving practices had been modified (and by implication, were producing fewer cases of Type II DCS) since then. The Department of Energy has therefore commissioned a continuation of this study, to bring the data up to date to the time of the DSM (August 1986) and to continue to monitor the effects of the DSM and of any dispensations given in terms of particular decompression schedules, for a further three-year period.

A large part of the information for 1984 and 1985 has now been collected, and a preliminary analysis of the 1984 data has been completed. At the date of preparing this paper, however, we are not at liberty to present these results publicly; the industry through the AODC Council has made strong representation to the Department of Energy that the results of the updated studies should not be revealed until there has been an opportunity for private discussion between the industry and the Department of Energy.

Analysis of data for 1985 will be completed by the end of 1987 and of those for 1986—up to the date of the DSM—shortly afterwards. These will again be discussed privately with the diving industry prior to any public disclosure, but hopefully the complete analysis from 1982 to August 1986, together with the initial results from the current year, will be available by mid-1988.

REFERENCES

Imbert, J. P. and Bontoux, M. Safety performances of French 1974 air decompression tables, UHMS Workshop on Decompression in Surface-based Diving. Tokyo, Sept. 1986.

Shields, T. G. and Lee, W. B. The incidence of decompression sickness arising from commercial offshore air-diving operations in the UK sector of the North Sea during 1982/83. Hyperbaric Medicine Unit Report submitted to Department of Energy under Contract No. TA 93/22/147.

21

Clinical Presentations and Significance of Decompression Sickness: an Analysis of Incidents from Commercial Diving Operations 1977–87

C. M. Childs, Occupational Health Limited, and University of Aberdeen, UK

INTRODUCTION

Pathophysiology of Decompression Sickness

A review of the symptomatology of decompression sickness must make reference to present understanding of the cause of the condition. Since Boyle[1,2] was able to produce decompression sickness (DCS) in animals by changes in ambient pressure, it has been held that at least the initial stimulus to the development of the condition is the formation of bubbles of inert gas within tissues and small blood vessels. The converse of this process, achieving elimination of these bubbles by recompression, was first realized in 1863.[3] The mechanical effects of undissolved gas liberated from solution within tissues and the bloodstream are not, however, sufficient to explain the variable symptomatology, and cannot account for variations in individual susceptibility, time of onset, duration, and the nature of the response to treatment. Explanations of the pathophysiology of decompression sickness are not complete, but it has been known since 1937[4] that activity at the surface of undissolved gas has the potential in plasma for causing alterations in the structure and reactivity of substances within the plasma. It is currently thought[5] that these alterations may lead to the denaturing of plasma proteins and the formation of lipid micelles from triglycerides, sterols and fatty acids normally contained within blood lipoproteins. These changes in the macromolecular constitution of blood will lead to changes in the flow characteristics of blood,[5] with increased viscosity and decreased local flow. It is suggested that there will then be intravascular aggregation followed by microembolism, platelet involvement and impaired tissue perfusion, leading to pulmonary and central nervous system damage.[5–7] Other changes thought to occur in blood following exposure to a gas surface, as would occur around bubbles, are many and will affect the coagulation, complement and fibrinolytic systems, and it is suggested[5] that

Advances in Underwater Technology, Ocean Science and Offshore Engineering, Volume 14: Submersible Technology

minor pre-dive abnormalities in the levels of inhibitors may allow the alterations in coagulation, fibrinolytic and complement activities to overwhelm any available inhibitors. Extensive local propagation or dissemination of an otherwise well-contained process may then occur.

Clinical Terminology of Decompression Sickness

A satisfactory clinical description which matches this interpretation of the cause of decompression sickness has not yet been achieved. Golding *et al.*[8] used the terms 'Type I' and 'Type II' decompression sickness in 1960 as a means of distinguishing between what they described as 'simple', and 'more serious and more complicated' cases. Opinion has varied since then about the validity of these two categories,[9] and clinical experience suggests that the terms do not accurately indicate the clinical nature and significance of symptoms they are used to describe. Further, differences in the syndrome related to the nature of the precipitating dive[9] and use of the same terms to describe the condition at stages varying from immediate post-decompression to some hours or days later do not take into account differences in the pathology of the condition which may have occurred because of the progression between initial bubble appearance and the sequence of intravascular and extravascular events outlined above.

Relation of Decompression Sickness to Dive Type and to Irregular Dives

It has not been the practice in most reviews and analyses of decompression sickness to examine incidents of decompression sickness in relation to the type of diving during which they occurred, although Doll[10] in his survey pointed out that dives carried out in the US Navy Experimental Diving Unit were often experimental in nature, and decompression sickness occurring as a result of such dives cannot be taken as typical. Incidents of decompression sickness from such military diving and from sports and semiprofessional diving in which procedures and decompression tables are not standardized, may give an impression of the clinical presentation and significance of decompression sickness which is not representative of all diving. It has also not been the practice in reviews of decompression sickness to analyse the causative dive for any problems, difficulties or incorrect practice, and to assess the treatment of early symptoms for promptness and adequacy.

Analysis of Decompression Sickness Risk

It is accepted that some episodes of decompression sickness will occur no matter how safe the procedures or how fit the diver,[11] because knowledge of the causation of the condition is still limited. Exposure to the risk of injury or illness from diving must be avoided to the extent that measures necessary to accomplish this can be expected to succeed without depriving the adventurous of the satisfaction of diving as a recreation, or professional divers from a safe living. No attempt has yet been made fully to quantify the implications of decompression sickness for the individual by long-term health assessment and to suggest from this information the level of risk of decompression sickness which can be tolerated in relation to different types of diving. Levels have been suggested[12,13] which would be acceptable during the development of decompression tables, but these criteria only deal with a small and specialized field. Any attempt to make an analysis of risk for other diving activities will depend upon accurate assessment of the seriousness of decompression incidents arising from them. It will be necessary not only that the full circumstances surrounding each incident are known, so that external precipitating factors such as incorrect procedures can be understood, but also that any terminology used to describe incidents that occur reliably represents the clinical significance of these incidents. Any categorization must take account of whether incidents are described immediately post-decompression or at some later stage when effects other than the presence of simple bubbles may have significant consequences.

Use of the Terms 'Type I' and 'Type II'

In recent years there has been concern about the sequelae of Type II decompression sickness arising from commercial diving operations, although little information about these incidents is available. For their report Shields and Lee[14] were not able to examine divers or their full case histories and could not therefore determine the accuracy of the diagnosis, the full circumstances surrounding incidents of decompression sickness, or the consequences to individual divers. The criterion they used to determine the seriousness or otherwise of incidents was the classification Type I or Type II, often originally decided upon by non-medical personnel in the diving team offshore. Any incident described as Type II decompression sickness was taken by these authors to represent a serious threat to health. In the United Kingdom this has resulted in a restriction on air diving enforced by the Department of Energy,[15] but based on the conclusion by Shields and Lee that 'the only acceptable incidence for Type II DCS in an operational situation (other than the exceedingly rare fortuitous event for which no decompression procedure can cater) is zero'.

There is a danger that assumptions will be made about the seriousness of decompression sickness based on a classification rather than on clinical findings, and on a misunderstanding of the significance of incidents, caused by lack of differentiation between diving techniques. It is proposed in this chapter that external factors such as use of a decompression table inappropriate for the dive, injury and intercurrent illness cannot be ignored in assessments of the causation of the condition. Experience in the author's clinical practice of patients with decompression sickness from commercial diving does not support the use of the term 'Type II' DCS as a reliable guide to the clinical significance of decompression sickness or to an accurate understanding of the circumstances that have led to the development or aggravation of the condition.

This chapter examines clinical data from 103 cases of decompression sickness to analyse the patterns of symptoms from air diving compared with heliox saturation diving, and assesses the seriousness of various groups of symptoms, judged by clinical recovery. The study then investigates the differences in resolution after treatment of different groups of symptoms, and the effects of influences on the dive and diver which may provoke or encourage the development of decompression sickness.

METHODS

The cases were all referred to the author for medical advice following episodes of decompression sickness while the patients were working as commercial offshore divers in the United Kingdom or abroad. The number of cases of mild decompression sickness is not great as these generally do not require medical attention and will have been treated offshore. So that the results truly represent decompression sickness from commercial diving, no patients from experimental dives are included and no cases of sports divers or semiprofessional divers, such as clam divers, are in the series. Only cases are included for which sufficient clinical information, especially descriptions and timing of symptoms, was recorded, as well as information about associated diving which would at least provide information of depth and duration.

All patients with neurological and respiratory symptoms were seen and examined by the author, as were the majority of patients with skin or joint symptoms. Records of interviews and examinations were in the first instance made in the normal format of patients' notes, and the details subsequently transferred to files in a standard Apricot computer program which allowed sorting of data by any chosen parameter.

The number of cases available with decompression sickness did not render data from them suitable for statistical analysis, and this was not considered to be appropriate in a largely descriptive clinical review. Where appropriate, the data are grouped and percentages of the components of the groups described, with the total in each group being given. The cases do not represent a statistically valid sample for the purposes of calculating the incidence or prevalence of decompression sickness, because no information was available about the total numbers of dives or divers active during the period of the study.

RESULTS

The signs and symptoms in 103 cases of decompression sickness are described. All presented to the author during the years 1977 to 1987. The patients had a mean age of 33 years with a maximum of 42 years and a minimum of 23 years.

Cases from a total of 83 divers are included, 69 of whom suffered one incident of decompression sickness. Two episodes of decompression sickness were recorded for 10 divers, four episodes for three divers and four episodes for one diver. Of the 69 divers with only one episode of decompression sickness, 56 were carrying out air diving and 13 were working on heliox saturation diving. Details of the symptoms shown by the divers with multiple episodes are described later.

Analysis of Symptoms

In this analysis of presenting symptoms and symptoms which developed subsequently, all the sites of decompression sickness from each case are included in Tables 1 and 2, with the number of patients involved in each category noted. Fifty-one cases of decompression sickness demonstrating mild symptoms were recorded from air-diving operations. Fifty of these arose during or after surface decompression diving. In these 51 cases, 75 different manifes-

TABLE 1 *Skin and joint symptoms[a] of decompression sickness (air diving and heliox saturation diving)*

	Number of cases with symptoms in the skin		*Number of cases with symptoms in or around joints*	
Site of symptoms	*Air diving (25 patients)*	*Heliox diving (2 patients)*	*Air diving (29 patients)*	*Heliox diving (10 patients)*
Shoulder	1		6	
Arm	7		4	
Elbow			11	
Forearm	1		1	
Wrist			2	
Hand	1		1	
Hip	3		5	
Thigh	8			
Knee			4	10
Shin	1			
Thorax	8			
Abdomen	6	1		
Back	5	1	2	
Total no. of presentations	41	2	36	10

[a] Symptoms may occur in more than one site in the same patient.

tations were recorded and are shown in Table 1. A total of 55% of the manifestations were in the skin, with the remainder around or close to joints. Twelve cases of mild decompression sickness from heliox saturation diving are reported (Table 1), showing 12 manifestations of the condition. All but two of these were in or close to the knee. Skin decompression sickness from both air and saturation diving occurred most often over the thorax, abdomen and back. Next most frequently it occurred in the lower limb. Pain-only (joint) decompression sickness from air diving was seen most often in the elbow and other parts of the upper limb. Pain-only (joint) decompression sickness from heliox saturation diving presented exclusively in the knee. Cases of decompression sickness seen in the skin are shown in relation to the depths of the preceding dives in Fig. 1, with incidence of pain-only (joint) decompression sickness shown separately in the same figure.

Fifty-one cases are also reported from air diving of decompression sickness which demonstrated neurological or respiratory symptoms (Table 2). Forty-four of these arose during or after surface decompression diving, the others occurring in association with in-water decompression diving, chamber dives breathing air, or SCUBA diving. A further eight cases are reported, with 12 manifestations of neurological and/or respiratory manifestations from heliox saturation diving (Table 2). Cases of decompression sickness showing respiratory and/or neurological symptoms are shown in relation to the depths of the preceding dives in Fig. 1. In association with air diving, dizziness and non-specific cerebral symptoms occurred most commonly (36% of all neurological

TABLE 2 *Neurological and respiratory symptoms[a] of decompression sickness (air diving and heliox saturation diving)*

Type of symptoms	*Number of cases with neurological or respiratory symptoms* — *Air diving (51 patients)*	*Heliox diving (8 patients)*
Dizziness	20	3
Confusion	1	
Disorientation	3	
Tiredness	3	
Restlessness	2	
Lightheadedness	3	
Incomprehension	1	
Anxiety	1	
Loss of concentration	1	
Eye signs		
nystagmus	2	
tunnel vision	5	
blurring	10	1
scotomata	6	
flashing	4	
mixed	1	1
Sensory impairment		
hyperaesthesia	1	
paraesthesia	7	2
numbness	4	1
heaviness	1	1
Motor weakness or loss	3	1
Dysarthria	4	
Nausea and vomiting	9	2
Respiratory		
dyspnoea	6	
coughing	4	
Total number of presentations	102	12

[a] Symptoms may be of more than one type in the same patient.

and respiratory symptoms), with visual symptoms presenting as 26% of all manifestations. Sensory or motor impairment, usually only occurring as paraesthesia, hyperaesthesia or weakness in some muscle groups, constituted 20% of all reported neurological and respiratory symptoms. Cases from heliox saturation diving showed a reversal of this frequency, with sensory or motor impairment being most common (41%), dizziness occurring as 25% of symptoms, and visual disturbance as 17%.

The times of onset of symptoms in relation to the associated dives are shown for surface decompression air diving in Fig. 2 (other forms of air diving having been excluded in this instance). Fourteen of the 18 cases of decompression sickness associated with heliox saturation diving presented during decompres-

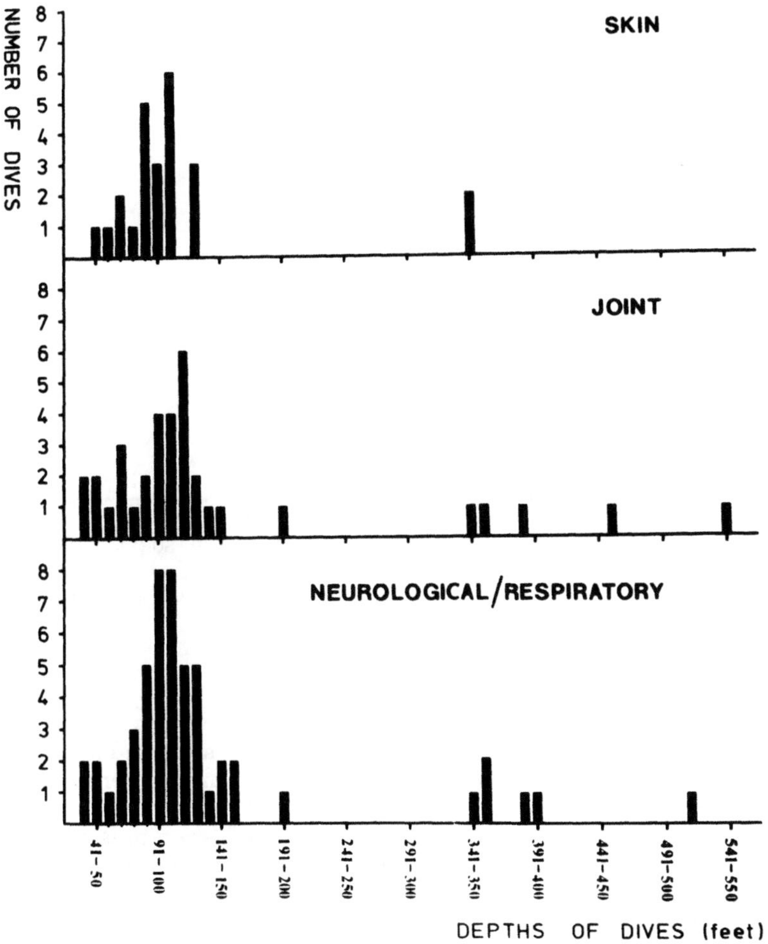

Fig. 1 Number of dives in the study developing skin, joint and neurological/respiratory symptoms plotted against the depths of the preceding dives.

sion. The four remaining cases, all exhibiting neurological symptoms, first presented at 0, 6, 10 and 14 hours after surfacing.

IDENTIFICATION OF DIRECT CAUSES OR CONTRIBUTORY FACTORS

The incidents of decompression sickness which were recorded have been analysed to determine as far as possible direct causes or other factors which might have provoked decompression sickness in divers who, under normal circumstances, would not be expected to have developed the condition. The incidents have also been analysed to determine whether the actions taken

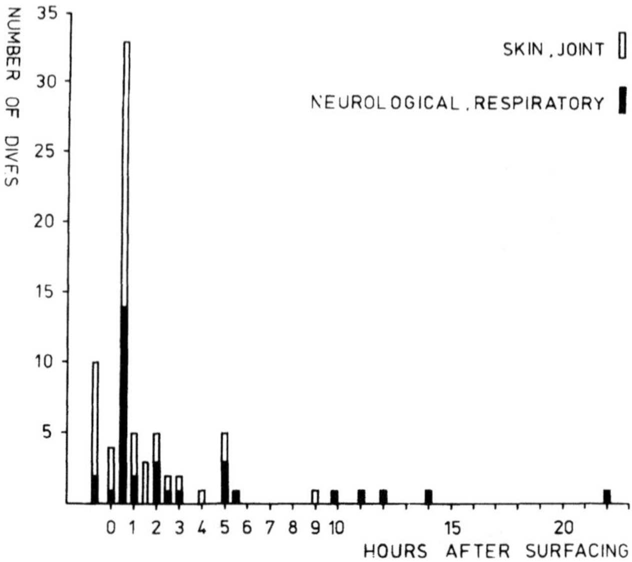

Fig. 2 Time of onset of symptoms as reported by divers during medical consultations.

following initial development of symptoms would now be regarded as appropriate and adequate. The incidents have been further analysed for the response to the treatment given, with particular attention to any patients in which the response was not immediate or who were left with lasting symptoms. Finally, the cases have been analysed for repeated episodes of decompression sickness in the same diver, to investigate the possibility that there is increased susceptibility to the condition following an incident. Evident was also sought for continuing effects upon health, or for idiosyncratic sensitivity to decompression sickness in certain divers.

In the 103 cases, 27 showed evidence of difficulties or incorrect practice during the dive. These are summarized in Table 3. Twenty-six of these cases occurred during air-diving operations, the other during heliox saturation diving. Of the 27 cases, 17 showed signs or symptoms of pulmonary or neurological decompression sickness.

Injury during the dive in four of the 103 cases was followed by the development of decompression sickness, the site of decompression sickness being close to an area of local trauma in two of the four cases. The types of injury followed by decompression sickness are summarized in Table 3.

In eight of the cases there was evidence of intercurrent illness or injury prior to the dive. Six of these cases showed signs or symptoms of pulmonary or neurological decompression sickness. These cases are summarized in Table 3.

In three of the cases it was clear that surface decompression diving had been used repetitively and that decompression sickness occurred after a period of at least one week, during which dives had been carried out each day. In one further case the diver was clearly obese and was forbidden to dive until he had lost weight to bring him into the normal range. In another three cases the symptoms

TABLE 3 *Cases demonstrating causative and/or provocative factors and other external factors*

Factor	*Cases affected*	*Details*
Dive	8	Incorrect decompression table
	6	Stops omitted by supervisor error
	3	Diver decompressed despite symptoms developing
	2	Ascent rate too fast
	2	Bottom time exceeded
	2	Trapped at depth leading to delay
	2	Bottom depth varied
	1	Heavy swell during water stop at 30 feet
	1	Irregular ascents to and from surface during dive
	1	Delay in use of oxygen during surface decompression
Injury during dive	1	Underwater explosion and inhalation of seawater
	1	Blow on hand from hammer→symptoms close to injury site
	1	Muscular strain left arm→symptoms close to injury site
	1	Electrical leak through arm during work underwater
Intercurrent illness/injury	1	Gastroenteritis
	1	Influenza
	1	Cerebral space-occupying lesion
	1	Pre-existing spinal pathology
	1	Previous local trauma→symptoms close to injury site
	1	Chronic respiratory condition, ? sarcoidosis
	2	Marked psychological disturbance
Other external factors	3	Repetitive diving
	1	Obesity
	3	Hot shower after dive

were recorded to have begun while the diver was taking a hot shower immediately after a dive. These three further factors are summarized in Table 3.

It is known that in 10 cases significant delay (20 minutes to 5 days) occurred between the onset of symptoms and the institution of treatment. This was because the diver failed to report his symptoms, the supervisor failed to take prompt action, or mechanical malfunction prevented immediate treatment. Eight of these cases originally or ultimately involved neurological or respiratory symptoms.

In five cases the treatment table or procedure would now be regarded as inadequate, either because of the incorporation into decompression tables in recent years by diving contractors of extra margins to increase safety after adverse experience with particular tables, or because USN Table 5 is now considered inappropriate for the treatment of serious symptoms. All five cases originally or ultimately involved respiratory or neurological symptoms.

Records show that in 57 of the 103 cases one or more of these causative or provocative factors was present, and that in 46 none was reported. In 28 of these, neurological or respiratory symptoms were evident.

Clinical Outcome in Cases Showing Evidence of Contributory Factors

To investigate the outcome of cases in which the various adverse factors have been identified, the cases were analysed according to the effect these factors had on the immediate response to treatment. Of the 86 cases in which full resolution of significant signs and symptoms occurred during immediate on-site treatment, there were no records of the wrong treatment table having been used (Table 4). In six cases (7%) there was a delay before treatment, and in 22 (26%) there had been difficulties or incorrect practice during the dive. In the 17 cases where full early resolution was not obtained, five cases (29%) had received treatment on an incorrect table or by incorrect procedures. Four cases (24%) had experienced a significant delay between the onset of symptoms and the start of treatment, and in five cases (29%) there had been difficulties or incorrect practice during the dive.

TABLE 4 *Effect of adverse factors on resolution of symptoms*

	Incorrect treatment table used		*Delay before treatment*		*Problems during dive*	
Total 103 cases	*No.*	*%*	*No.*	*%*	*No.*	*%*
86 with full resolution during immediate treatment	0	0	6	7%	22	26%
17 without full resolution during immediate treatment	5	29%	4	24%	5	29%

Effect of Treatment Irrespective of Causes and Contributory Factors

In an effort to avoid judgement of the effects of what have been described earlier as causes of decompression sickness, or provocative factors, all 103 cases were analysed for the effectiveness of treatment and the presence of lasting signs and/or symptoms, irrespective of supposed causes or provocative factors.

Five cases out of 103 showed signs and/or symptoms whose outcome is uncertain because they were lost to follow-up (Table 5). Improvement was continuing in one at the three-week stage, but after this he returned home to the United States. It is not known whether a doctor who suffered decompression sickness after a chamber exposure still has signs or symptoms. The remaining three were also lost to review, but their symptoms at the time they were last seen would be expected to have recovered and it is likely that they could have returned to diving.

Decompression sickness made necessary the withdrawal of Certificates of Medical Fitness from divers in the following four cases (Table 5). Two of these four divers had suffered repeated episodes of decompression sickness (four episodes and three episodes) and are discussed later. The other two divers had symptoms which were persistent but resolved after two years.

In a further eight cases (Table 5) a return to diving was allowed after a period of time. In five of these, ENG abnormalities had been detected and had returned to normal within a maximum of 36 weeks, but more often within nine weeks, after which time a return to diving was allowed. One further case showed neurological signs which resolved in eight weeks and the diver was allowed to return to diving. One case was asymptomatic after four weeks and returned to diving, but the diver suffered from decompression sickness again three months later and his Certificate of Medical Fitness was withdrawn. This case is further discussed later. The final case of these eight was asymptomatic after one year but did not need to return to diving because he was an engineer and had been compressed in a chamber only as part of a course on non-destructive testing.

The remaining 86 cases had shown full recovery from their signs and symptoms of decompression sickness as a result of immediate treatment by recompression and, as far as is known, have returned to diving.

Comparative Response to Treatment of Different Symptom Types

In a separate analysis of the effects of treatment on cases grouped into cases with visual disturbance and/or dizziness, and motor and/or sensory impairment, the response to treatment was compared in the two groups. After exclusion from both groups of all cases whose treatment was delayed or inadequate, the cases in this study show that 86% (25 out of 29) of cases with visual disturbance and/or dizziness showed full resolution or symptoms after recompression treatment, whereas only 50% of cases with motor and/or sensory impairment (seven out of 14) showed such a response.

Patients with More Than One Recorded Episode of Decompression Sickness

The final analysis of the 103 cases was for divers with more than one episode of decompression sickness for which they had sought medical advice. Only cases where medical records are held were included in the study, so divers with repeated episodes of mild decompression sickness will generally not show in this analysis. The cases are summarized in Table 6. One diver suffered four incidents of decompression sickness (three of which produced respiratory or neurological symptoms) and was declared unfit to dive. Four suffered from three episodes and the two of these, who had shown neurological or respiratory symptoms, were declared unfit to dive. Ten divers had two episodes of decompression sickness each, and in five divers at least one episode showed respiratory or neurological symptoms. None of these divers was declared unfit, although one who suffered two episodes of neurological decompression sickness gave up diving after being recommended to do so. There was no repeated pattern of symptoms in any of the divers with more than one recorded incident of decompression sickness except in one who was subsequently found to have an unrelated intracerebral lesion.

TABLE 5 *Cases which did not resolve after on-site treatment*

Case No.	*Year*	*Dive type*	*Symptoms during incident*	*Subsequent result*	*Circumstances of dive or treatment*
			Cases which did not attend for review or were lost to review		
6	1978	IWD	Neurological	recovering at 3 wk; lost to review	SCUBA dive; uncontrolled ascent rate; incorrect treatment
28	1982	SD	Neurological (ENG)	failed to attend for review	nil
43	1983	chbr	Neurological	lost to review	symptoms in doctor attending diver in chamber; incorrect treatment
72	1985	SD	Neurological	failed to attend for review	nil
74	1985	sat	Neurological (ENG)	failed to attend for review	nil
			Cases after which the diver's Certificate of Medical Fitness was withdrawn		
16	1981	SD	Neurological, respiratory	recovery after 2 yr, but Certificate withdrawn	injury during dive; incorrect treatment
2	1982	SD	Neurological	asymptomatic but Certificate withdrawn because of repeated incidents	nil
17	1982	SD	Neurological (ENG)	asymptomatic but Certificate withdrawn because of repeated incidents	nil
46	1983	SD	Neurological (ENG)	asymptomatic after 2 yr, but certificate withdrawn	badly supervised dive; treatment wrong and delayed

Cases in which return to diving was allowed or not contra-indicated					
4	1977	sat	Neurological (ENG)	normal after 36 wk, returned to diving	Abnormality found at diving medical
8	1979	SD	Neurological (ENG)	normal after 9 wk, returned to diving	no dive details available
26	1982	SD	Neurological (ENG)	normal after 20 wk, returned to diving	missed stop during dive
50	1984	SD	Neurological (ENG) and respiratory	normal after 8 wk, returned to diving	trapped underwater; inhaled seawater
71	1985	sat	Neurological (ENG)	normal after 8 wk, returned to diving	nil
15	1981	SD	Neurological	asymptomatic after 4 wk, returned to diving but subsequent incident (case no. 17).	nil
7	1979	SD	Neurological	asymptomatic after 8 wk, returned to diving	nil
13	1981	chbr	Neurological	asymptomatic after 1 yr, return to diving unnecessary because an engineer	dive badly managed; treatment delayed

TABLE 6 *Divers with more than one recorded incident[a] of decompression sickness*

Case number	*Episode A Symptoms of DCS*	*Days*	*Episode B Symptoms of DCS*	*Days*	*Episode C Symptoms of DCS*	*Days*	*Episode D Symptoms of DCS*
25	Sk—thighs	4	Weakness in legs				
100	Sk—thighs, buttocks Pn—L and R hips	7	Sk—thighs, buttocks Pn—L and R hips				
86	Sk—trunk, arms	8	Pn—R knee				
87	Pn—R elbow	15	Pn—L elbow				
15	Sk—thorax	120	Chest pain	120	Dizziness	75	Sk—trunk
33	Dizziness, vis. dist.	217	Vis. dist.	210	Vis. dist.		Weakness in legs
49	Sk—thorax, shoulders Tunnel vision	279	Pn—L elbow Dizziness				
73	Pn—L knee	305	Pn—L elbow				
26	Pn—R shoulder Dizziness, weakness R hand	425	Pn—L knee	7	Pn—L knee		
40	Chest, pain, Paraesthesia hands and feet	445	Pn—L knee, hip & shoulder				
66	Pn—L wrist	485	Dizziness, nausea				
60	Sk—abdomen	640	Chest pain vis. dist.				
38	Sk—legs	670	Pn—R shoulder	128	Pn—R elbow		
30	Pn—R knee Vis. dist.	1520	Paraesthesia R arm				
2	Dizziness (2 episodes)	1825	Vis. dist.				

[a] Episodes are denoted *A, B, C, D*. Sk indicates DCS in the skin. Pn indicates pain in or close to a joint.

DISCUSSION

Eighty-five of the cases arose during air diving, predominantly using surface decompression (78 cases), and 18 arose as a result of heliox saturation diving. No conclusion can be drawn from the greater number of incidents reported from air diving than from saturation diving or from the predominance of incidents from surface decompression diving. The number of cases in the series with only mild symptoms in the skin or around joints (45 cases) is only slightly greater than the number of cases with neurological or respiratory symptoms (40 cases). It is the author's recent experience from commercial diving that the ratio of mild to more serious cases is around 4:1. The difference between this and the ratio found in the study reflects the fact that cases with mild symptoms are usually dealt with offshore and not referred for medical opinion.

In this present study a further 18 cases showed both mild and severe symptoms of decompression sickness. The observation that symptoms were also present in the skin or around joints in 18 cases with neurological or respiratory manifestations is important, because the former symptoms may in reality be more painful and therefore noticeable, and distract attention from the treatment of other potentially more serious but less obvious symptoms.

Although no conclusions about prevalence related to exposure can be drawn from the data given here, analysis of the incidents themselves reveals much information about the symptomatology and prognosis of individual cases and of the condition overall, and about the frequency of factors thought to influence the development of decompression sickness. The widespread symptomatology of decompression sickness is confirmed by the study.

Pain-Only (Joint) Decompression Sickness

In joint manifestations of the condition, data in this study from air diving show the commonest site to be the elbow, with the shoulder second. The study confirms the observation elsewhere that, in compressed-air workers, the upper limbs are affected approximately three times more frequently than the lower limbs.[16,17] In the present study, however, pain-only decompression sickness from heliox saturation diving was only seen in the knee (10 cases). The relative frequency of joint (pain) and skin manifestations in the present study was 45% and 55%, whereas Elliott and Kindwall[9] reported 70% and 30%. This may be due to differences between the techniques of the causative dives in the two series. The depths of dives which were followed by pain-only (joint) decompression sickness varied from 40 feet to 550 feet, with most occurring in dives of between 100 and 120 feet. This range was similar to that seen for cases involving respiratory and/or neurological symptoms but was wider than the range of dives associated with skin decompression sickness.

Skin Decompression Sickness

The distinctions drawn by other authors[18,19] between subdivisions of skin decompression sickness were not found to be of great use in predicting the results of treatment, and it may be that skin decompression sickness is a single condition which can be manifest in a number of ways of varying severity. The maximum number of cases of skin decompression sickness arose after dives to between 100 and 120 feet, with only two cases presenting after dives to more than 130 feet, confirming that the condition more often follows shallower diving[9] than deep diving.[20] Skin decompression sickness in this study was much less likely to be seen after dives in excess of 130 feet than any other type of decompression sickness. Elliott and Kindwall[9] report that the most common site for skin decompression sickness is in the shoulders and upper limbs, but in the present series it was more often seen over the thorax, abdomen and back, and more often in the lower limb than the upper limb.

Neurological and Respiratory Symptoms

The range of symptoms recorded in the neurological and respiratory systems was as wide as has been reported elsewhere.[17] Dizziness, visual difficulties and non-specific cerebral symptoms were by far the most common, at 62% of the total number of neurological and respiratory symptoms, with sensory and motor impairment representing only 20% of the total number. Although the same proportion of cases complained of dizziness as in the series reported by Elliott and Kindwall,[9] many more exhibited visual disturbance and many fewer showed sensory or motor impairment. Respiratory symptoms and nausea were about the same in the two series. Incidents in this study from heliox saturation diving showed a pattern quite different from air diving, with motor and sensory impairment being much more frequent than visual disturbance or dizziness. Whatever the causative dive, visual symptoms and dizziness were usually found to resolve fully with prompt and adequate recompression (86% full resolution), whereas motor and sensory impairment were more likely to persist (50% full resolution). Further, when dizziness was a symptom and electronystagmography was carried out, resolution of the condition could easily be monitored by this test and always occurred within a period of weeks. The cases in this study suggest strongly that any perception which regards all the symptoms usually classified as 'Type II' as being of equal and sinister clinical significance will be incorrect.

Time of onset of symptoms

The symptoms of 10 cases of decompression sickness from air diving presented before surfacing, but the majority occurred within the first hour and decreasingly up to 22 hours later. No cases of decompression sickness occurred in the water during surface decompression diving unless circumstances during the dive had precipitated the condition. With heliox saturation diving most cases were reported before decompression had been completed, but four cases presented at times up to 14 hours later. The times of presentation of decompression sickness differ from those described by Elliott and Kindwall,[9] particularly after air diving where the majority in the present study occurred after surfacing. In the study, data from air diving and heliox saturation diving have been clearly separated and it may be that consideration of information from all types of diving together does not permit an appreciation of different patterns of presentation.

Decompression Sickness in Relation to Adverse Factors During Dive or Treatment

Forty-three of the cases showed that there had been injury, coincidental illness or problems during the dive, and that in some cases diving techniques had been improperly used. It is suggested that illness or injury factors could lead, because of interference with tissue perfusion and inert-gas elimination either systemically or locally, to the development of decompression sickness in divers who under normal conditions would not be susceptible. The nature of problems recorded as occurring during dives is analysed and includes use of an incorrect decompression table, over-running of bottom time and omission of decompression stops. These factors mean that decompression for the dive actually carried out has not been properly completed and decompression sickness could be expected. Further factors such as obesity and taking a hot shower immediately after a dive were also found in some of the cases, and it is further suggested that these factors are also likely to encourage the development of the condition because of increased body fat in the first instance and alteration in tissue perfusion in the second.

After symptoms have developed it is generally accepted that treatment by recompression using widely accepted tables and procedures should be undertaken at once. The data in this study show that in 12 cases such immediate and adequate treatment was not carried out. In seven, a significant delay is known to have occurred before treatment; in twos, the treatment was inadequate; and in a further three cases treatment was both delayed and inadequate. In the study overall, 82% of cases which did not resolve during immediate treatment were associated with problems during the dive, inadequate treatment or other factors, but only 33% of cases which resolved fully had this association. This part of the study clearly shows that decompression sickness is less likely to follow properly conducted dives by fit individuals, and that if the condition does occur, prompt and adequate treatment will give the best chance of full resolution of symptoms. The requirement for early and adequate treatment in order to minimize lasting symptoms will mean that diving carried out within immediate reach of a pressure chamber will result in less morbidity from decompression sickness than, for example, sports or semiprofessional diving, where recompression facilities are seldom available close to the site of diving. Incomplete resolution resulting from delayed or inadequate treatment may indicate that when symptoms persist for any length of time, decompression sickness has advanced beyond the simple presence of discrete gas bubbles in the microvasculature and tissues. This change from a simply physical cause of symptoms to a continuing pathological process of microthrombosis and more intractable ischaemia is confirmed by the finding in very serious decompression sickness of a fall in platelet count accompanied later by a rise in fibrin degradation products and other indices of intravascular coagulation.[21]

If the incidence of decompression sickness from any one form of diving is to be assessed, it is essential that details are known of all the circumstances of the dives and recompression treatments, to avoid the misinterpretation that persisting symptoms of the condition result from a properly executed dive rather than from a badly conducted dive, perhaps with improper treatment of early decompression sickness. This study shows that the latter sequence of events is more likely to be the cause of persisting symptoms of decompression sickness.

Divers with Repeated Episodes of Decompression Sickness

Fifteen divers who suffered between two and four episodes of decompression sickness are reported, the time between episodes varying between four days and

1520 days. Repeated episodes of neurological symptoms of decompression sickness are recognized as a reason for the withdrawal of a diver's Certificate of Medical Fitness, although opinion varies on the number of incidents that can be tolerated and the period of time during which incidents should be regarded as associated. Patterns of symptoms were not generally repeated in the same diver. In one diver, however, all three episodes caused visual symptoms but it was subsequently found that he had a cerebral lesion not associated with diving which was likely to have been responsible for particular susceptibility to decompression sickness. In the cases leading to the withdrawal of Certificates of Medical Fitness, the reason was generally that the divers were thought to show idiosyncratic sensitivity to decompression sickness rather than there being clinical evidence of lasting damage in the central nervous system from decompression sickness. They did not reliably show repeated similar patterns of symptoms in repeated episodes, the condition did not always arise from the same type of dive, and the period between incidents was unpredictable. At the time of withdrawal of Certificates none of the divers showed any clinical sequelae of decompression sickness, and any ENG traces which had shown abnormalities had returned to normal.

CONCLUSION

The pattern of symptomatology from commercial diving operations is shown by this study to differ from that reported from some other types of diving. Symptoms of neurological decompression sickness suggesting initial severe damage or lasting damage after treatment are not frequent, and are even less so when dives are properly conducted and adequate early treatment of any symptoms is carried out.

Standard practice in commercial diving operations following the development of decompression sickness is to carry out recompression treatment at once in a chamber at the site of the diving operation. This practice may be of primary importance in the differences in symptom patterns and morbidity. A further reason is likely to be that commercial dives should be carried out according to strict procedures, and the study shows that when these are not followed, decompression sickness is more likely to occur and to be more resistant to treatment. Resistance to treatment in such cases when dives have been improperly carried out, or the persistence of initially minor symptoms of decompression sickness which have been treated incorrectly, would fit the hypothesis that immediate recompression is able to remove the causative bubbles but that any factor in the dive or treatment which allows bubbles to persist and their surface activity to begin the processes causing aggregation, microembolism and microvascular-flow impairment may lead to the production of substances within the microvasculature and to vascular change which would not be dispelled by the physical effect of recompression.

Divers who suffer from repeated episodes of decompression sickness do not generally show consistent patterns of symptoms, and there is no clinical evidence from examination suggesting that they have any lasting local damage from previous decompression sickness. This evidence suggests individual susceptibility and does not indicate that, for all divers, resistance to decompression sickness is decreased by an episode of the condition.

The types and patterns of symptoms found in this study confirm that the range of symptoms to which the term 'Type II' decompression sickness is applied is wide but that, in patients from commercial diving, the symptoms are more of minor visual disturbance of dizziness with cerebral origin than motor or sensory, suggesting cord damage. The frequency with which these symptoms are dispelled fully with recompression treatment, the absence of findings on subsequent clinical examination, the return of ENG abnormalities to normal, and the absence of consistent patterns of symptoms in divers with more than one episode of decompression sickness, suggest no evidence of persistent local damage.

In some sets of diving procedures there is provision for an abbreviated form of recompression for skin and pain-only decompression sickness, and it may be that the terms 'Type I' and 'Type II' are justifiable, to be used, for example, by diving supervisors, in distinguishing between cases to be given this treatment and others who should be given full treatment. In all other respects the terms will not constitute an adequate clinical description, and they are not appropriate as categories for studying the clinical significance and long-term morbidity of decompression sickness.

REFERENCES

1. Boyle, R. New pneumatical observations about respiration, *Phil. Trans. R. Soc. London* 1670; **5,** 2011–2031.
2. Boyle, R. Continuation of observations concerning respiration, *Phil. Trans. R. Soc. London* 1670; **5,** 2035–2056.
3. Foley, A. E. *Du Travail dans l'Air Comprimé. Étude Médicale, Hygiénique et Biologique Faite au Pont d'Argenteuil*, Paris: Bert, 1863.
4. Bull, H. B. and Neurath. H. The denaturation and hydration of proteins. II. Surface denaturation of egg albumin, *J. Biol. Chem.* 1937; **119,** 163–175.
5. Hallenbeck, J. M. and Andersen, J. C. Pathogenesis of the decompression disorders, in P. B. Bennett and D. H. Elliott (Eds), *The Physiology and Medicine of Diving*, London: Baillière Tindall, 1982; pp. 435–460.
6. Swindle, P. F. Occlusion of blood vessels by agglutinated red cells, mainly as seen in tadpoles and very young kangaroos, *Am. J. Physiol.* 1937; **120,** 59–74.
7. Lee, W. H. and Hairston, P. Structural effects on blood proteins at the gas–blood interface, *Fedn. Proc.* 1971; **30,** 1615–1622.
8. Golding, F. C., Griffiths, P., Hempleman, H. V., Paton, W. D. M. and Walder, D. N. Decompression sickness during construction of the Dartford Tunnel, *Br. J. Ind. Med.* 1960; **17,** 167–180.
9. Elliott, D. H. and Kindwall, E. P. Manifestations of the decompression disorders, in P. B. Bennett and D. H. Elliott (Eds), *The Physiology and Medicine of Diving*, London: Baillière Tindall, 1982; pp. 461–472.
10. Doll, R. E. Decompression sickness among US Navy operational divers: an estimate of incidence using air-decompression tables, *Navy Experimental Diving Unit, 1965; Research Report* 4–64.
11. Ciria Underwater Engineering Group. *RNPL Metric Air Diving Tables, 1976*; Report UR7.
12. Hanson, R. de G. and Sowden R. R. Trials for testing diving tables, Admiralty Marine Technology Establishment, 1979; *Tech Memo AMTE(E) TM,* **79,** 403.
13. Hennessy, T. R. *Decompression Diving: Which Tables to Use?* South African Underwater Union, 1975.
14. Shields, T. G. and Lee, W. B. The incidence of decompression sickness arising from commercial offshore air diving operations in the UK sector of the North Sea during 1982–83. Department of Energy, London, 1986.

15. Department of Energy. *Diving Safety Memorandum No. 7/86.* Department of Energy, London, 1986.
16. Slark, A. G. Treatment of 137 cases of decompression sickness, Medical Research Council, RN Personnel Research Committee Report 1962; 63/1030.
17. Rivera, J. C. Decompression sickness among divers: an analysis of 935 cases, *Milit. Med.* 1964; **129:** 314–334.
18. Behnke, A. R. Decompression sickness following exposures to high pressures, in J. F. Fulton (Ed.), *Decompression Sickness*, London, Philadelphia: Saunders, 1951.
19. Rashbass, C. Aetiology of itching on decompression, Medical Research Council, RN Personnel Research Committee Report 1957; UPS 167.
20. Miles, S. *Underwater Medicine*, London: Staples Press, 1966.
21. Philp, R. B., Inwood, M. J. and Warren, B. A. Interaction between gas bubbles and components of the blood: implications in decompression sickness, *Aerospace Med.* 1972; **43,** 946–956.

Part V

Developments in Underwater Intervention (including Robotic Systems)

22

Design, Installation and Operation of an ROV-operated Subsea Production System in 450 msw

F. R. Frisbie, E. W. Hughes and C. E. Brown,
Ocean Systems Engineering Inc., USA

The first large-scale template-based subsea production system to be installed beyond the practical depth for diver intervention was set in February 1987 by Placid Oil Company in Green Canyon, Block 29, in the Gulf of Mexico at a depth of 450+ m. The installation relies heavily on remote-work system intervention and support for both operation and maintenance. The system consists of a 24-slot main production template producing through a fixed buoyant riser to a floating drilling/production semisubmersible permanently anchored overhead. Future satellite templates are planned which will produce back to the main template.

The remote work-system which will be used will be a fully founded, dual-manipulator, electrohydraulic, work-capable remotely operated vehicle (ROV). In some cases the remote work-system is used as a back-up/secondary means of operation. In many instances, however, it provides the primary and often the only facility for operation of the subsea components. The concept of using the remote work-system as the delivery vehicle for subsea valve operators allows the operators to be easily removed to the surface for maintenance and repair. It also eliminates the costly and often complex task of designing components for a 20+ year life in a fully submerged condition. The design parameters which were used for this development are summarized below:

- Gulf of Mexico
- Water depth 500–1000 m
- Semisubmersible work platform
- Fixed buoyant riser
- 24-slot main template
- Pipelines to shallow-water platform
- Four ea. 4-slot satellite templates
- ROV support—HYDRA™ 1000

INTERVENTION TASKS

The intervention tasks to be performed by the remote work-system evolved over a period of months. The tasks cover the spectrum of underwater activities from installation through long-term operations. Some of the work tasks required that hardware be developed. Others required procedure development. A list of the

Advances in Underwater Technology, Ocean Science and Offshore Engineering, Volume 14: Submersible Technology

intervention tasks by major subsea component follows. This list outlines only work tasks. Inspection tasks—of which there are many—are not included.

Main Template

- Lift sling disconnection and removal*
- Pipeline ball valve operations*
- Pig trap isolation valve operations*
- Pig trap launching valve operations*
- Ball-valve stem lubrication*
- Ball-valve sealant injection*
- Guidepost replacement*
- Guidewire establishment*
- Shuttle-valve operation and replacement*
- Anode replacement*
- Pipeline/flowline pull-in

---

*denotes ROV used as primary method

Template/Satellite Trees

- Hydraulic gate-valve override
  production master valves
  annulus master valves
  wing valves
- Needle manual gate-valve operations*
  isolation
  chemical injection
  testing
  tree cap
- Sealing-ring replacement*
- Connector override/release

Production Riser
- Air-valve replacement*
- Control umbilical installation/replacement
- Installation guidewire removal*

Control Pods

- Needle/ball-valve actuation*

---

* denotes ROV used as a primary method

HARDWARE FOR REMOTE WORK-SYSTEM INTERVENTION

To enable the remote work-system to perform some of these tasks required that certain hardware be developed. Most of this hardware took the form of valve actuators for various types of valves and widely varying torques. A list of the hardware designed and produced is shown below.

- Needle/manual gate-valve actuator
- Guidepost replacement tool
- Pig-launching valve actuator
- Pig-isolation valve actuator
- Production-pipeline valve actuator
- Hydraulic main-block gate-valve actuator
- Control umbilical riser clamps
- Replaceable hydraulic shuttle valve and panel
- ROV docking/interfaces
- Ocean System's Engineering (OSE) patented pressure-balanced connectors
- Launch-sling release

DESIGN DETAILS

Details of each of the tool designs are given in the sections which follow. The smaller, lower-force requirements were designed to utilize the manipulators. Larger, higher-force requirements eliminate the need for manipulators wherever possible. Where practical the remote work-system is docked next to the component which it is actuating, to eliminate relative motion between. This simplifies the task for the operator.

Needle/Manual Gate-Valve Actuator

The needle/manual gate-valve actuator is a hydraulically powered rotary tool providing 0–108 N m if torque. Shown in Fig. 1, the actuator is 580 mm × 240 mm × 330 mm (l × w × h) and weighs 14 kg in air. The tool is used for operating $\frac{1}{2}''$ manually operated needle valves and gate valves on the trees and control pods. The actuator is provided with a mechanical turns-counter to allow the operator to monitor visually the number and direction of revolutions through which the tool has turned.

Fig. 1

Guidepost Replacement Tool

The guidepost replacement tool is used for removing and installing an annular pin which retains the guidepost. The tool can be set and operated from two sides of each guidepost to minimize the possibility of interference with a bent guidepost prohibiting the tool from being set. The guidepost replacement tool is shown in Fig. 2. The tool is 400 mm × 115 mm × 320 mm and weighs 12 kg in air. It can provide up to 250 kgf in either the pushing or pulling mode. The tool is designed to capture fully the pin prior to removing the pin from the guidepost socket, to minimize the possibility of dropping the pin during installation or retrieval. The template was set without the guideposts installed, so one of the first tasks which the remote work-system must perform prior to commencement of drilling operations is guidepost installation.

Fig. 2

Pig-Launching Valve Actuator

The pig-launching valve actuator, shown in Fig. 3, is used to rotate a modified ball valve 180° for launching ball pigs. There are two pig launchers on the template—one for each sales line. Pig launchers holding multiple pigs will be lowered and connected using mechanical connectors. The remote work-system will then operate either the 20-inch launching valve or the 16-inch launching valve to release individual pigs as required. The pig-launching valve actuator is 430 mm × 330 mm × 610 mm and weighs 25 kg in air. It is capable of developing up to 4750 N m of torque through multiple rotations.

Pig-Isolation Valve Actuator

Two isolation ball valves are installed downstream of each of the pig launchers for pressure integrity and control. These are located inside the template structure and cannot be accessed by the remote work-system. Rather than permanently install actuators on the ball valves, yoke and pull-rod assemblies were installed on each of the ball valves. The actuator for operating the isolation valves is shown in Fig. 4. An overview of two of the actuator docking assemblies is shown in Fig. 5. The pull-rods have sufficient columnar stability to allow actuation of the ball valves by either pushing or pulling, but the primary

Fig. 3

Fig. 4

Fig. 5

mode of operation is to pull the rod. The actuator is 330 mm × 330 mm × 1375 mm and weighs 70 kg in air. It can develop 32 tonnes of force, thereby imparting 95 000 N m of torque to the valve.

Production Pipeline Valve Actuator

The template is fitted for the future connection of up to four import pipelines. Each of these lines is fitted with a ball valve. The actuator for operating these 14-inch and 16-inch ball valves is shown mated to one of the valves during testing in Fig. 6. The actuator is 940 mm × 660 mm × 510 mm and weighs 60 kg in air. It is capable of developing up to 40 500 N m.

Fig. 6

Hydraulic Main-Block Gate Valve

A linear actuator was built for operating and locking open the hydraulically operated gate valves on the trees. The actuator assembly consists of two discrete components—the actuator and the locking assembly. In this manner, only one actuator must be built. As many locking assemblies as may be necessary can be built, set, operated, and left on the trees using only one actuator. The locking assembly is uniquely configured to provide precise full-open locking of the gate to allow passage of a wireline plug. The locking device can tolerate manufacturing tolerances of up to ±6 mm between the location of the valve stem with respect to the valve body as it varies from valve to valve. This was necessary due to the actual manufacturing tolerances associated with the construction of these subsea valves. The actuator is 450 mm × 180 mm (l × dia), weighs 28 kg, and can develop 30 tonnes of force. The locking actuators each weigh 20 kg, are 350 mm × 180 mm (l × dia), and are designed to lock and hold 30 tonnes on the valve stem.

Control Umbilical Riser Clamps

Clamps for attaching the control umbilicals to the production riser, which can be opened and closed by the remote work-system, were designed. The control umbilicals are 200 mm in diameter. The riser clamps are frequently located the length of the production umbilical to eliminate the possibility of the umbilical billowing out in the current and increasing the drag on the riser.

Replaceable Hydraulic Shuttle-Valve Panel

Shuttle valves which are operated by the remote-work system are installed on the template. These interface with the control system to provide selection between the control pods for certain functions. The design of the shuttle valves is based on OSE's highly successful, patented, ROV-operable, pressure-balanced connector. The shuttle-valve panel is shown in Fig. 7. The unique feature of the valve design is that all of the seals are installed on the male stab. Should any of the seals fail, the male can be removed from the valve body, returned to the surface, the seals replaced, and the male reinserted into the valve body by the

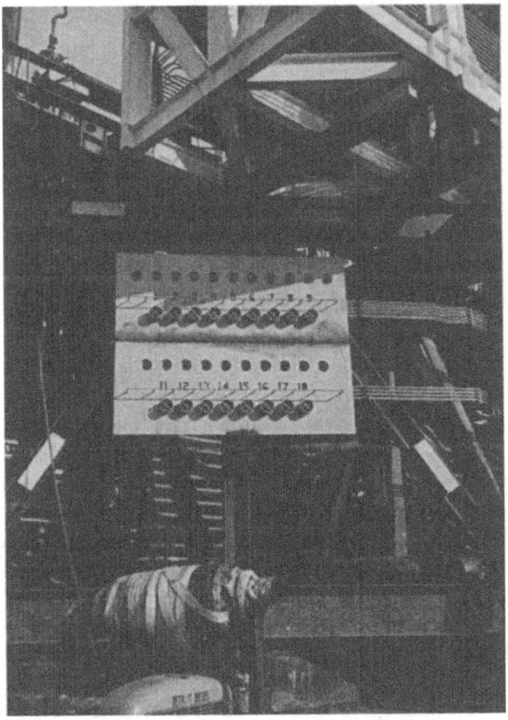

Fig. 7

remote work-system. Each of the shuttle-valve assemblies is 355 mm × 180 mm (l × dia) and is rated for 5000 p.s.i. operations.

ROV Docking/Interfaces

Each of the tools and actuators required special mating assemblies on the subsea hardware. Wherever possible, the remote work-system is docked prior to engaging the tool. Docking ports are evident on the shuttle-valve panel shown in Fig. 7. The docking configuration and valve interfaces on the tree are shown in Fig. 8.

Fig. 8

OSE Patented Pressure-Balanced Hydraulic Connector

Ocean Systems Engineering's patented, pressure-balanced hydraulic connector was used throughout the installation. The connectors are used in several places for applications other than for hydraulics. The connectors are also utilized as the means for sealant injection and stem lubrication on the ball valves on the template.

Launch-Sling Release

Removal of the launch slings during the installation required that 200-mm diameter shackle pins be retracted. Hydraulic cylinder assemblies were built which would allow the remote work-system to retract the pins. Hydraulics were provided from the Hydra^{Tm} 1000 using another of OSE's pressure-balanced hydraulic connectors. This operation was successfully completed by the remote work-system in a matter of minutes.

CONCLUSION

A subsea template has been installed beyond the range of divers which relies heavily on remote work-systems for operation and support. A wide range of tools and actuators has been designed and built to ensure the long-term success of this operation.

23

The Role of Two New Autonomous Diving Systems in Underwater Intervention

G. S. Hawkes and S. A. Earle,
Deep Ocean Engineering Incorporated, USA

Underwater vehicles are tethered or autonomous, manned or unmanned. Tethered work systems, whether remotely operated or manned, have the advantage of unlimited surface-supplied power, but are limited in range by the length of their cables to a support vessel or surface platform. Such systems include nearly 70 vehicles that have been operated successfully during the past decade, including 15 Jims, 18 Wasps, 30 Mantis submersibles, two Wranglers and two Spiders. While autonomous in terms of life support, all typically rely on a tether for launch and recovery as well as for primary power and communications.

Complete autonomy is normal for large manned submersibles, but it is largely atypical for atmospheric diving systems (ADS), microsubmersibles such as Mantis, and for most remotely operated vehicles (ROVs). This changed with the advent of the autonomous one-person microsubmersible, Deep Rover (Fig. 1), and subsequently, the smaller, wider-ranging Deep Flight (Fig. 2). A description follows of the development and application of these systems.

DEEP ROVER

Deep Rover was designed to incorporate desirable features of large submersibles, including autonomous operation and increased range and depth capability, in a small, transportable system that, like an ADS, could be piloted by diverse operators with relatively little training at a cost significantly lower than that required for larger, more complex systems.

The senior author, designer of Wasp, Mantis and various ROVs, set out to produce a submersible that is user friendly so that the operator can concentrate on the work at hand while comfortably managing and operating the submersible. Success in achieving this goal can be measured in part by the time required to master basic operational procedures—about 30 minutes. In a seated position, the pilot exerts slight forward motion on one or both arm-rests. Microswitches in the arm-rests activate appropriate thrusters and the sub moves forward; reverse is triggered by leaning back. Such movements soon become instinctive and most of the pilot's attention can therefore be concen-

Advances in Underwater Technology, Ocean Science and Offshore Engineering, Volume 14: Submersible Technology

Fig. 1 Deep Rover.

trated on working. The effect is in some ways comparable to SCUBA diving in terms of spontaneity of action, dexterity and manoeuverability—but without the burden of decompression nor with the complexities typically involved with piloting a small submersible. Such ease of operation is taken for granted by astronauts with moon-walking or space-walking suits, as well as by divers, but it is not often considered in autonomous submersibles. Most require a dedicated pilot, with work performed primarily by one or more crew.

Deep Rover's man/machine interfacing relates directly to its ADS heritage, where the concept of the vehicle behaving as an extension of the operator to perform work is fundamental to the commercial success of the system.

One indication of the ease of operation is the number of people who have served as pilots. Within the first two years of operation, Can-Dive Services Limited (Deep Rover's owners and operators), logged 165 working dives involving 80 different operators. An additional 100+ short, shallow dives were made by individuals ranging from the son and daughter of the designer (at ages 13 and 14) to a retired British admiral.

Deep Rover was originally intended to be used for subsea intervention on oil rigs in eastern Canadian waters. PetroCanada awarded Can-Dive a two-year contract for Deep Rover when it was in the design phase of development, but the rapid decline of the offshore oil industry caused the application of Deep Rover to rig work to be delayed indefinitely. Instead, the vehicle has been used for various survey and inspection assignments and has attracted considerable notoriety through use for scientific research and exploration projects.

The first research applications took place in Monterey Canyon, offshore from

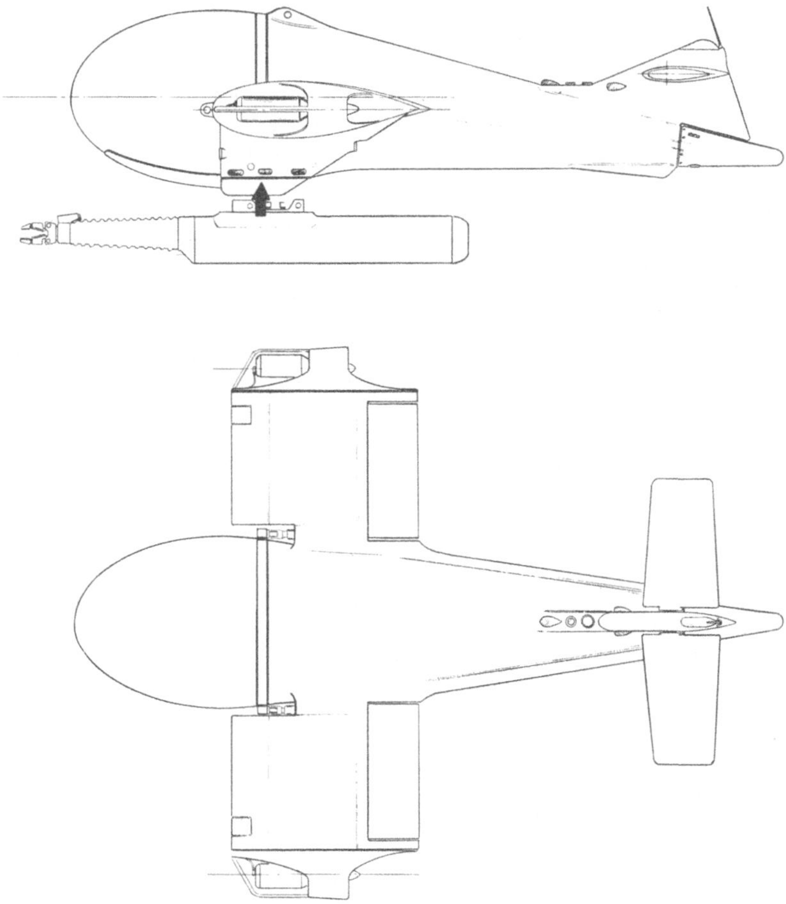

Fig. 2 Deep Flight.

California during August and September, 1986. Dr Bruce Robison, zoologist from the University of California in Santa Barbara, with support from the National Science Foundation, organized a project to explore and document life in the canyon. He and four other scientists were given standard operations instruction for Deep Rover and subsequently routinely dived Deep Rover to 2000 feet.

During 30 days, these scientists piloted Deep Rover for 51 dives averaging four hours per dive, resulting in hundreds of new observations about life in the deep sea. Concurrently, more than 60 hours of underwater video film of marine life was recorded by the scientist/pilots, both for scientific documentation and for use by Sea Studios Incorporated and the British Broadcasting Corporation.

In November 1985, Deep Rover returned to California from its home base in Vancouver, British Columbia, for a series of training and research dives near San Diego. Thirteen people trained in Deep Rover and four record-making descents were made to 3000 feet. The first, made by the senior author, extended the solo depth record in the sea by approximately 1000 feet.

From November 1985 through August 1986, Deep Rover remained in Canada and was engaged in commercial operations, including pipeline inspection, salvage, and several film projects. It was mobilized in late August for transport to the Bahamas for a month of work at the Caribbean Marine Research Center on Lee Stocking Island. During 28 days there were 66 dives including 21 working research dives. Maximum depth attained was 3000 feet and the longest dive was 7.5 hours.

Currently, Deep Rover is being prepared for a series of dives in Ontario, Canada in connection with inspection of dam gates, and additional research dives are being planned for late 1987 and 1988.

Originally designed for oil-field operations, Deep Rover has demonstrated versatility for other subsea applications. Use of manned systems generally has declined as the use of ROVs has gained favour, but Deep Rover continues to provide effective special-purpose working access under water.

DEEP ROVER FEATURES

Basic features of Deep Rover include:

- a pressure hull 5 inches thick, 5 feet in diameter, constructed of two acrylic hemispheres joined by an aluminium ring, affording 360° vision, rated to 3000 feet with a 4 to 1 safety factor, certified by Lloyds;
- two modular sensory seven-function manipulators with integrated control system that incorporates vehicle movement with manipulator controls;
- autonomous operation with power for a normal six- to eight-hour mission; emergency power and life support for 150 hours. Optional tether/cage deployment system for special applications;
- air weight, 7000 pounds;
- safety provisions including drop weights, air ballast, battery drop, payload drop;
- sensors and instruments including depth, altitude, magnetic heading, temperature, cabin pressure, oxygen and carbon dioxide sensors, colour sonar, external video and still cameras.

DEEP FLIGHT

Autonomous manned submersibles such as Deep Rover, and multipassenger submersibles including the Alvin, Pisces, and Johnson-Sea-Link I and II, avoid the problems inherent in tethered ROVs and ADS, but, despite their relative freedom, such systems are basically dirigibles relying mostly on ballast and buoyancy for slow vertical ascent and descent, and on battery power for lateral mobility. Speed for small manned vehicles rarely exceeds one to two knots, and battery capacity usually lasts for six to eight hours of normal operation. Range, therefore, is limited to a small working area. It is generally not regarded as practical to use such vehicles for search, survey, or any task requiring wide-ranging activity, or speed.

Deep Flight, a new vehicle designed by the senior author, is intended to bridge the gap between towed-array survey systems and site-specific vehicles. Built for rapid deployment and autonomous operations in depths to 4000 feet, Deep Flight is a relatively far-ranging fast vehicle more related to fixed-wing aircraft than to typical dirigible-style submersibles. Although basically a manned system, Deep Flight is designed to be adapted for future computer-controlled autonomous manned operation and has the potential for using a combination of artificial intelligence (AI) and manned control.

Development of an effective search vehicle, with survey, identification, and manipulative capability, has been made feasible by newly developed forward-looking, scanning-sonar systems. Compared to side-scan sonar, the new forward-looking systems have a shorter range but, in general, greater effective resolution because of the host vehicle's ability to fly close to the sea bed, giving a greater relief. The forward-looking aspect enables the vehicle's operator to acquire information on possible targets, then home in immediately for definitive visual identification. A forward-scanning sonar, like side-scan, must be used in conjunction with a separate camera system that can effect positive identification of likely targets. Combining these two systems in one vehicle that has freedom of movement creates a powerful tool for subsea search, since it has the capability of being able to acquire, locate and identify in one operation.

Specifications for a search vehicle capable of locating small targets such as wreckage or debris would be as follows:

1. The system should carry forward-scanning sonar with a maximum range consistent with the required resolution (approximately 300 feet).
2. The system should carry suitable target-identification equipment such as video cameras or specialized sensors.
3. The vehicle must be able to navigate with positional accuracy that is within the range of ± 5% of the sonar range in order to minimize search overlap.
4. The vehicle should cruise at the optimum search speed of the sonar. The optimum speed is the fastest consistent with the number of target scans for reliable operator recognition. With current sonar, this is approximately 5 knots, but increased speeds should be possible with future rapid electronic scanning.
5. The system should permit the immediate physical investigation of possible targets. Ideally, this means the ability to deviate 90° from a search pattern for a distance equal to the sonar range, stay long enough to effect identification, then return to the point of departure from the pattern and continue the search.
6. At all times the vehicle must hold pitch/roll/heading translations within limits that enable the sonar to maintain high resolution.
7. The system should operate to the depth required for the assigned task.

REQUIREMENTS FOR AN ROV SEARCH VEHICLE

The conceptual design for an ROV to meet the above specifications runs into seemingly fundamental design barriers. There is a trade-off between depth and speed imposed by cable limits of strength and/or power transmission and cable drag. An analysis by the senior author suggests that the conceptual limits of current materials technology are such that if priority is given to item 4, a search rate over the bottom of 3.8 ft/s (2.28 knots) is the maximum feasible for a 4000 ft free ROV system. Streamlining of the vehicle is almost irrelevant as the cable drag is overwhelmingly dominant. Use of depressor weights or other means of cable control are not feasible if item 5 is to remain uncompromised. Table 1 indicates current practical limits given surface-supplied power and present cable materials technology. It assumes a streamlined (tear-drop shaped) cable

TABLE 1

Operational depth (ft)	*Maximum feasible speed*	
	ft/s	*knots*
4000	3.8	2.28
2000	4.94	2.96
1000	6.42	3.21
500	8.35	5.00

cross-section with advanced-fibre structural members, fibre optics for data and ultra-high voltage power transmission.

REQUIREMENTS FOR AN AUTONOMOUS UNMANNED SEARCH VEHICLE

A successful autonomous unmanned search vehicle would require AI capable of coping with item 5, or would have to be controlled from the surface in real time. At present, AI systems have not been developed that can accomplish such tasks, nor is data transmission feasible over the required range at the data rate required.

It is anticipated that Deep Flight will be useful in helping to assess and solve problems associated with autonomous unmanned systems and that fully autonomous unmanned systems and AI-assisted manned vehicles will be practical in the near future.

REQUIREMENTS FOR AN AUTONOMOUS MANNED SEARCH VEHICLE

The operational limits for an autonomous manned vehicle are free of cable constraints. The payload available for energy storage versus depth (pressure hull weight) gives a trade-off in terms of speed or endurance versus depth. For example, within the limits of current materials technology, it is possible for a manned autonomous vehicle to operate at 4000 feet for five hours at a speed of 5 knots, with a total displacement of 3300 pounds.

Comparing the tethered ROV to the autonomous manned vehicle, the autonomous system operates more efficiently below 500 feet, but practical considerations, principally endurance, favour the ROV in depths to 2000 feet. The decision to develop a manned autonomous vehicle streamlined for search/survey/identification was based on the relative complexity of an ROV that could accomplish such tasks and the need for a sophisticated, dynamically positioned support ship to operate such a system effectively.

DEEP FLIGHT DESCRIPTION AND FEATURES

Deep Flight is streamlined with minimal frontal area to reduce drag and provide an efficient vehicle for covering distances. The cruising speed, five knots, is currently regarded as the optimum rate of through-water advance for forward-looking sonar. The vehicle has provision for advanced navigation systems, video cameras, and sonar, modular, interchangeable-mission payload pods for manipulators, extra batteries, beacons, or other accessories.

The pilot in Deep Flight is positioned face-forward, as in the tethered microsubmersible, Mantis, with head and shoulders surrounded by clear acrylic. The system is designed to have an advanced fly-by-wire control system, enabling the pilot to set a fixed course controlled by on-board computers that provide the stability required for sonar. The system will automatically maintain optimum altitude and search pattern.

Deep Flight with a moderate payload weighs approximately 3500 pounds, has a depth capability to 4000 feet, and is 14 feet long. Deployment may take place from various platforms with appropriate lift capability, and provision is being made for a towing bridle and recovery system for rapid towing behind a small, fast vessel for appropriate shore-based operations. Under special conditions, helicopter deployment is feasible. Such deployment modes provide alternatives not only to large, dedicated surface-support ships required for many manned submersibles and large ROVs, but also to large vessels of opportunity. Deep Flight's small size and weight, coupled with its versatility, mean that the cost of search, survey and identification operations can be substantially lower than that of conventional submersibles and ROVs.

Basic features of Deep Flight include:

- a pressure hull constructed of a forward acrylic section joined by an aluminium ring to a composite aft cylindrical hull employing novel composite materials;
- two modular sensory manipulators, retractable within special-mission pods;
- autonomous operation with power for five hours of continuous operation at 5 knots and emergency power and life support for 96 hours;
- 4000 ft depth rating with a four to one safety factor for the pressure hull;
- safety provision including drop weights, air ballast, battery drop, payload drop;
- sensors: depth, altitude, magnetic heading, attitude (pitch and roll), rate gyro, speed indicator, temperature, forward obstacle sonar, forward scanning sonar, SIT and colour cameras, homing and tracking systems, dead-reckoning navigational computer.

Two Deep Flight systems are under construction by Deep Ocean Engineering for delivery in 1988 to Sea Hawk Deep Ocean Systems, Tampa, Florida, USA.

REFERENCES

Earle, S. Microsubmersibles: putting more scientists in deep water. *Sea Technology* **27** (12): 14–21, 1986.

Hawkes, G. and Earle, S. Deep Flight: a new approach to autonomous underwater search and survey vehicles. *Proceedings, Marine Technology Society*, ROV '87, 1987.

24

Wellman: the Subsea Intervention System

E. W. Nesse, H. Lindland and R. Husebye, Statoil, Norway
and K. Höglund, Asea Oil and Gas, Sweden

In 1984, the Norwegian oil company Statoil and the Asea Group of Sweden, entered into a general technical cooperation agreement. The purpose of this cooperation was to explore the possibilities for advancement within offshore technology by combining the competence and resources of the two companies.

Subsea technology was recognized as a major area of importance as there are plans for more than 100 subsea installations on the Norwegian Continental Shelf alone over the next 10 years. On a global basis, this number might exceed 500, of which more than 100 could be installed in water depths of 350-1000 m, thus the potential market and the need for cost-effective solutions was deemed to justify a product development effort. Obviously, improved techniques to install and service such subsea installations would in themselves boost the market by reducing costs and simplifying offshore operations. Within this context, a joint working group concluded that remotely operated intervention systems offered substantial cost savings and their development would comply with Asea's vast experience in robot technology and design of remotely operated systems. With this background, the two companies developed Wellman, a diverless, versatile easy-to-handle and cost-effective intervention system. The name Wellman (= *Well*head *Man*ipulator) was chosen to indicate the system's flexibility and versatility.

The initial development project started in 1985 with conceptual engineering; it proceeded through detailed engineering and manufacturing and was completed following sea tests in Norway during the winter of 1986/87. This paper will highlight the design philosophy, system configuration, tests and further development of Wellman. Its cost-effectiveness will also be illustrated.

DESIGN PHILOSOPHY

Based on previous studies, the joint development team reached the following conclusions at an early stage:

- The intervention system should be operated from an inexpensive monohull support vessel.
- To replace a subsea installed component a two-trip operation would allow substantial simplification of the intervention system, compared to a one-

Advances in Underwater Technology, Ocean Science and Offshore Engineering, Volume 14: Submersible Technology

trip design. The additional time required was judged to be marginal, and in many cases the service crew would prefer to inspect the component at the surface before installing a new one.
- The intervention system should be landed and docked on the subsea installation before any subsea operation starts.
- Guidelines and dedicated docking posts were the preferred choice.
- The tool packages should be accurately positioned around the component to be replaced by a three-point docking arrangement. Reaction forces should be transferred to the subsea structure.

The project team further concluded that:

- Modification of an existing ROV to a common workbase providing power, communication, navigation and handling manipulators, should be the preferred approach.
- Development efforts should concentrate on tools, tool package deployment and system handling.
- The modified ROV with its replaceable tool packages should be easy to handle and operate. Fast exchange of tool packages was considered essential.
- Use of insert subsea production components provides a very suitable approach for the remotely operated replacement.
- The intervention system should replace components without disturbing the proper functioning of nearby stabs, seals, valves, etc.

After completion of initial conceptual studies, a development project was initiated with the objective to: 'Verify the soundness of the Wellman Intervention System by developing, manufacturing and testing the intervention system (by replacing a 3″ insert valve) designed for application down to 500 metres depth'.

The rationale behind this objective was twofold: (1) Relating the test to replacement of an already existing insert valve would allow the project team to concentrate on the design of the intervention system itself. (2) Replacement of the 3″ valve was, in terms of complexity and system performance required, considered representative for other subsea tasks envisaged.

Throughout the development work, a careful balance was kept between the new concepts and existing, well-proven equipment. This would help to minimize unpredictable difficulties during fabrication and offshore operation.

WELLMAN CONFIGURATION

Wellman, as tested in 1987, is a guideline based, suspended, remotely operated tool system to be handled from inexpensive support vessels. The guideline deployment system minimizes handling and avoids navigation-related problems. The present design can, however, be modified to accommodate thrusters or a winch system for applications where the use of guidelines is not recommended. The suspended concept eliminates the large and bulky characteristics associated with neutrally or positively buoyant systems.

The total intervention system consists of:

- ROV (tool package carrier vehicle)
- Tool packages and deployment devices
- Surface support vessels with launch- and handling system.

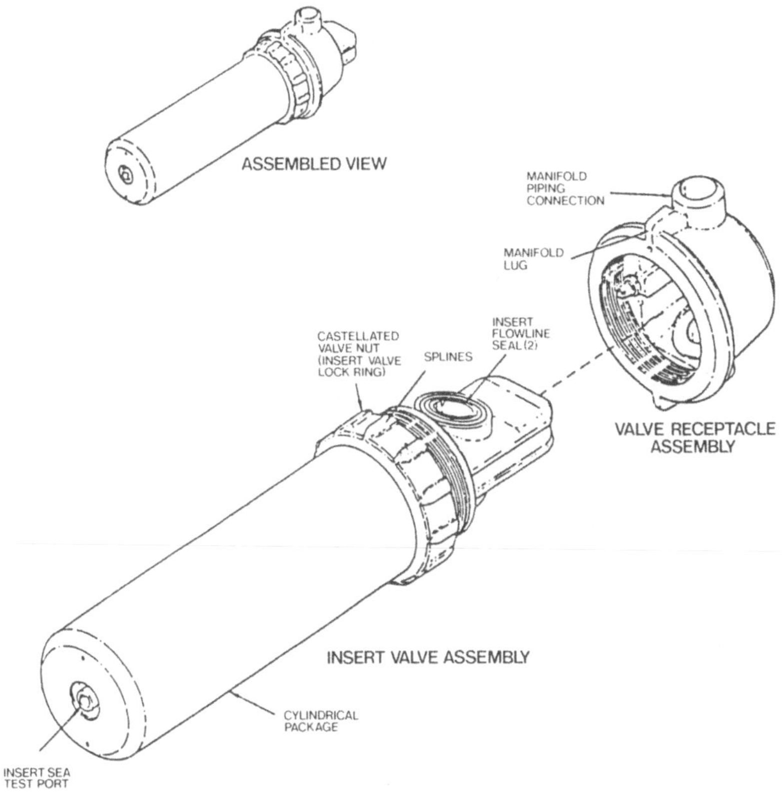

Fig. 1 The 3″ insert valve.

CARRIER VEHICLE

The carrier vehicle chosen for the intervention system is a modification of the Bandit drilling support vehicle, which is manufactured by Asea Oil and Gas under licence from Deep Ocean Technology. The Bandit's supporting frame was modified to accommodate the valve replacement tool package and its deployment system. The vehicle was also supplemented with a separate hydraulic system to provide the tool package with sufficient power for tightening/loosening the insert valve.

TOOL PACKAGE AND DEPLOYMENT DEVICES

A number of tool concepts were evaluated before it was decided to develop a tool which would enclose the valve and hydraulically tighten/loosen the castellated locking nut. The tool comprises a worm gear driven by two hydraulic motors.

The tool is suspended in a movable beam along which it can travel towards the valve. Deployment is accomplished by an initial docking of the movable beam to the Christmas tree or the manifold. The tool is then moved into position. Final alignment is effected by means of three hydraulically driven hook-shaped

Fig. 2 Overview of the intervention system and the test rig.

steering devices which attach themselves to fixed cones on the Christmas tree. The horizontal beam has an extended reach of 1 m and is capable of supporting a weight of approximately 1000 kg. It is designed to retract the tool package into the centre of the Wellman frame, thereby protecting the tool and valve from possible damage during handling and operation.

The Bandit's manipulators can be used as a secondary mean to move the tool back and forth.

Wellman is easily converted to vertical component replacement by substituting the horizontal beam with a vertical device. This device might facilitate 1.5 m of vertical movement and support up to 10 tons. It can also provide minor horizontal alignment of the tool package.

SURFACE SUPPORT AND HANDLING SYSTEM

Wellman can be launched and retrieved safely at sea states of 4.5 m significant wave height using DP-monohull vessels. The moonpool should preferably have a size of 5.0 × 3.5 m. In the case of a diving moonpool, a cursor guide system should be considered.

The handling system designed for Wellman offers two alternative solutions depending on the actual weather situation. At moderate sea states, a simple

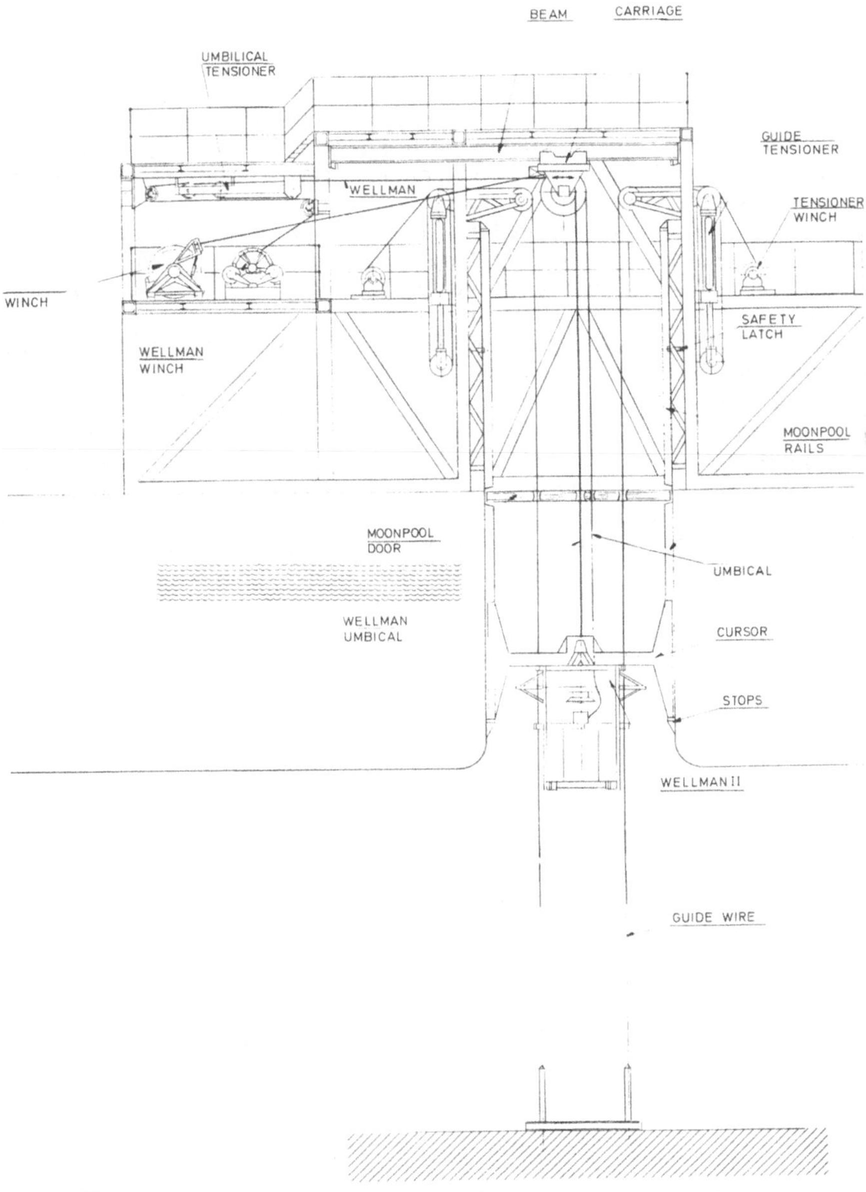

Fig. 3 Heave-compensated handling system for Wellman.

system without heave-compensation will be used. During periods of rough sea states, a cursor guided system with heave-compensation should be the preferred alternative.

TESTING

In the summer of 1986, Vetco, Aberdeen, delivered a dummy valve for use on a purpose-built test rig constructed by Asea. The complete test rig consisted of the 3″ dummy valve with its housing body, docking facilities for movable beam and two guideposts for Wellman.

The test programme was divided into a 3-month period of dry tests, followed by a 5-week period of sea tests. The dry tests were carried out at Asea in Sweden and allowed for necessary modifications prior to the sea tests. Following the successful completion of the dry tests, the Wellman system and the test rig were transported to NUTEC, Bergen, Norway, where the system was subjected to sea tests at a depth of 20 m. The subsea operations were supervised and controlled from a separate control container onshore. The operator had access to four video cameras, which provided an excellent overview of the work area.

It was earlier found by workshop tests that a torque of 20 000 N m was required to achieve sufficient sealing of the valve at 350 bar. During the replacement operations, the tool produced a torque of 20 000 N m for tightening and up to 40 000 N m of loosening the valve from its housing. Torsional moments were measured by means of force transducers situated in the main tool. A total of only

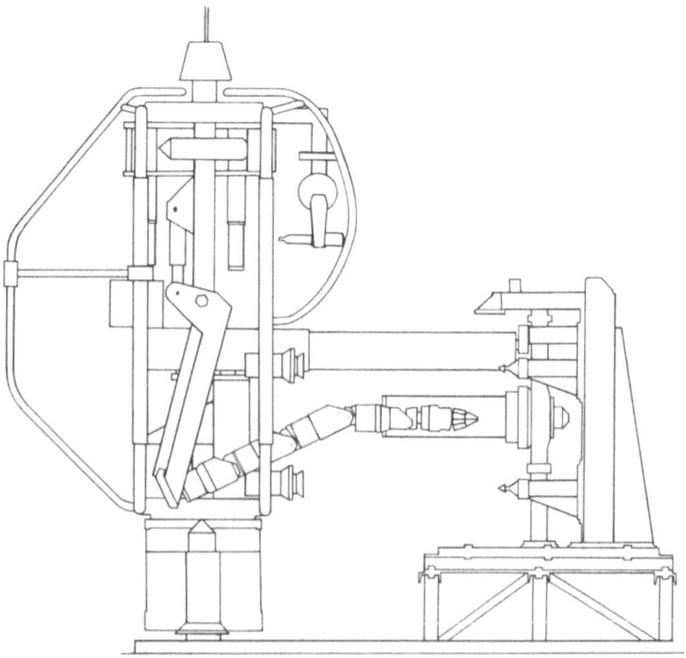

Fig. 4 Wellman after docking to the test rig.

three transducers was needed to measure the hydraulic pressure in the complete system.

A large number of valve replacement operations clearly proved that the system works according to expectations. The ease with which Wellman could be handled on the surface vessel proved to be a considerable advantage over traditional intervention systems.

FURTHER DEVELOPMENT OF WELLMAN

Parallel to the Wellman prototype manufacturing and testing, several studies were conducted to assess the application of the Wellman tool concept to carry out other subsea maintenance tasks. The conclusion of the studies was that the basic concept was very attractive. Some of the tasks would, however, require larger and heavier work packages than accommodated by the Bandit. Consequently, a completely new carrier vehicle has been designed that provides for a large open frame in which the tool packages can be placed.

This Wellman frame is approximately 5.0 × 2.5 × 2.0 m in size and weighs 6–8 tons. It is capable of handling tool packages varying in size up to a height of 2.8 m, and a width of 2.3 m or more and weighing up to 15 tons. The total assembly is negatively buoyant and can, as an option, be equipped with two rate-controlled 7-function manipulators.

Wellman utilizes common hydraulic and electrical supplies for all functions. The control system can accommodate not only the required functions of the carrier vehicle, but also those of the tool packages added to the frame for intervention purposes. The use of a common control system to operate all functions gives an efficient system that requires little time for shifting between different tool packages or modules.

The main frame assembly is equipped with several standard subsystem components, such as video cameras and powerful lights. Attached to each side of the main frame assembly are guidepost funnels, which are utilized to deploy the Wellman on standard API centre guidelines and guideposts. For non-standard API situations, modified guidepost funnel positions are available. The guidepost funnels may be positioned to deploy Wellman inside, outside, diagonally to, or between designated guideposts.

Several specialized tool packages have been conceptually designed. The final design of each package will depend on the components finally selected by the operator for the field development.

COST-EFFECTIVENESS

Wellman can easily be adapted to various subsea installations. Experience from application studies shows that only minor design modifications are necessary on a typical North Sea template in order to accommodate Wellman.

The following tasks have been evaluated for intervention with the Wellman system:

- Gate valve replacement, 3–5″ horizontally and vertically mounted tree and manifold valves
- Choke replacement

- Control pod replacement
- Sensor replacement
- Replacement of larger gate and ball valves (6–20″)
- Valve operation
- Pig launching
- Flowline pull-in and connection

Further, major subsea component suppliers have confirmed that their corresponding products can be redesigned to be compatible with Wellman in 0.5–2 years' time, depending on the type of equipment in question. The internal integrity of the components will be maintained.

A comparison of Wellman to a traditional intervention system for the above-mentioned template (6 wells and manifold) indicated a yearly saving in operation costs of approximately NOK 20 million (Table 1). Further, use of Wellman might reduce estimated production down-time from 14 to 6 days on a yearly basis.

In summary, Wellman's cost reducing effects are explained by:

- Full template shut-down might be avoided for replacement of Christmas tree valves.
- Considerable time-saving for manifold valve replacement.
- Considerably reduced need for drilling rig based intervention.
- Reduced need for DSV.
- Increased production availability.

TABLE 1 *Intervention costs with/without Wellman*

	Traditional intervention (days/year)	*Wellman supported intervention days/year*	*Reduced operation costs (million NOK)*
Rig-utilization	53	31	−20
DSV-utilization	29	27	−1
Wellman-utilization	—	4	+4
Production availability	95.4	97.7	—
Production shut-down	14	6	NA

CONCLUSIONS

The Wellman intervention concept has been verified for replacement of a field-proven insert valve through full-scale sea tests. The system's easy handling and cost-effectiveness have been demonstrated. By increasing the present Wellman dimensions and load capacity, the intervention system becomes a very versatile tool for the subsea operator. Compatibility is achieved through acceptable modifications to subsea installations and components. Many already existing and proven components can easily be adapted to the Wellman system. Development and testing of components and tool packages can be done within 0.5–2 years. Existing running tools can be modified and integrated into the Wellman system (e.g. existing running tools for flowline pull-in and connection operations).

25

An Overview of Current Projects in the Field of Diverless Subsea Production Systems

P. E. Hadfield and S. M. Adamson
Furness Underwater Engineering Limited, UK

The potential need to produce hydrocarbons from deepwater offshore fields, has created a fundamental requirement to develop highly subsea equipment which can be simply and effectively maintained without the aid of divers.

Many approaches to diverless maintenance of subsea production facilities have been considered in recent years. Those addressed in the paper range from one-atmosphere systems to state-of-the-art deepwater systems currently under development for potential application in the hostile waters of the Northern North Sea, West of Shetlands and elsewhere. By overviewing the evolution of subsea production systems it is intended to provide a context into which some of the more highly focused technical papers presented at this session may be placed.

The paper is presented in four major sections. The first is a history of the evolution of subsea production systems up to the first major milestones in dry one atmosphere systems (Garoupa Field development) and in wet hyperbaric systems (Central Cormorant UMC). The second section catalogues the large number of current projects and outlines principal features and common objectives. The third and fourth sections concentrate on subsea manifolds/templates which form the hub of most proposed deepwater developments.

A methodology is proposed by which they may be designed in a logical way and the three key stages of the process are explored. These are the development of a basis of design, the selection of the maintenance intervention systems and the analysis of reliability and production availability. Representative subsea manifolds/templates are compared in the fourth section to illustrate the effects of the decisions taken during design on the template layout and methods of operation and maintenance.

EVOLUTION OF SUBSEA PRODUCTION SYSTEMS (FIG. 1)

Early History

The first underwater completion, using a land type Christmas tree, was made in the shallow waters of Lake Erie, Canada, in 1947. Drilling in coastal and lake

Advances in Underwater Technology, Ocean Science and Offshore Engineering, Volume 14: Submersible Technology

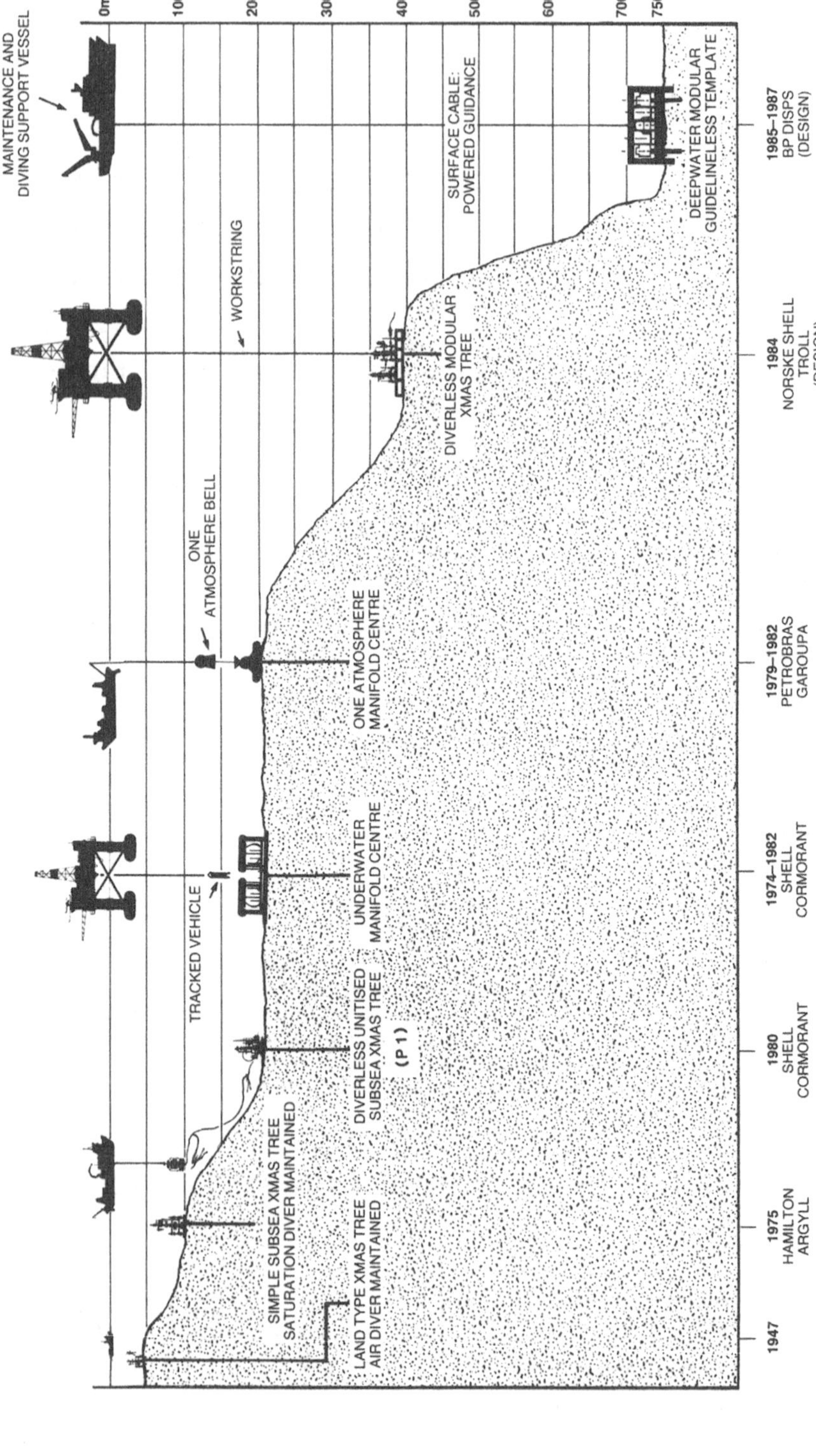

Fig. 1 Evolution of subsea production systems.

waters was commonplace by this time, having started at the Summerland Beach, California, as early as 1887. However the placement of the Christmas tree on the lake bed rather than on the drilling platform was a significant advance.

In the 1950s major breakthroughs were made in deepwater drilling technology, although the impact of these on oil and gas production was not felt until the early 1960s. Development programmes by Shell Oil Company and others resulted in the first semi-submersible drilling rig, dynamic positioning, heave compensation, underwater BOPs, and positioning systems for re-entry. These laid the foundations for the deepwater production technology in use today.

Offshore drilling and production continued throughout the decade, but primarily in shallow water from small platforms using land-based technology. In 1956 Shell once more pointed in the direction of things to come by building a prototype one-atmosphere chamber to house wellhead equipment on the sea floor for test purposes.

The increase in the use of jack-up drilling rigs in shallow water in the early 1960s, making the installation of a platform for each group of wells unnecessary, plus the ability to drill in deep water spurred on the development of subsea satellite well completions. Many of the features that are commonplace today were developed in these early years. For example the early Exxon satellite wells had remote flowline and riser connections, internal porting of all hydraulic control and electrical signal lines, a single casting for the tree block, metal-to-metal seals, through flowline servicing and electrical valve position indicators, pressure and temperature transducers.

The late 1960s saw the discovery of commercial oil reserves in the North Sea and elsewhere, within diver limits, then at around 120 m, and beyond. Viable alternatives had to be found to the costly construction of larger and larger platforms, with their long design, build and install times.

The steady improvement in reliability and confidence in subsea completions and the wide use of semi-submersibles for drilling led to one such alternative being used to produce the first oil from the UK sector of the North Sea in 1978. It is still in use today at the Argyll Field, operated by Hamilton Brothers. It employs a converted semi-submersible drilling platform for production from eight satellite wells connected to it by flowlines, tied into a base manifold, and a multiline riser. Oil is exported via the same riser system, through flowlines to a PLEM and shuttle tankers. The adjacent Duncan and Innes Fields have recently been tied in, and water injection has been added. The merits of the system have been technical simplicity, flexibility, operability, and above all low capital cost. The ultimate water depth for such systems is, however, limited by the use of the moored semi-submersible and riser. The concept has been used extensively throughout the world, most notably offshore Brazil. Here a significant difference has been the use of flexible risers in place of the rigid system in use in the North Sea.

The need for true deepwater, and hence diverless, production systems had been established, and numerous large and small scale research and development programmes were set up to investigate the problems and provide practical, cost-effective solutions.

Two of these are notable in that they resulted in successful working systems, albeit installed in diver depths, with very different approaches, viz: (1) dry atmosphere systems and (2) wet hyperbaric systems.

Dry One-Atmosphere Systems

The development of dry one-atmosphere systems was initiated by Lockheed of California in 1966. The initial project was to design and build a single well completion chamber, flowline pull-in system, service capsule and topside equipment. Successful qualification trials were held in both shallow (40 m) and deep water (273 m) off Vancouver Island in 1970. A group of oil companies led by Shell Oil Company entered a joint development programme with Lockheed Petroleum Services (LPS) (later to become CanOcean) which resulted in two one-atmosphere systems being installed in Shell fields in the Gulf of Mexico: a single well completion in 115 m water depth in 1972 (the deepest subsea well at the time) and a three slot manifold centre with one 'on board' well, two satellite well 'cellars' and a test separator in 74 metres water depth in 1976.

The programme culminated in 1978 in the design of an entire production system for 915 m water depth comprising a clustered well manifold centre for 11 through template wells and two satellites; a bundled flowline and control umbilical system, with axial pull first end connection and lateral pull second end connection; a riser base, riser and floating production facility; and a service system with a dedicated surface support vessel and three submersibles—one for transfer of five crew and 3400 kg payload, a smaller shuttle for crew only and a robot or unmanned version. Key features of the system were the use of guidelineless drilling techniques, commingled production, provision for TFL well servicing, gas or water injection, and gas lift. Export of processed fluids was to be by pipelines to a shallow water platform.

In parallel with the joint development programme LPS engineered three commercial single wells for Shell, Union and Tenneco which were installed in the Gulf in water depths between 37 and 76 m, and 14 pipeline/flowline riser connection chambers which were installed with Thistle 'A' in 159 m water in the North Sea and were used for three 16-inch pipeline tie-ins.

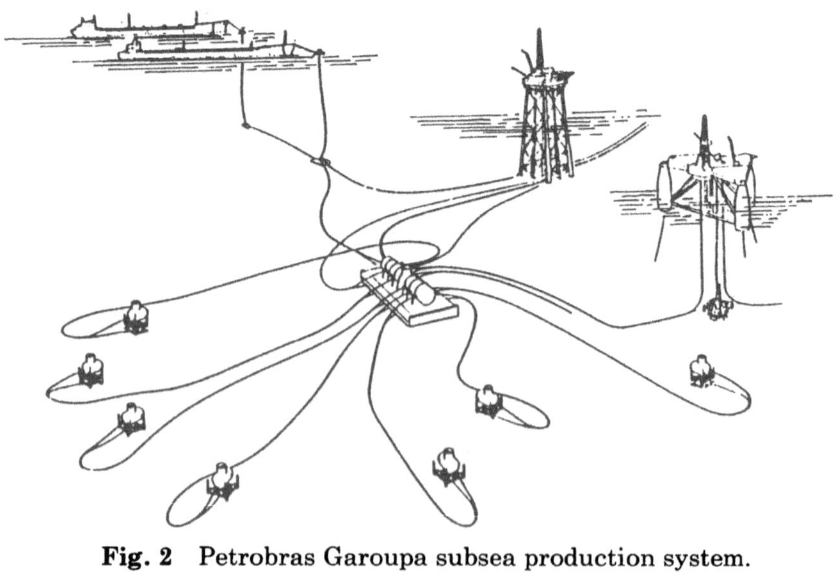

Fig. 2 Petrobras Garoupa subsea production system.

The culmination of commercial one atmosphere systems was the Garoupa Field development for Petrobras installed between 1977 and 1978. This comprised a one-atmosphere dry manifold centre configured to commingle produced oil from nine (although only seven were installed) outlying wellheads for export to a process tanker and later to a process platform in water depths ranging from 120 to 160 metres (Fig. 2).

The manifold centre incorporated production chokes, a test header, corrosion inhibitor injection arrangements in addition to provision for pigging and gas lift; it also served as a relay station for the tanker and later platform based control station and the subsea well completions.

A converted North Sea type supply vessel was used as a dedicated field support vessel: a service bell and its associated life support systems were installed along with an overstern 'A' frame and umbilical handling system for deployment (Fig. 3). Docking of the positively buoyant bell on to each wellhead cellar and the manifold centre was by a hauldown system to decouple it from surface vessel motions. In normal operational conditions the one-atmosphere enclosures were inerted with nitrogen in order to prevent the accumulation of a combustible atmosphere; during short maintenance intervention visits, crew transferring from the service bell to the seabed enclosures used a mask system to avoid lengthy cleaning and re-inerting the atmosphere.

The satellite wells were drilled at an early stage in field development and plugged live, i.e. with no stabilizing drilling mud, but with downhole safety valves and tubing hanger plugs. Having installed the manifold centre, landed

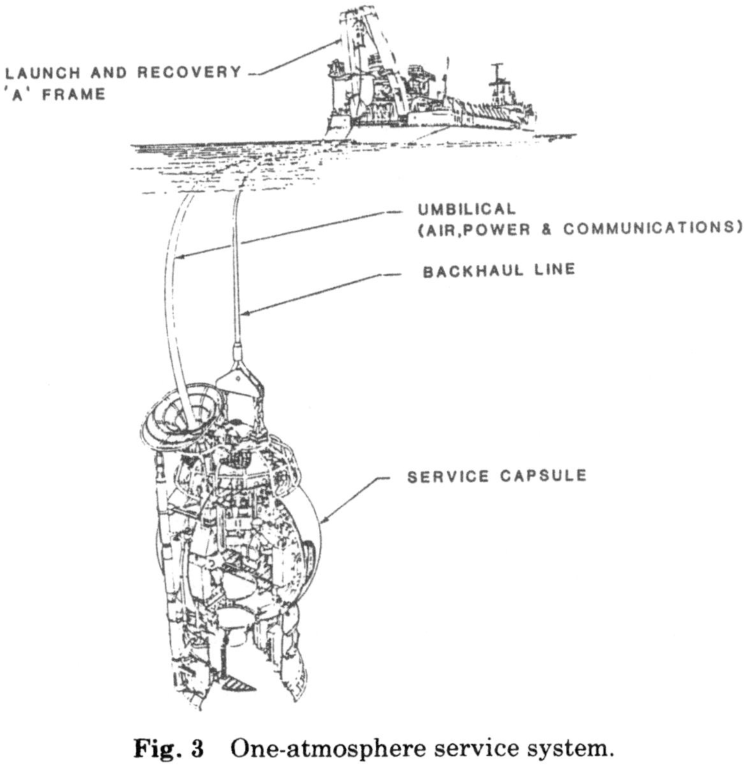

Fig. 3 One-atmosphere service system.

and attached the wellhead cellars to the satellite wells and completed flowline tie-ins, work started on hooking up the production equipment. At this stage some of the wells were found to be at pressure at the seabed, due to faulty downhole equipment, with the only safety barrier preventing blowout being the 180 mm blind flanges into the wellhead cellars. These had to be penetrated in order to reset the DHSVs and tubing hangers and render the wells safe. In a unique set of operations, the flanges were hot tapped, and a polished rod machine with BOP and lubricator used to remedy the downhole problems. The last of the seven wells was brought onstream in 1980; the one-atmosphere system operated efficiently throughout the production life of the field.

Although wet wells would have been completed differently, the ability to recover from a situation which could have so easily required abandoning the wells says a lot for the advantages of giving trained technicians access to production facilities in one-atmosphere chambers.

The Garoupa Field development for Petrobras has been the only major application of one atmosphere production systems, but their development has continued. The subsea atmospheric system (SAS) promoted by Mobil/Statoil/Kvaerner is described below in the section entitled 'Projects in the field of diverless production systems'.

Wet Hyperbaric Systems

A major programme was set up by Exxon USA in 1968 to develop subsea production capabilities for the Santa Barbara Channel off California. An integrated production system was devised based around a template capable of accommodating: 20 wet wells with modular completions; valves and control system (electro-hydraulic from the surface facility); and commingled production through flowlines to platform, production process tanker or shore. Template maintenance was by a combination of module and component replacement. Both this and flowline connections were entirely diverless. Well servicing was by TFL techniques. A three-well prototype of the system template was installed and tested in 1974 in 52 m water depth, off Louisiana.

In the same year Shell and Esso combined resources to design an underwater production system for the North Sea—the Underwater Manifold Centre (UMC) for Central Cormorant (Fig. 4). The UMC was designed to accommodate nine wells, but these could be either through template or satellite completions, unlike the SPS which was designed for template completions only. Other design features were practical improvements of those adopted in the SPS, e.g. the remote maintenance vehicle (Fig. 5), although since installation was to be in 150 m of water and within diver depths, maintenance activities were adapted for diver intervention if required.

In 1981, a prototype version of the UMC wells was installed as a satellite (P1) and tied into the South Cormorant platform in 155 m of water. The successful field operation of this diverless unitised (modular) subsea Christmas tree provided confidence to Shell/Esso prior to the installation of the UMC in the following year. Production from the UMC started in 1983, and operating experience has endorsed the design as a viable alternative to 'conventional' offshore developments. It has proved to be flexible, allowing freedom to change production strategies as the field has been delineated, operable, maintainable, reliable and safe.

The wet hyperbaric approach to deepwater subsea development has been

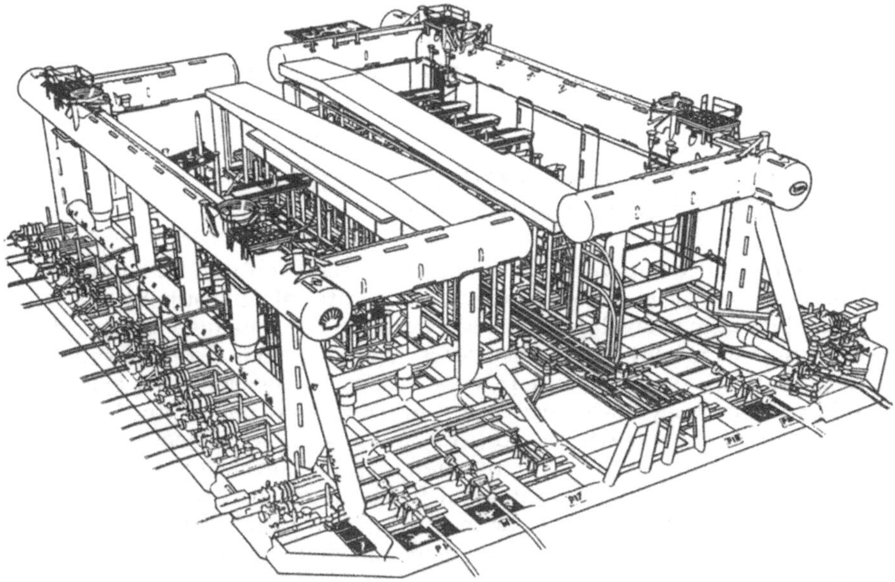

Fig. 4 Central Cormorant underwater manifold centre.

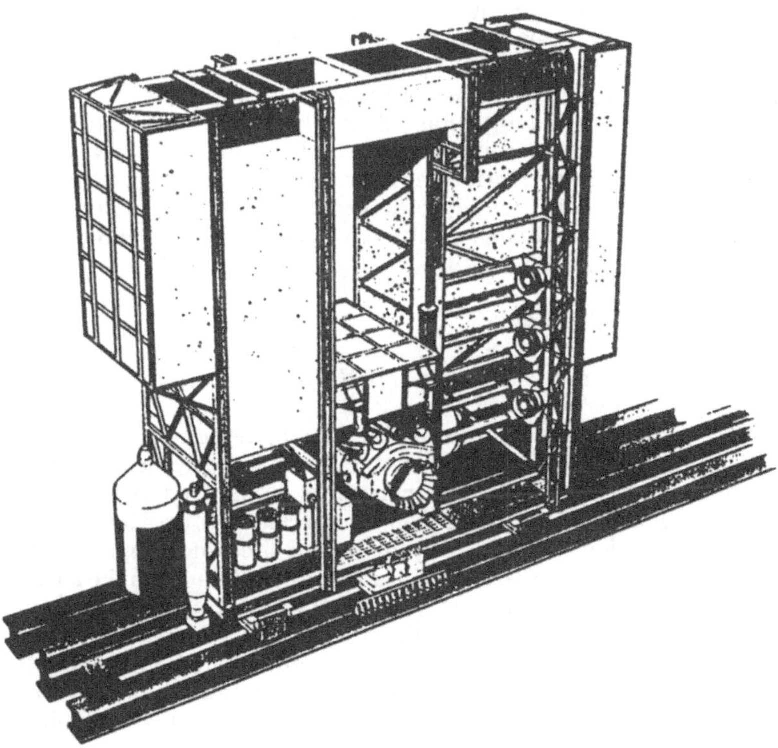

Fig. 5 Remote maintenance vehicle.

adopted by most of the oil majors. The wide variety of technical options that are being explored by them, and the methodologies by which the solutions are derived, are discussed below in the section entitled 'An overview of the design of subsea manifolds/templates'.

PROJECTS IN THE FIELD OF DIVERLESS SUBSEA PRODUCTION SYSTEMS

Several major oil companies are involved in research and development into diverless subsea production systems. The rate of investment has varied enormously: in the early days large expenditures were made in order to develop the basic new technologies. Twenty years on, the rate has slowed down because the engineering concepts, and in some cases hardware, have been proven feasible, and the immediate need to develop deepwater fields has been pushed back because of the vagaries of oil price and government fiscal policies.

Development in the UK has continued at a low level during recent years, either to fill gaps in the technological jigsaw (such as Esso's EDIPS, which builds on the UMC technology), or to develop new capability for the company (such as BP's DISPS), but with no field-specific developments targetted. These are now perceived to belong to the mid to late 1990s. In Norway, there is more activity on field-specific projects, some of which are likely to be installed in the early 1990s. Currently there is an extremely high level of activity in Brazil, where there is an immediate need to develop deepwater oil and gas fields to decrease the country's dependence on imported hydrocarbons. Independent manufacturing companies are also pursuing diverless components in their own fields with, in the majority of cases, oil company backing. The most significant of these developments is multiphase flow which has the potential to significantly reduce deepwater production costs.

For convenience of description, current projects are split into four groups: long term R & D projects, field specific projects, multiphase flow projects and others. Due to the large number of current projects, not all can be described in this paper.

Long Term R & D Projects

Diverless Subsea Production System (DISPS) (Figs 6 and 7)

BP's DISPS project centres around the development of the technology of a multislot template for commingled production from template or satellite wells. An eight-slot system for 400 m water depth is being used as a basis of design, although conceptual work is also being carried out on the implications of increasing the number of slots to eighteen and the water depth to 750 m.

The system is modular to allow maintenance by retrieval to the surface. Valve and control modules are retrieved guidelineless by a lift wire deployed from a DSV. The effect of vessel heave on the modules is reduced by a compensation system on the DSV, whilst lateral position and azimuth are controlled by an underwater thruster package attached to the modules via a running tool. Drilling and wireline BOPs and the Christmas trees are currently planned to be run using guidelines from a semi-submersible.

The Christmas tree has a standard set of valves: master, swab and wing for

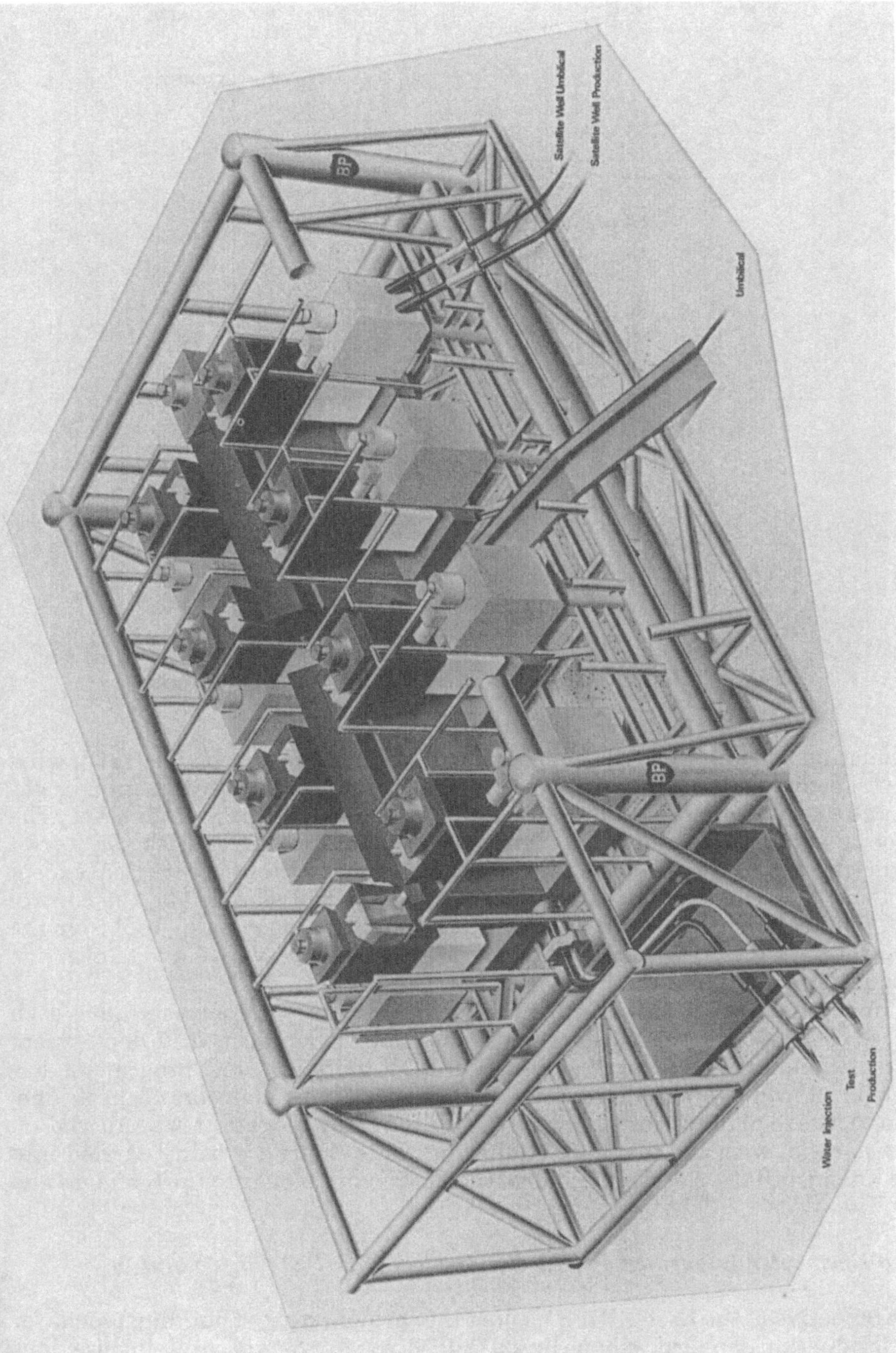

Fig. 6 BP diverless subsea production system.

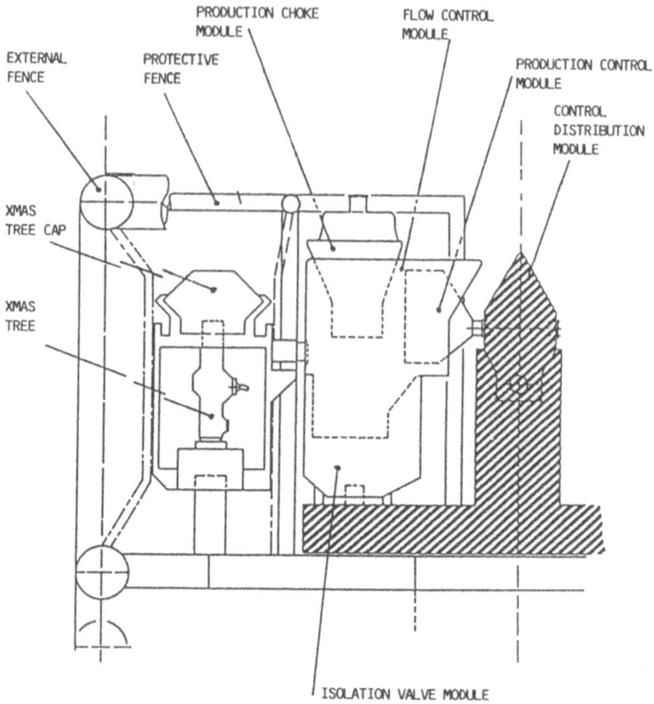

Fig. 7 Schematic of BP DISPS module arrangement.

production and annulus. The adjacent module stack has a secondary set of wing valves, a choke, flow control valves for production, test, water injection or gas lift and a set of valves to isolate the manifold header at the base of the stack. The stack is arranged so the most commonly worked and valves with the lowest mean time to failure are at the top: intermodule connections are vertical except for those between the Christmas tree, the flow control module and the control pod. Connectors have *in situ* hydraulics operated via the running tool from the thruster package. ROVs are used to operate the isolation valves and to override production valves and the horizontal connectors.

Conceptual design is well advanced, and a number of contracts have been placed to develop critical items such as the control system, module deployment system, connectors, compact valves, and ROV tooling. A test programme has been developed to prove these items and the underlying concepts of the design: land tests are planned for summer 1988, followed by in-water tests in 1989.

In parallel with the template studies, investigations are being carried into flowline installation at depth, and methods of connection into the template and satellite wells.

Esso Deepwater Integrated Production System (EDIPS) (Figs 8 and 9)

The objective of the Esso EDIPS project is to provide a set of building blocks for the production of hydrocarbons in 600 to 1050 m water depths, utilizing previous experience, including that of the UMC. A field scenario that identified the major

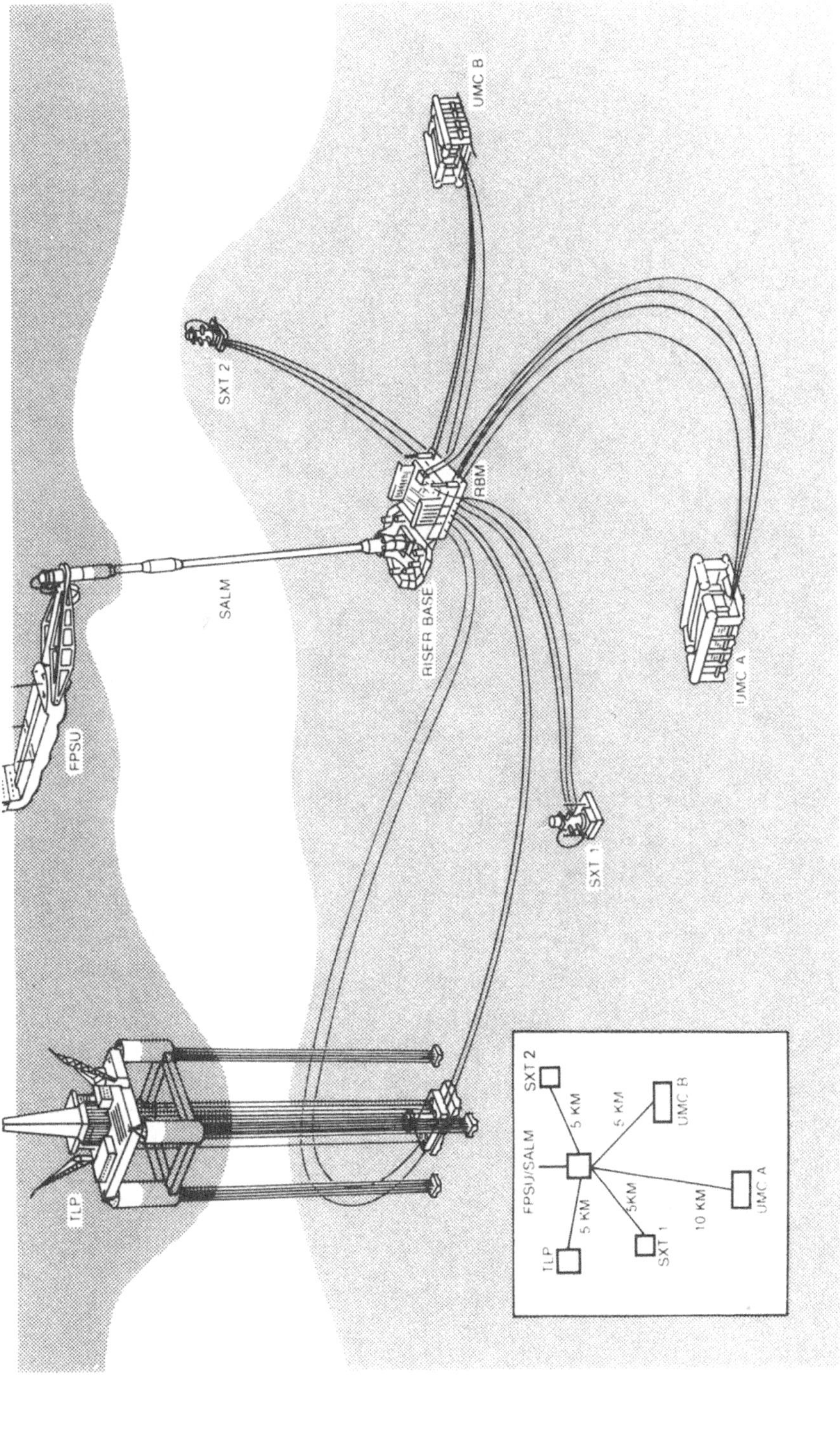

Fig. 8 Esso deepwater integrated production system.

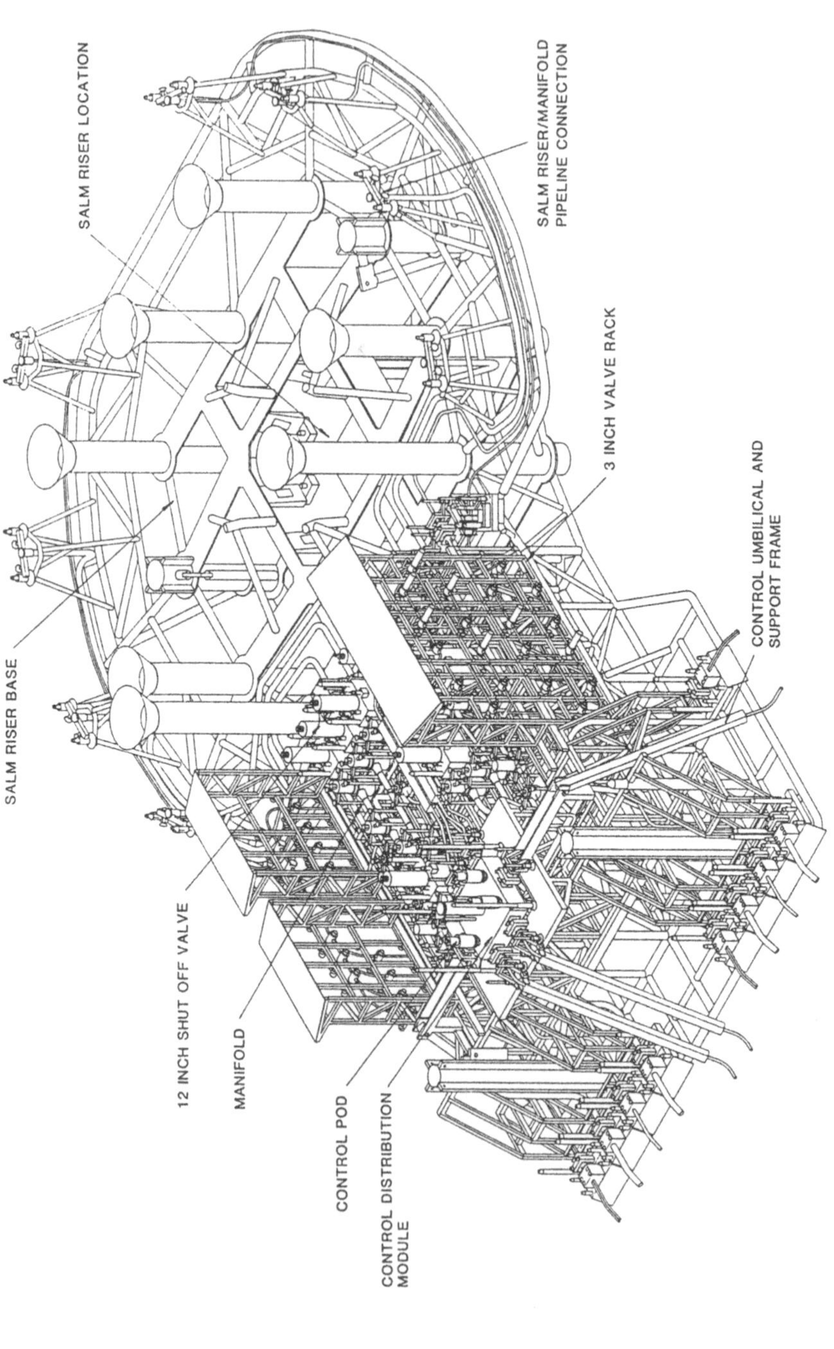

Fig. 9 EDIPS riser base manifold.

components was drawn up early in the project. It comprises a TLP, two underwater manifold centres, two satellite wells, a single anchor leg mooring (SALM) with a production manifold at the base and a floating production and storage unit (FPSU). TLP production is processed on the TLP and transported to the FPSU via the riser base manifold. Raw production from the subsea system is commingled at the riser base manifold and produced to the FPSU for processing and storage. The riser base manifold contains 3, 6 and 12 inch insert valves, diverters and isolation valves and four electro-hydraulic control pods that are mounted on a control distribution module. The 3 and 6 inch valves and control pods are maintained by ROV tool packages, whilst the 12 inch valves and the control distribution module are maintained by surface retrieval using running tools deployed on sloping guidewires from the FPSU. The flexible pipes and pipeline connections between the SALM riser and the manifold are replaceable by a combination of guideline deployed tools with ROV assistance.

The conceptual design stage, completed in June 1986, concluded that there were no major gaps in technology, but that key areas required development of existing technology in order to improve levels of confidence. Areas currently being worked on are: diverless maintenance methods including the design of insert valves, ROV tooling, and replacement of the SALM flexibles; model testing of the SALM itself; and hydrate control.

Subsea Atmospheric System (SAS) (Mobil/Statoil/Kvaerner)

The SAS has a long history, starting out as the SEAL atmospheric system designed in 1969 and funded by BP, Mobil, CFP, and others. A prototype was built, tested onshore and installed in the Gulf of Mexico in 1972 where it operated under test until 1976. The SAS forms part of an entire production system for deep water that includes a riser manifold, flowline bundles, a compliant production riser system and a floating production system. The SAS itself is a hybrid concept for 800 m water depths with critical production equipment located in a central one atmosphere enclosure and with the wellheads exposed to water.

The system has four major hardware elements; a base structure with well slots separated by guidance and protection structures; a central subsea work enclosure (SWE) with an entry 'teacup' for personnel transfer, an upper control section with breathable atmosphere where maintenance is performed, a lower service section with inert atmosphere which contains production equipment such as manifolds, valves and chokes, and a lower skirt through which flowline connections are made; a master valve assembly (MVA) which is essentially the lower part of a conventional Christmas tree re-packaged; and finally the wellhead connector assembly that joins the MVA to the SWE.

The system can accommodate nine wells, or satellites using a simplified wellhead connector assembly and a dummy master valve assembly. Equipment maintenance is primarily carried out by replacement of components in the one-atmosphere SWE; downhole maintenance is either by conventional workover using guideline or guidelineless techniques into the MVA (depending on installed depth) or by pumpdown tools. The MVA valves may be overridden by submersible mounted tools.

Statoil and Kvaerner joined the development in the early 1980s and, reportedly, have designed, built and tested all major components of a subsea system required for a North Sea installation. Although it is understood that no firm sites have been selected, a design for the Beta satellite well on the Statfjord Field is being considered.

Diverless Installable and Maintainable Oil Production System (DIMOS) (Norske Shell)

The field development scenario for DIMOS comprises satellite caisson wells producing via two modular manifold centres and a multibore riser tensioned either by a subsurface buoy or subsea boom to a surface floater based on the semi-spar design. No through template drilling is envisaged. Water and gas injection and production from dual zone reservoirs are provided for, as are downhole pumping or gas lift to deal with low reservoir pressures.

Maintenance of the satellite wellheads is by guidelineless techniques, whilst downhole maintenance is by TFL. Maintenance of the manifold centres is by surface recovery of modules: the initial concept was to deploy these using a tethered variable buoyancy maintenance vehicle launched from a semi-submersible. This has been altered to a smaller negatively buoyant vehicle that can provide lateral positioning ability to a heave compensated lift wire, deployed from a DSV. This is similar to the system being developed for DISPS and is considered a simpler and more cost effective solution.

Two items have been identified as critical to the concept and unavailable using existing technology. These are the multibore horizontal intermodule connector and the TFL selector that directs the TFL train to the appropriate satellite well; these are currently undergoing testing. Further development contracts have been placed for the design of the remote maintenance system and for optimisation of module layout to minimize system downtime.

Field-Specific Projects

The majority of field specific projects have been for application in Norwegian waters. These include the Troll template for Norske Shell, the TOGI scheme for Norsk Hydro, the Snorre template for Saga and the East Frigg development for Elf. Significant developments are also under way offshore Brazil for the Albacora and Marlim Fields.

The basic Troll oil production template design (Fig. 10) is a simple four well unit for 400 m water depths with separate production from each well. The Christmas tree is very simple, comprising only those valves necessary to ensure the pressure integrity of, and vertical access to, the well; the working valves are located in an adjacent valve module. Connections between the Christmas tree, the valve module and the piping module are horizontal. Spare wellslots are provided for use in the event that a well cannot be completed via the main wellslot. Primary maintenance is by surface retrieval of modules using workstring and guideline techniques, although the template layout is open to provide access by ROV for selected intervention tasks.

The Troll-Oseberg gas injection (TOGI) template is a compact six slot template using modular monoblock trees, a central manifold and valve module, an expansion module and pull-in and connection system for the pipelines and umbilicals. The control system is to be electro-hydraulic with fluid supplied from Oseberg 50 km distant and locally distributed by a multiplexed solenoid valve system. The unprocessed wellstream, gas with some water and conden-

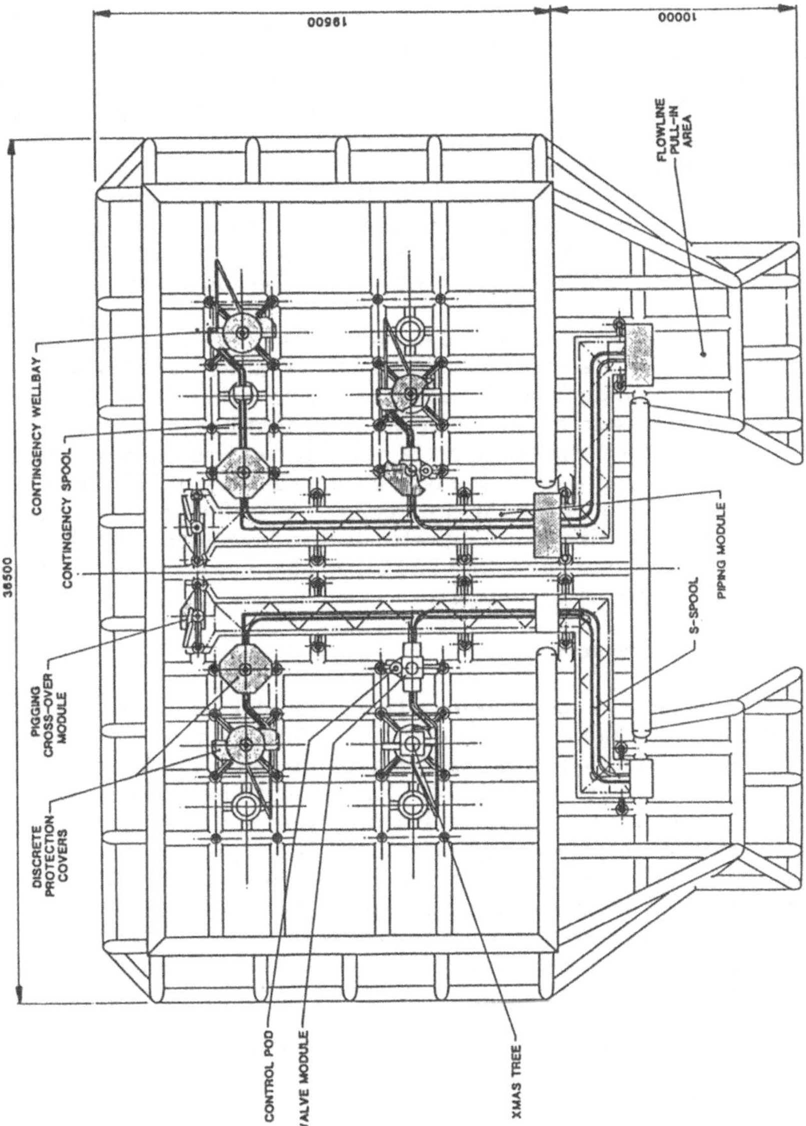

Fig. 10 Norske Shell Troll field template.

sates, is to be exported for injection into the Oseberg Field. Maintenance of the template is to be a mixture of retrieval of modules to the surface using guidelines, and intervention using ROV-carried tooling. If current plans are followed, the TOGI scheme will be the first Norwegian deepwater development (in 305 m water depth), operating in late 1991.

The Snorre template for Saga is at the conceptual stage, but is planned for installation in 1992. The current concept provides for two rows of five well slots either side of a central manifold. This will allow abandonment of the inner rows at the end of their short life and directional drilling of a further ten wells from the outer rows. Template maintenance will follow the UMC pattern with a remote maintenance vehicle that is currently under design, whilst downhole maintenance will use TFL techniques.

Elf is currently pursuing the Super-Skuld trial, postponed from 1986. Following land tests the template will be installed in 300 m water depth, modules will be run and retrieved and emergency procedures tested. The template modules are run using guidewires, for which a special connection tool has been developed.

The scale of Petrobras' Marlim and Albacora discoveries is staggering—still only partly defined, oil reserves are estimated to be more than 2.8 billion barrels in water depths extending from 300 to 2000 m. Exploitation will continue to use the step-by-step extension of current technology that has given Petrobras the record for a deepwater completion in 412 m water depth in the Marimba Field. Petrobras' favoured approach is the use of semi-submersibles for drilling and production, tied by flexibles to a range of template/manifolds and satellite wells. This is likely to see continued use in water depths down to around 1200 metres beyond which all options, including one atmosphere systems and multiphase flow and pumping, are being evaluated.

Multiphase Flow Projects

The ability to transport unprocessed wellstream product that may contain oils, gases, hydrates, and solids over large distances offers one means of significantly reducing subsea production costs. Research into the problems and development of viable solutions is very active: multiphase pumps have been developed for some time by the Institut Français du Petrole, Weir, and Stothert and Pitt; the problems of multiphase flow in pipelines, notably the growth of long oil slugs and hydrate formation are being investigated by Statoil and BHRA; existing subsea slug catcher designs are being improved by Norsk Hydro, and three phase metering systems are being developed by several companies.

Many of the above programmes are at the hardware testing stage, and it seems likely that multiphase pumping will be a real possibility within the next five years.

An alternative approach to essentially the same problem is being studied in a three year European Community funded project to develop a 30 000 b/d subsea processing unit for 150 to 400 m water depths up to 30 km from the field centre.

Other Projects

Two other groups of projects may make worthwhile contributions to deepwater subsea technology. The first is in the development of satellite wells: increasing depth capability, adaptation for ROV intervention, and changing the method of control from direct or electro-hydraulic to an integral power pack and acoustic commands in order to remove the need for controls umbilical. Work on this approach is being carried out in a UK joint industry study and by Agip.

The second is in pipeline technology, by a step-by-step increase in depth capability, improvements in diverless installation and connection, and in emergency repair techniques. Of note is the branch barrier valve developed for Total/Statoil as part of their project to develop controllable 'tee' and 'wye' pipeline connections. The valve will allow production to continue (although not to be shut off) if the valve itself needs to be repaired. Also of interest is the re-awakening of interest amongst some parties of autonomous mini-submarines such as surface independent subsea intervention system (SISIS) for pipeline repair and even subsea completion work.

Common Objectives of Current Research and Development

Although the various current projects vary widely in approach and detail, the objectives are largely the same. These are:

- Provide the ability to recover oil and gas from deep water
- To do so safely; with minimum risk to life and the environment
- Provide the most cost-effective overall production scenario (i.e. one that has the lowest combination of research and development, capital and operational expenditure)
- Minimize technical risk of failure as far as possible

Some of the means by which these may be achieved are as follows:

- Adoption of a systems approach
- Use of the simplest technical solutions where possible
- Application of existing technology with known characteristics where possible
- Testing of novel technologies and equipment in order to reduce technical risk
- Balancing redundancy and reliability throughout the system
- Developing diverless technologies that can be applied in diver-accessible depths in order to reduce both costs and risk to life
- Application of a rigorous QA/QC regime during construction followed by land testing prior to installation

AN OVERVIEW OF THE DESIGN OF SUBSEA MANIFOLDS/TEMPLATES

Design Methodology

Subsea manifolds/templates are complex and have many subsystems which are highly interactive. There are a large number of options for each of these subsystems and, therefore, an even greater number of decisions that must be made before the design is frozen for detailing. A sound methodology is needed to ensure that all options are carefully evaluated and that key decisions are taken at the right time.

The methodology presented in Figure 11 has been pursued with success in

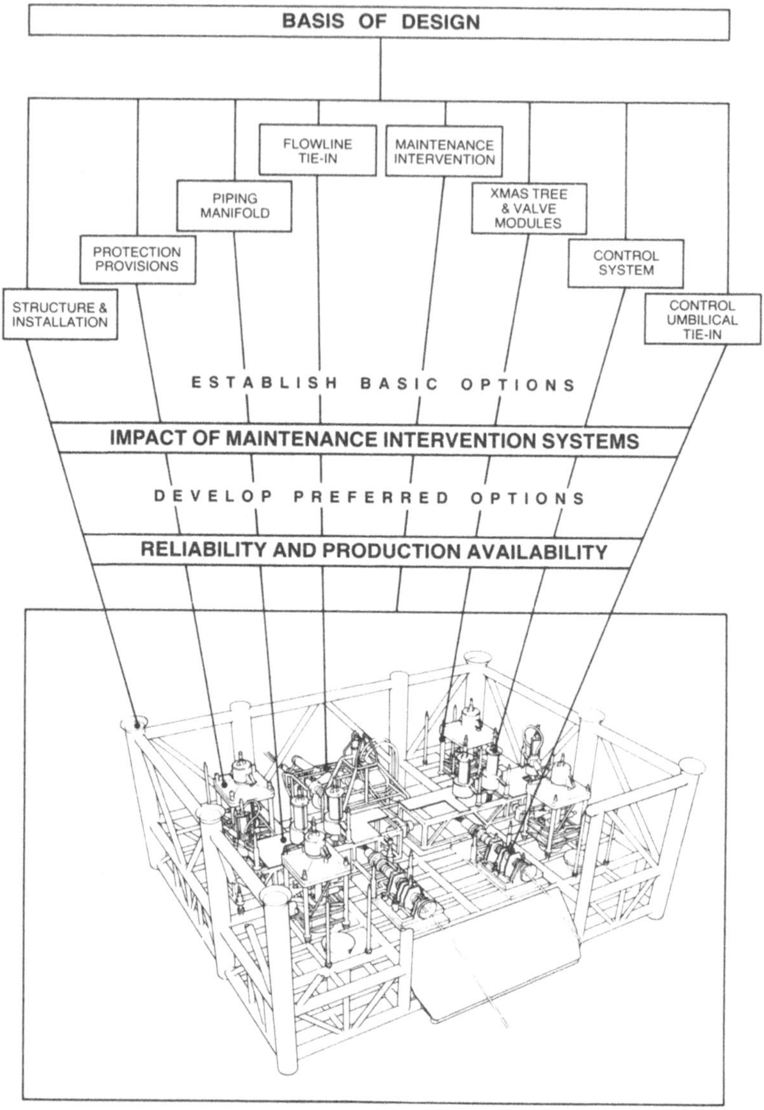

Fig. 11 Template design methodology.

several subsea template design programmes. The methodology comprises: the development of a basis of design, establishment of the basic options, evaluation of the impact of maintenance systems, development of the preferred options and analysis of reliability and production availability as part of a system performance analysis.

Development of a Basis of Design

A sound basis of design in which fundamental decisions on equipment design and operational philosophies have been made is essential to the progress of the design. Many of the decisions will be affected by field characteristics, and the oil companies' own previous experience and established policies.

Some of the fundamental issues that need to be addressed are listed below.

- *Subsea Well Completion/Manifolding* Encapsulation in one-atmosphere chambers or direct exposure to seabed ambient wet conditions
- *Installation and Maintenance* Diverless systems with or without the use of guidelines
- *Well Grouping Options* Vertically drilled single satellite wells or deviated wells, in clusters or drilled through the template
- *Well Servicing Philosophy* Wireline servicing from a surface vessel, subsea machining, or by TFL techniques from a field processing facility
- *Flowline Arrangements* Separate or commingled flow carried by single or bundled flowlines
- *Well Control and Monitoring* Simple hydraulic or electro-hydraulic operation, multiplexed or acoustic telemetry
- *Production of Subsea Equipment* A trade-off between the need to protect against dragged anchors and fishing equipment and dropped objects and the provision of unimpeded access for maintenance intervention.

Once a basis of design has been developed, the basic options for each of the template subsystems are established. This work is followed by an assessment of the impact of maintenance intervention systems and the selection of the most appropriate.

The Selection of Maintenance Intervention Systems

The ability to maintain any subsea production is the cornerstone to its success. Careful design of the system and its components may reduce the number of areas requiring maintenance; better component reliability may make maintenance less frequent; built-in redundancy may make it possible to postpone maintenance once a failure occurs and continue production; but at some stage during the life of a subsea production system maintenance *will* be required.

Some of the fundamental decisions on maintenance intervention have to be made in development of a basis of design: where to perform the maintenance—*in situ* or on the surface—and how to access or retrieve the equipment. The range of available intervention systems is great, and each has its limitations which have a strong influence on template design. A firm methodology is once more required to rationalize the selection process (Fig. 12).

The first stage is to identify all areas requiring maintenance intervention, and to list all the tasks that must be performed. For a typical multiwell subsea template/manifold thc areas would include:

- Christmas tree and associated equipment
- Production equipment and/or modules
- Piping system/module
- Connectors
- Control pods
- Flowline connection arrangements
- Control umbilical connection arrangements
- Main template structure
- Damage protection provisions

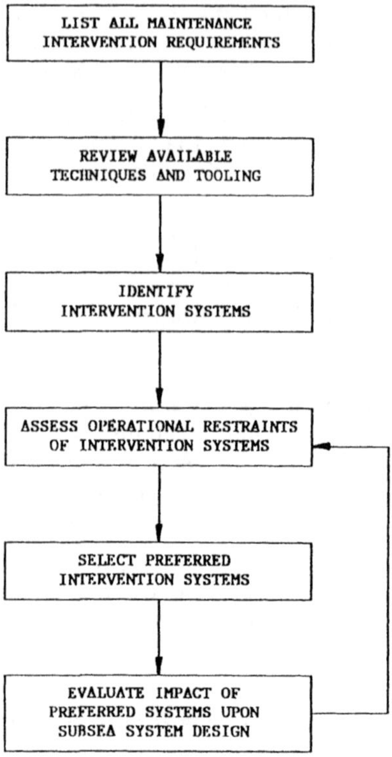

Fig. 12 Methodology for selection of the preferred maintenance intervention system.

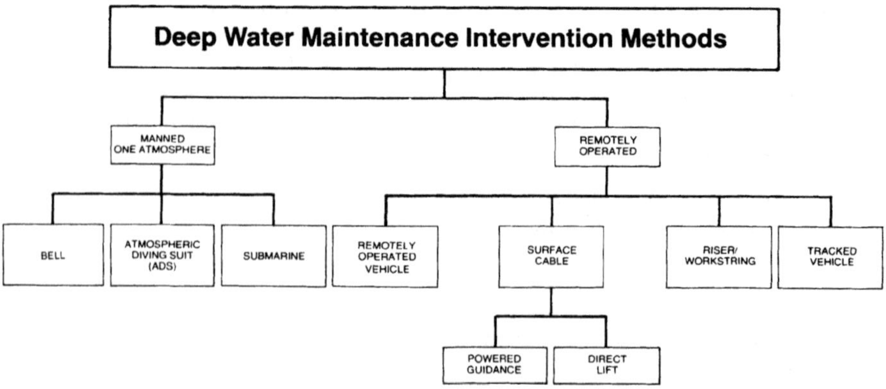

Fig. 13 Range of deep water intervention systems.

The individual maintenance tasks are classified into groups which require similar intervention techniques. A commonly used set of classifications is *observation, manipulation* and *intervention and retrieval*. Typical observation tasks would range from anode inspection to checking valve position indicators; manipulation tasks would include operation of manual valves; installation and retrieval tasks would range from valve replacement through to Christmas tree retrieval.

The next stage is to identify the particular techniques and tooling required for each task (e.g. an existing design of end effector for valve operation, or a purpose designed tool for valve replacement) and a suitable intervention system (see Fig. 7). Each of these intervention systems has particular operational limitations which govern its suitability for each of the listed tasks. For example, an ROV has limited weight-carrying capacity, but good manoeuvrability and the ability to access vertically and horizontally. On the other hand a riser/workstring has extremely good weight-carrying capacity, but limited manoeuvrability and only vertical access. Each has different surface support requirements which may or may not be available when maintenance is required.

These operational restraints must be carefully considered prior to selection of the preferred intervention system against the following criteria:

- Ability to execute tasks efficiently
- System manoeuvrability
- Topside support requirements and deployment techniques
- Weather dependency
- Potential risk of damage to adjacent equipment
- Safety to human life
- Operator preferences

Once the preferred intervention systems have been chosen their effect on the design of the overall template is evaluated, and a range of feasible alternatives generated. These preferred options are developed to a stage where reliability and production availability analysis may be carried out.

The Analysis of Reliability and Production Availability

Reliability and availability analysis form a central part of systems performance analysis which are used as tools in the optimization of template design and perhaps more importantly of field layout and operating methods. The ultimate aim is to produce oil and gas in the most cost-effective manner, so all elements of cost throughout the life of the field must be determined and compared. These include capital, development, installation, operations, and shutdown costs. Reliability and production availability analysis helps provide operations costs.

In simple terms, production availability is an integration of the analysis of reliability, leading to a knowledge of the mean time to failure of each component, and of maintainability, leading to a mean time for repair. Simple analysis may be performed manually, but a greater idea of the sensitivity of the system to changes in individual subsystems may be gained by computer modelling. A flow diagram of the performance analysis of a subsea template/manifold is presented in Figure 14.

The end products of the analysis will be knowledge of system availability, a plan for intervention operations, and a period-by-period operations cost for each of the preferred options. This will allow the most cost-effective overall system to be chosen with a reasonable degree of certainty.

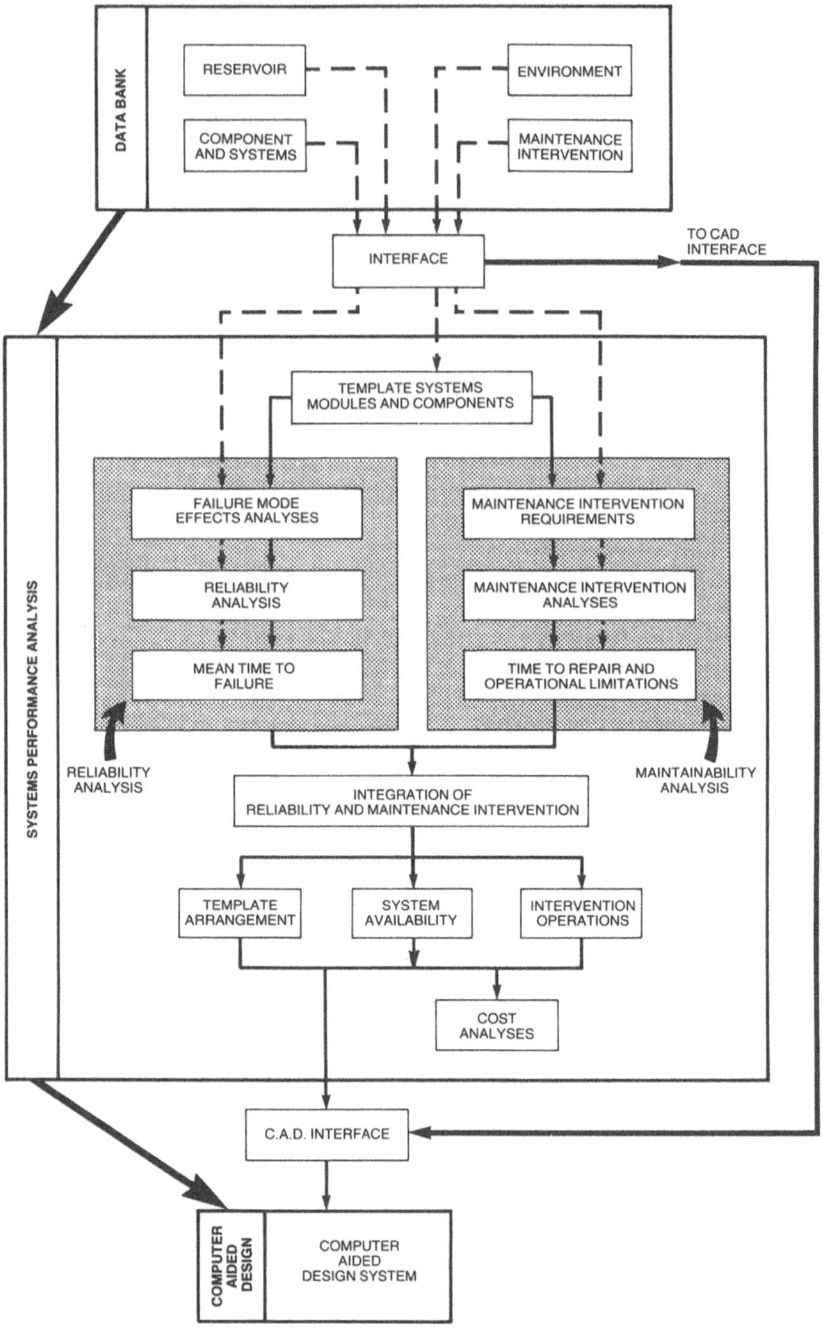

Fig. 14 Systems performance analysis of template/manifold.

COMPARISON OF REPRESENTATIVE SUBSEA MANIFOLD/TEMPLATES

The following comparisons between representative subsea manifold/templates will illustrate the extent of the influence of maintenance intervention systems on layout.

In the UMC, maintenance of the control modules and manifold header valves is performed *in situ* using the RMV, deployed from a diving support vessel. The RMV moves along a central corridor on tracks, so the control modules and valves are horizontally aligned and inwards facing. Downhole servicing is by TFL techniques so that vertical access is not needed for the Christmas trees.

In the EDIPS riser base manifold (RBM) which has no provision for drilling or completion, maintenance of the smaller valves and control pods is also performed *in situ*, but using ROV-carried tooling. This allows access for maintenance to be either vertical or horizontal: both are used, with the 3 inch valves being located horizontally in racks on the outside of the structure, and the 6 inch valves and control pods being vertically aligned towards the centre. Maintenance of larger components, such as the 12 inch shut off valves is by surface retrieval using liftwires from the FPSU. This requires clear vertical access, but the use of guidewires means that component spacing can be tight, and little protection between components is required. It is a philosophy of the EDIPS RBM to replace failed components *in situ*, as far as practicable, on the basis that it is preferable not to disturb other equipment which is functioning normally, and not to mount production equipment in modules.

The DISPS and TROLL templates are both designed against the philosophy of maintenance by surface retrieval, and both have a fundamental requirement to provide independent removal of the Christmas trees and valve modules. However, the different field scenarios result in two very different templates.

The TROLL template is only four well, has separate export from each well, and has very simple valving and control requirements. This results in only two modules being required for each wellbay which are arranged and interconnected horizontally to allow independent retrieval. Since guidewires are used, well spacing can be close and no significant protection between bays is necessary. This in turn allows clear access for inspection or valve override by ROV.

The DISPS template is designed in its basic form with eight wellslots; these may be for production, water or gas injection or replaced by satellite tie-ins. Production is also commingled so valve and choke requirements are more complex. Since the ultimate design depth is beyond current guideline capability, guidance and protection must be provided between each wellbay for guidelineless running (although drilling operations at the base case design depth of 400 metres could be carried out with guidelines). The modules would be handled by a lift wire from a DSV, so size and weight are restricted. If therefore all the modules were arranged and interconnected horizontally as in the case of Troll, the template would be much larger and potentially more difficult to install. So a predominantly vertical stacking of the modules between the Christmas tree and control pod has been adopted requiring both vertical and horizontal connections. The uppermost module contains the lowest reliability/highest utilization components that will require to be retrieved most frequently, whilst the bottom module serves only to isolate the valve modules from the manifold headers and is unlikely to require retrieval.

ROV access is necessary at the Christmas trees and module stacks for some intervention tasks including valve override, and the need for access corridors between the wellbay protection fences has a major influence on template size.

Thus the effect of the maintenance intervention philosophies and systems can be seen to have a far-reaching effect on template layout.

CONCLUSIONS

This paper has briefly charted the evolution of diverless subsea production systems, has described the projects, both conceptual and for identified field developments that are currently being worked on. It has attempted to demonstrate the complexity of the design process for manifold/templates and the need for a strong methodology to carry out this work efficiently. It has shown the strong influence of maintenance intervention requirements on template layout.

The next step is for the concepts to become reality, and the preparation of a paper describing the success of a number of diverless subsea production systems early in the next decade.

26

Advanced Robotics under Water: the DTI Initiative

R. W. Barrett, UEG, UK

The feasibility study on the application of advanced robotics (AR) to underwater work was undertaken by UEG, the Offshore and Underwater Engineering Group of CIRIA, between May 1986 and July 1987 on behalf of the UK Department of Trade and Industry (DTI). It was one of a number of DTI studies in separate application areas forming part of a nationally coordinated initiative to develop UK capabilities in AR through collaborative ventures. For the purpose of the initiative, the term 'advanced robotics' covers devices which are independently mobile and are able to make decisions about how to proceed with the tasks required.

The feasibility study report completes stage one of a three-stage process envisaged by the DTI for the development of an autonomous device (or devices) within each industry area.

Stage 1 A feasibility study fully funded by DTI to identify and assess potential applications and to recommend a preferred project on the basis of technical and market factors.

Stage 2 A project definition study (PDS), with funding shared by DTI and interested non-government organizations, to produce a proposal for an industrial development project including R and D requirements, design schematics, a detailed market analysis, work share proposals, cost estimates, and a detailed project plan.

Stage 3 An industrial development project, also with shared DTI/Industry funding, to develop, through one or more stages, an advanced robotic device (or devices).

This study (Stage 1) has been managed and coordinated by UEG with main subcontractor assistance being provided by Slingsby Engineering Limited, on technical appraisal and concept development, and Hollobone Hibbert & Associates Limited for the market related aspects. In addition, specialist input on specific areas has been provided by Heriot-Watt University, W. D. Loth and Co. Ltd and Subsea Developments Limited.

The study was steered by a national Underwater Collaborative Group comprising industrial, academic and government representatives of the underwater field and representatives from other technology fields relevant to the development of an advanced robotic underwater device (e.g. robotics, computing, artificial intelligence). Input has been provided by group members both at formal progress meetings and individually by interview and correspondence.

Advances in Underwater Technology, Ocean Science and Offshore Engineering, Volume 14: Submersible Technology

IDENTIFICATION AND ASSESSMENT OF POTENTIAL APPLICATIONS

The initial approach to the study was to identify all conceivable underwater operations based upon operational areas agreed by the Underwater Collaborative Group. These areas are:

> Oil and gas production system, pipelines, salvage, ships and floating objects, underwater food resources, mid-water operations, general seabed operations, inshore/inland operations, and contaminated environments.

The main operations identified within these areas, some of which are common to a number of them, were:

> Exploration, construction/installation/hook-up, intervention/cleaning and inspection, maintenance and repair, demolition/removal of structures, survey, search, hull/cargo recovery, research/survey/catch/processing/farming of underwater food resources, environmental monitoring, marine sciences research, waste disposal and monitoring, mineral extraction, and excavation/dredging.

Each operation, 31 in all, was then defined in terms of the specific tasks involved plus the additional aspects of special equipment used, mission duration, power requirement, task complexity and working environment each of which needs to be taken into consideration in determining operational suitability for AR development. From subsequent analysis of these parameters, it became clear that many of the tasks and equipment were common to a number of identified operations. A further task group was therefore made so that commonalities of task could be coalesced under 'generic activity areas'. Fourteen of these 'generic areas' were identified:

> Bottom survey, bottom sampling, seismic survey, cleaning and inspection, cleaning, coating and painting, low volume dredging, high volume dredging, repair and maintenance, demolition, dismantling and salvage, environmental monitoring, fishing, aquaculture, mineral mining, and drilling and production support.

Companies directly involved in each of these areas were contacted to reassess the tasks and, more particularly, to establish the estimated current and future market size for each. This activity, involving the use of a market-based questionnaire, included visits to at least one organization experienced in each area and was backed up by telephone discussions with others. The fourteen areas were then analysed in terms of their overall suitability to AR by a technical and market appraisal by specifically considering the tasks involved, the current approach, current shortcomings, current and future market, and potential developments.

The three areas considered most promising for AR, selected for further detailed analysis were:

- Bottom survey and environmental monitoring
- Structural inspection and its associated cleaning
- Structural maintenance and repair

In parallel with this work, data searches and literature surveys were undertaken to assess the state-of-the-art of AR developments, both in the underwater field and in other areas where the technology requirements may be similar.

For each of the three selected areas, an outline specification was produced for an AR device capable of performing the identified tasks. The 'enabling technologies' for each application were considered in some detail and the developments required were identified.

At the same time a further, more detailed, market survey was undertaken for each of the three proposed applications. Companies approached fell into the categories of client, contractor and manufacturer. The survey included an assessment of current and future markets, the economics for AR devices, and identifying the potential benefits and spin-offs of AR.

ADVANCED ROBOTICS DEVELOPMENTS AND REVIEW OF ENABLING TECHNOLOGIES

Currently, the most sophisticated underwater vehicles are untethered (autonomous) vehicles (UAVs). These are self-powered vehicles, operating without a physical connection to the surface, acting according to a preprogrammed schedule or commands from the surface via an acoustic link. The majority of UAVs are being developed for search/survey (e.g. AUSS, Epaulard, ARCS) and pipeline/structural inspection (e.g. Eave-East, Elit). Additionally, there are a number of UAVs being developed for specialist applications (e.g. Telemine for vessel destruction, PLA2 for manganese nodule collection).

An autonomous underwater vehicle must carry its own energy supply, sufficient for a full mission plus any contingency requirement; energy requirements depend on the nature and duration of the mission. A cost reduction in high performance batteries would lead to their being a realistic energy source for an autonomous survey vehicle. Fuel cells represent a useful energy source if developed further. Progressively higher power is required for inspection and repair and maintenance operations. If internal combustion, Stirling, and, possibly, Rankine cycle engines are commercially developed for underwater use, hydrocarbon energy storage will present the most practical solution for missions with medium to high energy requirements.

Navigation requirements vary considerably according to mission type. A seabed survey vehicle operating within range of acoustic transponders can operate with a high positional accuracy and repeatability using existing technology. However, for long-distance navigation with large variations in depth and altitude, inertial navigation is the only method currently available. Substantial developments are required in this area before reliable systems become available. For inspection, maintenance and repair activities, long-distance navigation capability is not required and operations can be assumed to lie within the range of acoustic positioning systems. High positional accuracy is required, however, if working in close proximity to large complex structures; navigation within jacket structures using acoustic techniques is possible although difficult. The use of several on-board visual and acoustic imaging systems is likely to prove the most practical and effective means of navigating AR devices within complex environments.

One advantage of an autonomous robotic vehicle would be that high data-rate through-water links are rendered unnecessary by the level of on-board intelligence. However, reliable and effective data links would make operation easier, safer and more versatile and will play a significant part in the steps towards AR. Three basic means of cableless communication are possible in water: acoustic, electromagnetic and optical. The range of electromagnetic and optical transmission in sea water is very limited, even using range-gated laser systems and these methods are therefore not suited for survey type applications. Acoustic signalling is feasible for ranges up to 10–20 km and has been used for many years.

The problems of multipath reflections and high bandwidth transmission are the subject of much research, and high data-rate transmission using narrow beam transducers is under development. Further developments underway in data-rate reduction techniques will be of considerable benefit.

An AR device must function effectively as an underwater vehicle with weight, control and drag characteristics appropriate to the performance required. Many lessons have been, and will continue to be, learned from the use of ROVs and manned submersibles, and requirements in terms of subsea vehicle technologies are likely to be satisfied by the techniques available from these spheres.

Sensors of various types are required to provide data for navigation and communication, and for the performance of specific tasks. Numerous sensors are available for underwater use or could be adapted if necessary. Development work, particularly on vision systems, is occurring worldwide and research into acoustic imaging is continuing; further developments into tactile and proximity sensors will provide additional capabilities in manipulation. Interpretation of the raw data, however, presents more problems than data acquisition, so sensors themselves should not be a limiting factor in development.

Overall computer control of an AR device will be achieved by combining subsystems, in a hierarchical systems framework using artificial intelligence (AI) in the form of adaptive knowledge based systems. The vehicle control system must be fully integrated with other vehicle systems responsible for navigation, strategic mission planning and management, environmental perception and interpretation (e.g. obstacle avoidance) and task accomplishment. Additional subsystems will be required for communication management, selective data storage and man–machine interface (MMI) facilities.

Development of many individual subsystems is already well advanced and, being task specific, is expected to progress as requirements continue to arise in a variety of subsea areas. Distributed processing techniques are already used on conventional ROVs. Pattern recognition and 'low level' image processing techniques (e.g. edge detection, histogram equalization) are also well established. Knowledge-based image-processing techniques for sonar and video images are, however, at an early stage of development as far as subsea operations are concerned.

The technology for the control of an intelligent, autonomous AR device therefore exists at present, at least to allow the simpler type of mission to be performed. Systems capable of being continually updated to encompass more complex mission profiles will be assisted by the adoption of a common language for AI systems.

Data storage requirements depend on mission type with a full record of vehicle activities being necessary for survey missions, and video and still photography as well as NDT results being required for more interactive tasks. Various methods of storing data are available at present including floppy and hard disks, magnetic tape and solid state memories. Video tape currently provides the highest energy storage density at more than 1 Gbyte/hour. Compact and laser discs have very high storage capacities but cannot presently be used for the recording of data on-site. Solid-state memories and disks allow fast retrieval of information. Data storage will present a problem in terms of bulk, but data reduction techniques, developed mainly for image processing, are continually being improved and will ease data storage problems provided the processing power can be made available onboard.

In addition to the above-mentioned technologies which would be 'common' to all underwater AR devices, further developments in attachment methods, manipulator systems and tooling, cleaning systems and NDT methods will be required for AR devices undertaking manipulative and other non-survey type operations, such as close inspection and repair and maintenance activities. A number of underwater manipulators with closed loop control are commercially available, though they tend to be general purpose arms and lack the reach and dexterity essential for automated intervention. For this, the development of an appropriate intelligent manipulator subsystem would be required to incorporate force, tactile, proximity and vision monitoring, which in turn would be integrated with the computer control system.

MARKET ASSESSMENT

The detailed market assessments undertaken for the three selected areas show that the offshore hydrocarbon industry is the major driving force, accounting for the largest potential commercial market in each area.

In the area of bottom survey, there are further primary markets for marine sciences and hydrographic activities. The adoption of AR in these roles will depend upon the availability of suitable, reliable devices at competitive prices and, in some cases, the requirement to carry out surveys in very deep water or under ice.

Estimates of the worldwide market value for a bottom survey AR device show steady increases to the end of the century in both the hydrocarbon and marine sciences fields, with an estimated combined value rising to over £200 M per year by the year 2000.

Although not considered as part of this study, spin-offs into military applications including mine countermeasures, search and surveillance will form large additional markets in the future.

Other markets for a bottom survey device may include video recording for leisure market purposes and fishing industry support in the form of search, monitoring and herding of shoals.

The hydrocarbon extraction industry is presently the major potential market for an AR cleaning and inspection device. The non-hydrocarbon sector, which largely covers the civil engineering industry, offers a separate market such as inspection of bridge and harbour works, seawater intakes, sewage outfalls and containment tanks. If such devices were developed, it is likely that the technology could find other markets, including aquaculture equipment inspection, detailed inspection of wrecks, prior to salvage and various military uses. In addition, developments towards a machine which can be mobile whilst attached to a structure (in other words capable of climbing) could have spin-offs in spheres other than underwater. In the next ten to fifteen years, it is unlikely that hydrocarbons will be produced in water depths significantly greater than at present. Accordingly, in the short term, it is necessary to look at a scenario in which AR inspection devices will compete with existing means of inspection, and thus, for the development to be justified, the technology they incorporate must have applications in other markets.

Estimates of the primary worldwide markets for AR underwater inspection devices show a steady increase in the ten years up to the year 2000, when the value is estimated to be over £100 M per year.

At present, the repair and maintenance (R & M) market is principally centred on the offshore hydrocarbon industry, where it encompasses such activities as structural repair and replacement of components. However, there could be a large future market in platform dismantling and decommissioning and many of the features of an R & M device would also apply here. As with the bottom survey and cleaning and inspection devices, the technology required to develop suitable AR devices for these markets could find outlets in several others, not just in the underwater field. Assessment of market values is particularly difficult as levels of R & M vary considerably from year to year. However, the estimates made for the size of the R & M market indicate a steadily increasing value reaching almost £400 M per year by the year 2000. AR technology required for devices designed for the R & M market would require to have a high degree of sophistication. Devices developed for this area would thus provide the greatest technological spin-off across the underwater field and into other areas.

The estimates of the worldwide market values for bottom survey, cleaning and inspection, and repair and maintenance of AR devices given above represent only the values of the predicted major markets. The technology which would be developed for AR devices to be used in these markets would find spin-off applications in many other areas, the value of which could significantly increase these figures.

REVIEW OF THE THREE SELECTED AREAS

Bottom Survey

The specification prepared for an AR device in this area covered conventional bottom survey, but also included some environmental monitoring and marine sciences activities. These include visual and sonar search for inspection of pipelines, cables, miscellaneous objects, minerals and underwater food resources; monitoring of salinity levels, temperature, current and tides, pollution and specific measurements (e.g. oxygen); study of water conditions and marine life in deep water and under ice; and, charting surveys for subsea structure sites, pipeline and cable routes, waste disposal sites and plotting of lake bed contours.

Considerable cost savings would be achieved by an AR survey device by the requirement for only intermittent use of a small 'support' vessel, in comparison with the continuous requirement for surface support vessels currently required when using towed fish, ROVs or manned submersibles. Removing the need for an umbilical cable would provide further benefits in manoeuvrability, enable under-ice surveys to be conducted, and remove the costs of complex cable and winch systems.

The major developments required for an AR survey device are in the areas of advanced computer control, environmental interpretation, power/energy systems and long-distance navigation. Advances in these technologies will lead to spin-off, not only for further underwater developments but also for many other terrestrial and aerospace applications; a free-swimming autonomous vehicle would also provide an idea test bed for more complex developments.

Cleaning and Inspection

Cleaning and inspection consist of operations which are performed within a structured environment and which currently involve, to a large extent, the intervention of man under water. Cleaning in this context is associated only with the area to be inspected and does not include cleaning of large areas. All potential AR applications necessitate the accurate positioning of a tool or instrument together with precise control over its movement. Tasks identified include general visual inspection, crack detection, marine growth removal, detection and measurement of porosity, and assessment of collision and abrasion damage.

At present, divers and ROVs are used for structural inspection work. The amount of close inspection work that can be carried out by a diver is limited, and diving operations are expensive and carry an element of risk to the divers. ROVs can work for longer periods, but are currently limited in terms of access and manipulative performance. An AR device could go a long way towards solving these problems and reducing 'manual' costs.

The major developments required for an operational AR device involve intelligent manipulator systems, crack detection techniques, navigation within structures, and the application of knowledge-based systems to a fully operational device. Developments towards a mobile device with good manipulative capability and which could navigate through a hazardous area, would also find uses in other spheres for example, the nuclear industry, bomb disposal, and the fire fighting and rescue areas.

Repair and Maintenance

Repair and maintenance (R and M) cover an extremely varied range of tasks, including observation, installation/replacement of structural members, debris clearance, cable deburial, replacement of various valves and sensors, cutting and welding, marking structures, and recharging subsea accumulators. The repair tasks in particular are both complex and unpredictable. Taking these factors into consideration, it is extremely unlikely that a single device would be able to perform all of them.

The replacement of divers, particularly for dangerous tasks like repair and maintenance work, is a major incentive for the use of AR. Removal of the umbilical cable is also of great benefit in this type of work. The incentives to produce a fully autonomous R and M device with no human intervention are not clear cut, and operation with a 'man in the loop', even for monitoring purposes, would often be desirable.

The technical requirements of an R and M robot are by far the most demanding of the three areas considered. Considerable developments in manipulative techniques and the application of the knowledge-based systems to complex, non-repetitive tasks are essential to this field. Components of the vehicle and subsystems, which would need to be developed for a repair and maintenance device, would form the basis of a virtually universal AR device. The number of potential spin-offs is therefore enormous and would include the nuclear industry, construction, military, fire fighting, space and many more.

EVALUATION OF THE THREE APPLICATION AREAS

Each of the proposed developments was examined in detail and evaluated in terms of separate technical and market factors, each contributing to the decision about the suitability of each area for AR development. The resulting evaluation is shown in Table 1 with factors being equally weighted in all but two cases, where considerable subjectivity was involved.

TABLE 1

Factor	*Weighting*	*Bottom survey*	*Cleaning and inspection*	*Maintenance and repair*
Technical feasibility	(10)	7	6	3
Development required	(10)	7	7	2
Benefits over non-AR	(10)	8	5	5
Market size	(10)	5	3	7
Economic feasibility	(5)	4	1	2
Potential users' views	(5)	4	2	3
Spin-off markets	(10)	6	6	8
Total		41	30	30

The above totals provide a numerical indication of the relative merits of the three applications, but other considerations must also be taken into account in selecting the most promising area for development. One concerns the progression of the development of AR technology. There are clearly benefits in adopting a staged approach to AR thereby capitalizing on existing technology.

The development of a bottom survey device would meet many of the market requirements using largely existing technology. Furthermore, the majority of 'enabling technologies' developed for such a device would be common to almost any AR underwater device and hence would considerably assist AR developments for other applications.

Another consideration concerns spin-off into the defence markets. Although not considered as part of this study, military applications such as search and surveillance offer potentially large home and overseas markets for AR devices. An AR bottom survey type vehicle is also seen as offering the greatest potential for spin-off into these areas.

OVERALL CONCLUSIONS

The feasibility study has led to the following conclusions:

(1) There are sufficient world markets, showing steady increases up to the

year 2000, to provide a stimulus for AR developments in each of the three selected areas.

(2) There are sufficient benefits to be gained by introducing AR into underwater work, including removing man from hazardous environments, eliminating the need for an umbilical, and improving operational procedures.

(3) 'Advanced robotics' is a long-term objective. Any developments should adopt a step-wise or 'staged' approach towards AR and embody as much existing technology as possible.

(4) The development of any underwater AR device would provide a stimulus for many 'common' AR enabling technologies (e.g. navigation, sensors, data storage) and would lead to spin-offs into other underwater developments and across many other fields.

(5) Based on the technical and market assessments undertaken and other considerations, the area of bottom survey has been identified as the most promising for the initial application of AR to underwater work. From the technical point of view, many of the 'enabling technologies' are sufficiently advanced to enable a prototype AR device to be realised in the medium term (up to 1995). Cleaning and inspection, and repair and maintenance would also benefit from the application of robotics and ultimately AR, and many of the technological advances and sub-assemblies resulting from a bottom survey development would be applicable for use in these and other areas.

THE PROJECT DEFINITION STUDY

An outline plan for a project definition study (PDS) on the recommended area 'bottom survey' has been prepared, but the approach adopted would also apply for a study in either of the other two selected areas.

The outline proposal developed for the PDS covers the following project aspects: objectives and scope of work; areas of work to be covered; project management and methodology; costs estimates; funding and timescale. The development and promotion of a proposal for the PDS are also addressed.

It is envisaged that the 'product' of the PDS would be a detailed report presenting a proposal for an industrial development project and would include research and development requirements, design schematics, detailed market analysis, work share proposals, cost estimates, and a detailed project plan.

The initial cost and time estimates for the PDS are £380K and 18 months respectively.

27

New Developments in Flooded Member Detection

A. M. Leonard, BP Petroleum Development Limited, UK

The improvement in our understanding of the fatigue lives of offshore welded tabular structures has led to a change in our inspection philosophy. For the older platforms we are able to predict very accurately where and when trouble can be expected. The newer structures have been designed to ensure that all nodal welds have fatigue lives in excess of three times that of the required working life of the platform, so there are no particular areas of concern. Any propagating defects in the new platforms will be due to fabrication faults which evaded detection during construction. Because the locations of these faults cannot be predicted, it is not possible to direct inspection programmes at any specific areas of high risk. Therefore a rapid method is required for checking primary members in the newer and older platforms for propagating defects. ROVs can check for gross damage, but it is desirable to detect defects as early as possible. The philosophy behind flooded member detection is that because all jacket members are seal welded in air at one atmosphere during construction, water within a member must indicate a through-thickness defect. The challenge has been to find an accurate, rapid and cheap method of checking jacket members for flooding and it has proved remarkably difficult.

A number of techniques have been used for flooded member detection. All have their advantages and disadvantages, most of which I do not intend to discuss in this paper because I wish to take a positive approach to some new developments.

ULTRASONICS

The technique currently favoured by the industry uses ultrasonics. The principle of operation (Fig. 1) is that if a member contains water, ultrasound will be transmitted through the water, reflect at the water/air or water/steel interface on the opposite side, and return to give a signal on an A-scan screen which is unmistakable. Unfortunately, badly corroded internal or external surfaces, or internal debris, will scatter the ultrasound (Fig. 2), giving a 'non-flooded' signal which may be false. To detect horizontal members that are less than half full, the probe has to be held at the 6 o'clock position. This is where debris and loose corrosion products will collect, so the problem of ultrasound attenuation is particularly severe. The quality of the reflected signal will also

Advances in Underwater Technology, Ocean Science and Offshore Engineering, Volume 14: Submersible Technology

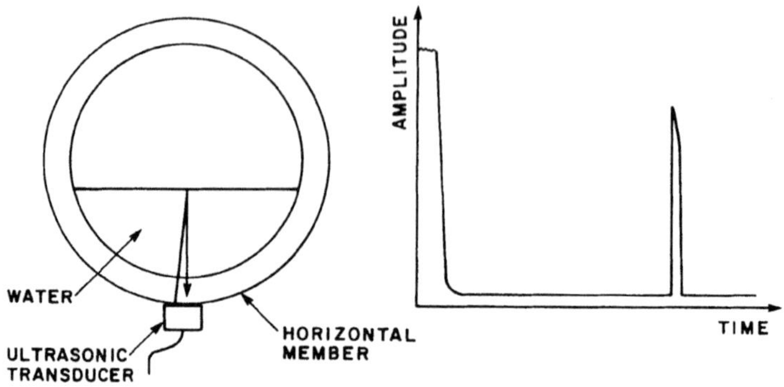

Fig. 1 Ultrasonic FMD. The ultrasound reflects off the internal water surface and the A-scan displays the echo.

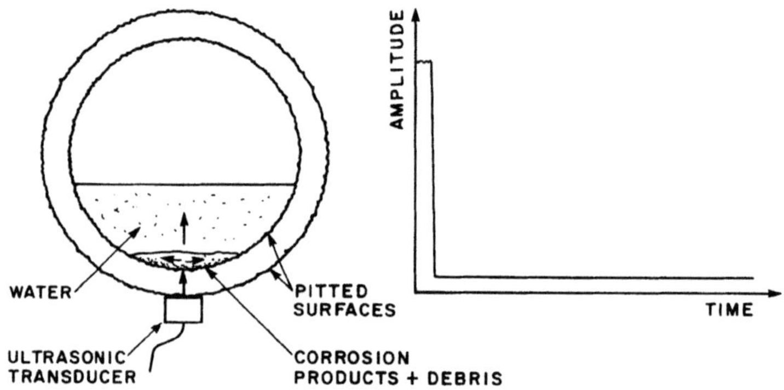

Fig. 2 In corroded members, the ultrasound is scattered. No echo is received, and the A-scan indicates an empty member.

depend on how close to the 6 o'clock position the probe is placed, on the angle of the member to the horizontal, and on any ripple on the water surface inside the member.

In some recent offshore trials conducted by BP, the 'hit rate' was indeed found to be variable. Tests on samples on the surface and on the wet bell staging under water gave a high detection rate approaching 100%, and it was discovered that the signal quality was considerably improved by cleaning the steel surface where the probe was applied to bright shining metal (BSM). However, when tried on two known flooded members, cleaned locally to BSM, the detection rate reduced markedly. On the horizontal member, no positive indications could be obtained at the 6 o'clock position, and the hit rate for the 3, 9 and 12 o'clock positions was approximately 40%. On a vertical section of flooded redundant riser, the hit rate was approximately 60%.

Care must be taken when attempting to draw conclusions from such a limited trial. A more rigorous probability of detection evaluation is required which would take a statistically valid number of readings on many different-sized members of various ages in various fields. In the absence of such an evaluation, we have concluded from our trials that if a large number of readings are taken on BSM at both ends of a horizontal member and at 3, 6, 9 and 12 o'clock positions, there is a reasonably high probability that flooding will be detected. For non-horizontal members the number of readings can be reduced because the water will obviously collect at the bottom, reducing the problems associated with the detection of partial flooding. However, there is still concern that ultrasonics is not completely reliable and that flooded members may go undetected. BP are therefore actively encouraging the development of new techniques.

GAMMA RADIOGRAPHY

The basis for the technique of gamma radiography is that if a source of gamma radiation and a detector are positioned at 6 and 12 o'clock respectively on a member, the number of counts will be less for a flooded member than for an empty one. Chevron and BP have co-funded the development by Sub Sea Offshore Ltd of a practical system which is deployed on an ROV.

A large claw mounted on the ROV has a gamma source in the bottom half and a detector in the top. There are rotating brushes which remove the marine growth in both areas and an ultrasonic device in the bottom arm measures wall thickness and inspects for flooding. The bottom arm is fixed, and a moving trolley ensures that first the brush, then the ultrasonic device, then the gamma radiation source are always presented at the 6 o'clock position on a horizontal member. The top arm is hinged, and the diameter of the member is calculated from the angle at the hinge when clamped. All sensor information is fed to a computer which compares the actual count rate for the member under inspection with that calculated for the measured diameter and wall thickness. A

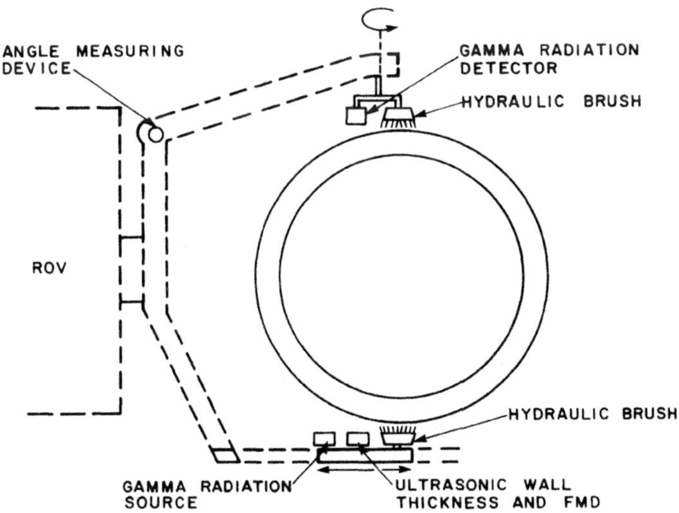

Fig. 3 Gamma radiographic FMD.

significantly lower count rate than calculated may indicate flooding or debris. The system is suitable for members of diameters 0.55–1.8 m, and the whole claw assembly rotates about its axis to allow clamping in any orientation.

The fundamental advantage of the gamma radiographic system is that it fails positive. It will always detect a flooded member (workshop trials indicate that it may be able to detect as little as 50 mm depth of water within a member), but it may gave false alarms because low counts due to the presence of debris cannot be distinguished from low counts due to flooding (or a combination). However, this is considered preferable to failing negative. Until some field experience has been gained, it is not possible to quantify the number of false positives that may be expected and the consequential cost of verification. Again, it is too soon to quantify the financial advantage of an ROV-mounted system, which is preferable in principle to a diver-held system but which make take longer to complete each inspection. The disadvantages of this system are that it requires a large vehicle to deploy it, and a radiographer is required to take responsibility for the gamma source.

THERMAL SYSTEMS

BP are now sponsoring the development of a new technique which detects flooded members by applying a heat source and measuring the resulting temperature profile. Owing to the different heat transfer characteristics of steel/air and steel/water interfaces, an empty member will heat up more quickly and to a higher equilibrium temperature than one containing water. The present prototype design (Fig. 4) uses an annular electrical resistance heater which is thermally insulated on the outer side, and protected by a thin steel mesh on the inner side. The inner side is placed against the member (at 6 o'clock for horizontal members) and eight thermocouples monitor the temperature profile across the diameter.

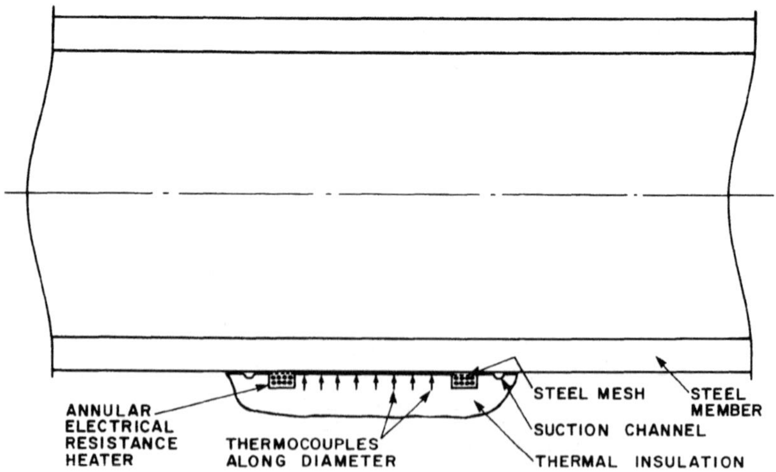

Fig. 4 Thermal FMD.

A computer model developed by Development Engineering International Ltd showed that detection of water in contact with steel could be achieved with relatively low power inputs, and in a series of trials using a number of prototypes, it was shown that the technique could be used to differentiate between dry and flooded members, even when filled with sand to a depth of 10–20 mm.

The first prototype was bonded directly to a section of 762 mm (30 inch) diameter, 41.5 mm ($1\frac{5}{8}$ inch) wall thickness pipe. Figure 5 shows that temperature profiles for various cases after equilibrium had been achieved with a 230 watt input. It can be seen that there is a clear distinction between the flooded and non-flooded cases. The presence of sand reduces the discrimination of the system, but the difference between a dry member and one containing 20 mm of water remains clear. The sand will tend to conduct some heat away from the internal surface of the steel, which has the effect of lowering the measured temperature for the air-filled case. But when water is added, the sand restricts convention in the water, so the measured temperature will be higher than for the case with no sand. From the graph it may be concluded that the minimum detectable water depth is approximately 20 mm, but it should be noted that this is a function of the geometry of the prototype.

The second prototype was the same size as the first, but was formed around a moulded pad which could be applied to the surface of the member and held in place by suction. The suction is achieved by evacuating the annular channel

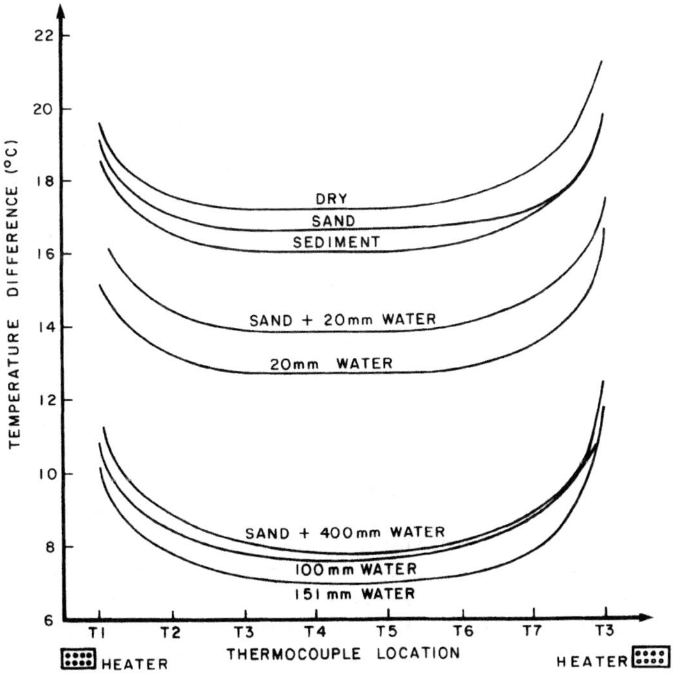

Fig. 5 Plots of steady state temperature difference against thermocouple location for dry and flooded members with and without contaminants.

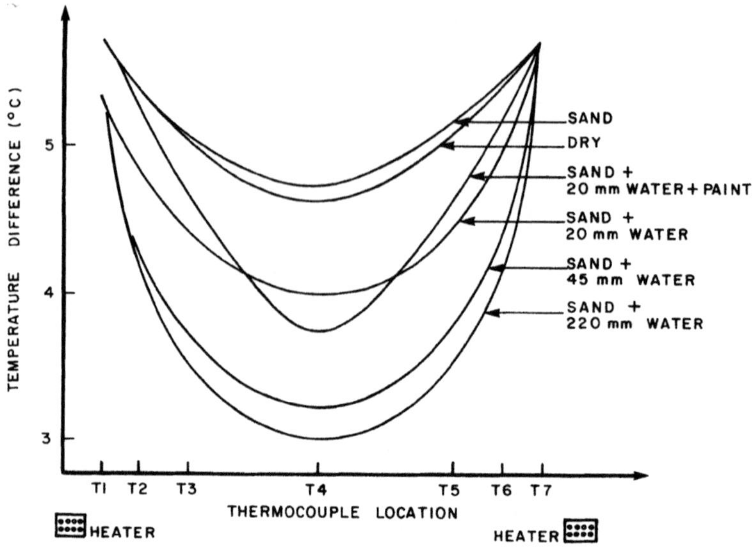

Fig. 6 Plots of steady state temperature difference against thermocouple location for a constant temperature difference of 5.7°C at T7.

outside the heating element. To reduce the time required to achieve equilibrium, tests were carried out maintaining a constant temperature at the thermocouple closest to the heater coil rather than constant power input. By this means. equilibrium was achieved in approximately 20 minutes compared with 1–$1\frac{1}{2}$ hours for the first prototype. Figure 6 shows the temperature profiles for a constant temperature difference of 5.7°C between the nearest thermocouple to the heater and the ambient water. The discrimination between the flooded and non-flooded cases is quite clear, and again the minimum detectable depth of water appears to be about 20 mm. It can be seen from the graph that where the pad was applied to an area of pipe coated with about 2 mm of epoxy paint, the dip in the curve is more marked. This is to be expected: the paint coat will tend to restrict the transfer of heat into and out of the member, with a consequential drop in the measured temperatures. This effect can be countered to a certain extent by improving the insulation on the outer surface of the pad. Additionally, allowance can be made for the coatings by altering the calibration, or it may be preferred to remove the coating at the location of the test. The important point is that such an effect would give a 'fail positive' result, that is, the unit would indicate flooding where it did not exist rather than vice versa.

The temperature gradient with respect to time during the heating process is also dependent on the state of the member. The trials showed that an empty member heats up faster as well as to a higher temperature than one containing water. It is not yet possible to say whether measurement of temperature gradient or final temperature will be the more accurate. A combination of the two might give mutual back up and a higher confidence in both.

Work is now in progress to develop a working prototype for offshore use. The aim is to produce a small device, capable of deployment by diver or small ROV, suitable for any diameter in excess of 0.3 m and wall thicknesses up to 55 mm.

The size of the unit will be reduced as far as possible and the time required per reading will be minimised. The advantages of the resulting tool should be that

- It will not give false negatives
- It will give very few false positives
- It will not require specialized personnel in attendance
- It need only be applied to one side of the member
- It is not essential that it is located accurately at the 6 o'clock position on a horizontal member

It is hoped that this new development will soon be providing a highly reliable method of rapidly checking platform members for flooding.

28

Progress of Automation in Remote Intervention Tasks

D. J. Partridge,
Slingsby Engineering Limited, UK

This brief chapter will examine only a limited number of areas: those where significant progress in subsea automation has been made or where there is potential for further progress. It is hoped this will stimulate ideas and discussion for the future, for as all too often in our industry there is too ready an acceptance of past practice, although it may be less efficient or more costly than what is achievable today because 'it worked last time'. So it may have, but we must ask 'Was it time efficient and was it cost effective?'

VEHICLE-CONTROL AUTOMATION

Pilot Comfort

Our aim must be to replace man from loop whenever it is possible. If this is not possible, then we must make him as comfortable as possible by removing monotonous tasks and by simplifying tiring or difficult tasks; although he must still be in control, but in a supervisory role if at all possible. We must use computers to simplify the controls for the pilot, never forgetting his need for communications to the outside world and to other operating staff. Finally, the provision of auto-fault monitoring takes the strain off the pilot, and above all significantly decreases system downtime. I would suggest that auto-diagnostics should be mandatory for any work-cable ROV or RIV today, and for any ROV, including observation vehicles, in the near future.

Ergonomics

All controls must be easy to see and to reach. The use of glass instrument panels reduces the size of a work station, and computer-generated graphics reduces both cost and complexity as well as reducing MTBF. Pilot strain and communications loads can be reduced by the use of remote cameras for monitoring external systems (winches, etc). Good and controllable internal

Advances in Underwater Technology, Ocean Science and Offshore Engineering, Volume 14: Submersible Technology

lighting in the control cabin is essential. Good video displays ease strain, and finally, environmental control of heat, light and noise, is essential.

Incorporation of sound ergonomics into the vehicle design stage is practical today and must become more widespread.

Removal of Individual Basic Pilot Tasks

We must take the load off the pilot to allow concentration on the real job at hand. Basic vehicle control is the obvious choice as this is quite simple but requires considerable concentration if done manually; examples which should be standard on all of today's work-capable ROVs are:

- auto-pitch;
- auto-roll;
- auto-depth;
- auto-altitude;
- auto-heading.

This list can be extended by providing computer control of thrusters during long transitions to the worksite, and to offset the effect of water currents; this is called *auto-cruise*.

Also, the subsea control of master–slave manipulators can be eased by getting the computer to calculate the position of the end effector continually in all three dimensions, and then by vectoring a camera onto this position; this is *auto-camera-follow*.

Replacement of Conventional Multiplexers

Using the experience of conventional data links on earlier ROVs, Slingsby Engineering Limited (SEL) developed the SELMEC modular data link system for the TROJAN and CIRRUS vehicles. Ten of these systems have now been in the field for some years and by the end of 1987 will be approaching 20 000 operating hours with no serious failure to date.

The SELMEC Concept

The SELMEC was designed to replace the conventional multiplexers which, although being in widespread use, suffer from several disadvantages. A major disadvantage is that the multiplexer system is highly complex and any component failure in the multiplexer causes immediate and total failure to the entire data link, which could lead to severe damage or indeed loss of any ROV or similar system to which it is connected.

The SELMEC system is based on the concepts of multiprocessing, multiterminal data links which are now common in the computer industry, but not in the ROV/RIV industry.

By dividing up the available time on the data link into blocks (or time slots) and allocating a pair of SELMEC cards, each of which has a microprocessor to each time slot, it becomes apparent that the data link will consist of multiple cards at each end of the data link, each talking to its opposite number at their allocated time. Consideration of this approach shows some of the advantages of such a system:

1. Modularity—each supplied system need only be as large as is required. Further capacity can easily be added by simply attaching the additional cards required.
2. Reliability—in the SELMEC design there is no central multiplexer; each pair of SELMEC cards is almost entirely autonomous from the other cards in the system. So, if a SELMEC card fails then the only effect will be that the data transmitted in its individual time slot will be lost. All other time slots will be unaffected and will continue to operate as before. This level of reliability is totally impossible to achieve with the conventional multiplexer. In addition, as a Manchester coding system is used the data buses are transformer isolated, thus preventing earth loops and offset voltages from disrupting the data link.
3. Security—the SELMEC system uses Manchester-encoded data with separate data checking for each time slot. Thus, even if some interference is present on the link, no defective data will be accepted by the receiving time-slot cards. In addition, any interference will only destroy the data in the single time slot during which it occurs, so maximizing the data-transfer rate on a noisy link. In a standard multiplexer system, any interference may destroy the entire data-update cycle for all channels.

Auto-diagnostics

Fault finding/repair and preventive maintenance is simplified and speeded by the use of real-time monitoring of vehicle systems and subsystems displayed at the pilot console.

By using dedicated data channels, we can monitor conditions within the vehicle. The computer can monitor all critical aspects of the vehicle's operation and warn the operator of impending problems without the operator needing to remind himself continually of the need to do this. Thus, continuous monitoring is removed from the pilot's workloads and the pilot has access to an accumulation of detailed information on which to base a retrieval decision should any faults arise, thus reducing the likelihood of an unnecessary retrieval. Any necessary retrieval and repair is speeded by knowing before recovery what parts, staff and tools must be available for an immediate mending. Anyone who has tried to find a fault in an ROV on a typical North Sea winter night would bless the use of auto-diagnostics. Even those of you now confined to a more comfortable office should welcome the reduction in downtime.

If a fault occurs and you know what is wrong you may be able to recover the vehicle without causing even more damage. An example is falling compensation volume which will cause auto-shutdown of hydraulics. But using diagnostics the operator can calculate how much time remains before water ingress and can plan recovery to ensure minimal damage to the hydraulics. Similarly, the diagnostics can warn of imminent failure, thus allowing fall-back positions to be taken.

IR System

Careful design, using modular, dedicated micro-systems, enables systems or parts of systems affected by water ingress to be isolated such that an operation may not need to be aborted and the vehicle remains operational and recoverable without total systems failure. In addition, individual failed systems can be isolated to prevent damage to other systems.

Real-time Optimization

Computer-based multiplexer/control systems have enabled on-line programming and optimization, that is, the vehicle control software can be modified and expanded on site/during operations to correct or add features such as new instrumentation or changed operational requirements.

Auto-dial Vehicles Systems

With on-line programming combined with a modem/comms package we can provide the vehicle system with the means to dial up the manufacturer if a fault occurs which is too complex for on-site personnel to analyse. By the same means the manufacturer can reprogram the system without needing to transfer skilled personnel to the site, with the inherent cost/delay/risk normally involved.

Computer-based Manuals and Maintenance

Obviously, if an industry-standard computer system is used then we can take advantage of the packages available and produce operation/repair manuals actually on the vehicle VDUs. Indeed, we can take this further and couple the repair manual to the diagnostics system such that the manual suggests the correct course of action when a fault occurs. Also the computer can data-log pertinent information continuously during the mission. Planned maintenance control can also be provided.

Self-diagnostics

By using the on-board system to control the existing auto-diagnostics facility we can get the computer to pinpoint the fault without operator input or knowledge.

If the system has in-built redundancy then the computer may even correct the fault without the need for operator intervention, although the operator will be warned.

WHERE TO IN VEHICLE CONTROL AUTOMATION?

We have described what has been done or is being done, in vehicle-control automation. Now we describe what we have not yet done but can do; and where we believe the industry should move to, given further development, for vehicle control. I would hesitate to predict timescales—as usual, economics will dictate where and how fast we move.

Expert Systems (or Knowledge-Based Systems)

We could apply these with our present state of knowledge. Relevant applications are discussed below.

Optimizing Autopilots

These are autopilot control loops which continually test the vehicle's response to its own control actions and learn the vehicle's characteristics by their own induction. The vehicle would then alter the auto-control loop parameters to gradually achieve stable and repeatable 'fly by wire' performance in previously unpredicted hydro-environmental conditions.

Work Envelope Awareness

The vehicle will learn about the immediate area around it and the position of all its tooling and manips, etc., such that it will perform clash detection and develop the necessary avoidance techniques to continue operations unaffected.

Auto-sonar piloting

This involves detecting an image on the sonar and moving the vehicle towards it or around it. Working out the best route to take and performing obstacle avoidance is extremely complex and probably beyond all but the most intelligent engineers.

Position Keeping

All present, auto-controls only keep the vehicle's attitude correct in a given water column; they do not correct movements of the vehicle from a given point over the seabed. This could be corrected by the use of an inertial guidance system or a Doppler sonar. This would be an improvement; however, neither system is perfect and neither positioning technique offers sufficient accuracy on its own for intervention positioning, and Doppler sonar may not work well in the acoustically cluttered environment of a subsea production template.

Route Following

Fairly simple acoustic positioning, coupled with reasonable Doppler sonar or inertial guidance or a cable/pipeline tracking system, will allow the following of a predetermined route or pipeline by closing the control loop with the auto-functions of the computer.

Fault Diagnosis/Prognosis

If the expert system is taught the details of the vehicle and its known and learnt reactions/symptoms during fault conditions and if the system is also allowed to learn from discoveries made during these conditions, then it will suggest courses of action to take when symptoms or faults are reported by the diagnostics. This will help if the fault is complex or unusual, in that the expert system can calculate probable outcomes from a combination of suspected failures. Obviously, this technique can be applied by simulation during the design of the vehicle to reduce the probability of such faults occurring and this will prevent small faults from causing disasters.

Vehicle Sensor Interpretation

No sensor is perfect in its operation. An expert system can learn, for example, the drift characteristics of a gyro and attempt to correct them. Similarly, faults of sensors, such as oscillation, can be 'learnt out'.

Vehicle Sensory Interpretation

This is a huge problem: basically it involves the combination of woolly information from several sensor systems to produce a hard model of the operating environment; in other words, to give the vehicle the spatial awareness normally provided by the human (who is by now asleep?).

Auto-work Site Acquisition

This is probably one of the most forbidding areas yet to be tackled. It's important, whether the vehicle is under manual control or is autonomous, that it gets to the work site reliably and quickly. This involves two major problems:

(i) *Knowing where you are.* Existing forms of navigation all suffer from problems when used within the confines of a modern subsea production environment, although they are adequate when used in an open environment such as cable tracking, etc.

Within the structure most standard acoustic systems fail due to poor progorgation: multipath, spurious reflections or scatter.

Poor visibility will probably cause visual systems to fail. Laser-range gated cameras are too expensive to be used at present. A simple solution would be to fit labels all over the subsea structure which can be read and then looked up on a reference drawing to give a POSITION. This will only work manually if a magnetic interrogation technique is used (which would only have a short range). Dare we hope that ROVs or RIVs will get what the NDT diver never got—although he always asked for it?

(ii) *Getting around.* It is important that the vehicle does not cause damage when in the confines of the structure. This particularly applies to the case of autonomy and is one reason why autonomous vehicles are a long way off, as the risk factor is too high. In particular, the pilot or computer must have a very good sense of spatial awareness and also must not forget where his umbilical is. To this end the industry would probably benefit from a research programme which combined solid modelling with the application of expert systems and vehicle-modelling packages.

I would now like to turn to an examination of task automation—as opposed to vehicle automation.

TASK AUTOMATION—INTRODUCTION

The reasons for task automation are:

- Potentially to simplify or reduce surface back-up;
- to take man out of loop as far as possible; *or*

- to reduce the load on the pilot;
- to simplify task performance (i.e. to make even difficult tasks more routine);
- and to increase task cost-effectiveness.

Teach-and-Learn Control of Manipulators

This is a current-day example of task automation. Repetitive manipulative tasks are those such as deploying pingers, and retrieving tooling from storage racks is a tricky but simple task, although it tends to be time consuming; these forms of tasks are better done by an 'electronic memory' such as the SEL teach-and-learn system. The system can also be used to teach action such as cleaning a node, as once this process begins visibility vanishes and effective manual control is impossible.

Auto-camera Follow

I mentioned this briefly earlier, but perhaps it would be more appropriate under Task Automation. Suffice to ask, 'Why should the pilot have to control a manipulator and also think of where he needs his camera to point so that he can see to exercise such control?' I believe that this is also a good current-day example of what can be offered to ease pilot fatigue and to speed task completion.

Requirements To Achieve Necessary Performance

Upgraded sensor technology will require sensors of high intelligence in order to allow sufficient data-gathering in order that a computer, not a human being, can make decisions. Advanced sensors and technologies may include pattern/image recognition from sonar(s) or TV cameras—complex technologies in their own right. Other sensors include normal navigation sensors (E/S, depth transducer, pingers, etc.) supplemented by 'distance-from-target' measuring systems and angle/attitude-measurement sensors.

The more data that are made available the more the task of decision-making without human beings is eased. Nevertheless, even the most powerful computer in the world (even if it could be mounted on a ROV) can't duplicate human overall intelligence, capacity for decision making and, more importantly, the act of performing a function *efficiently*.

The important enabling technologies, covering sensors, artificial intelligence, etc. are all, on paper, at a reasonable level of advancement, but incorporation into a real, functional piece of hardware is in its infancy, for the most part. It is difficult to have a universal, structured approach in a generally structured but variable scenario—it is the extra add-on in structures in the form of debris or anodes that present the major difficulties.

Discussion of Task Automation

There are many areas for automation—we shall look at only three as examples. There are obviously many others associated with remote intervention, but the three will serve to illustrate typical problems to be encountered, as well as showing that defining a standard approach to every automated task subsea is not likely to be the best approach—task-specific hardware may become the order of the day. No-one to date has yet produced a vehicle with the level of sophistication so far discussed, but several items of hardware address specific automated tasks such as: (a) deploy, survey in preprogrammed mode and return to start point; (b) pick and place items under automatic control using manipulators (teach-and-learn programs); or (c) prepared programs for task sequencing.

These types of approach ultimately may all be incorporated in a free-swimming, intelligent (decision-making), work-capable piece of subsea hardware that does not require a form of command intervention—human being—at all while it is subsea. Is it a dream? In the meantime, back to near-future practicality.

A major element in deciding how far to automate is the degree of predictability of task. The less the variation in the task and the higher the predictability, the more a task is suited to automation.

Let's look at three typical underwater tasks:

1. *Bottom-related tasks*

   Characteristics:
   - largely predictable;
   - long duration and very repetitious;
   - high percent of costs relate to the parent surface vessel;
   - large volume of information is generated, but may not require real-time data reduction.

   Examples: cable burial and pipeline survey.

2. *Inspection of underwater structures*
   - the task is predictable, but the variations are large so it becomes to an extent unpredictable;
   - repetitious;
   - light work and forces;
   - good worksite acquisition needed;
   - good mapping capabilities needed;
   - requires stable base or attachment;
   - accuracy and repeatability of a high order.

3. *Intervention on subsea production systems*
   - predictable;
   - often large forces, heavy modules, high torques and high sea states oppose the need for accuracy;
   - good worksite acquisition needed (probably with positive location).

In all three cases, task automation brings cost efficiency through the reduction of manual inputs to the loop (and perhaps the removal of the human being from the loop). Let's look at possibilities in these three areas.

In the case of (1), *bottom survey:*

- The tasks are very repetitious and of long duration. Therefore, it is clearly desirable to reduce manual input. As an example, current SEL technology provides for closed-loop cable tracking. Directional prediction and control of vehicle motion relative to the seabed and in cable inspection have materially reduced manual input. This technique could be, and I would suggest should be, applied to other bottom-related tasks.

Further developments could include:

(a) The recognition of faults or obstacles by the machine which would reduce the need to store or manually supervise the analysis of large amounts of data.
(b) The ability to store large amounts of data on board the vehicle could remove the need to analyse at the time and perhaps eventually remove the need for data transmission to the surface. Indeed, the provision of an on-board power source may negate the need for an umbilical in bottom survey altogether. I accept that autonomous vehicles will probably first emerge in non-oil and gas activities; but we should be aware of the potential for large cost savings by removal of a DP parent vessel in repetitious tasks.

In the case of (2), *inspection of underwater structures:*

- Current technology—high manpower input, expensive parent vessels, precise mapping is difficult, low repeatability, and less than adequate flaw-detection rates.
- The amount of automation at present is typically small; remotely controlled devices are beginning to appear, but we have a very long way to go.
- We need more control and dexterity on the delivery device (manipulator) than in any other remote intervention area.
- We must remove manual control of repetitious actions including cleaning, application of positional sensors, and finally of NDT sensors. On the PIC ROV, for instance, operator fatigue cycles for NDT sensor application were typically less than 30 minutes.
- Advances are required on getting to the worksite and identifying it, on mapping the area accurately and on being able to return to a particular point with precision. This requires the application of expert systems and manipulators with exceptional repeatability not so far developed.

Constant, high-intensity human monitoring may not be needed if we can provide:

(a) fault-recognition designed into the inspection tools or sensors;
(b) good mapping and data storage;
(c) auto-curvilinear path calculation, prediction and auto-tracking (application of spherical trigonometry).

In the case of (3), *planned maintenance of subsea production systems:*

- These tasks are very close to being wholly predictable and repetitious (or they should be).
- At present, intervention on subsea production systems is carried out in a very few cases by manned hyperbaric installations, but mainly by divers. Interest is centred on remote intervention onto new subsea structures specifically designed to interface with dedicated machines: for example the RGV for BP's DISPS, ROT for Statoil and UPS and ROVM for Saga. These are heavy-lift machines designed to change out valve modules or valve inserts, but essentially they rely on current ROV technology. Some of the systems under consideration require the use of ROVs for a variety of tasks, including observation, intervention during installation (operation of flooding valves, shackle release, pinger and level sensor installation, etc.) and operational intervention through life.

ROV operational intervention tasks on production templates, valve modules, production trees and flowline valves include tasks such as:

1. operation/over-ride of gate and choke valves;
2. hot stab into hydraulic lines;
3. preparation of valve modules prior to intervention by heavy-lift machine, i.e. removal of debris caps.

To accomplish these tasks the ROV will most likely be positively located to the subsea structure; at the very least, provision for this must be made at the structure design stage.

The ROV will be equipped with tool packages specifically designed to carry out these tasks on specific structures: unfortunately it seems that these packages will be different for each operator, and even perhaps for two fields of one operator.

A certain amount of automation to the level available in high-quality present-day usage may be incorporated in these packages, i.e. manipulator 'teach-and-learn' if a manipulator is used, camera 'auto-follow' and also simple preprogrammed sequencing of latching and tool deployment rams/actuators.

At present it is envisaged that these ROV intervention tasks will be carried out under manned control, but owing to the structured environment of the template structure (thus it is predictable) further automation could be accommodated; this is especially true if the template is designed from the outset for remote intervention. Perhaps the ROV or RIV will have an advantage the poor diver has never had!

Automation is required in:

(a) deployment through to template acquisition;
(b) positioning of the ROV or RIV within template—both positive and sensory;
(c) in performance of work tasks.

Extensive design content lies in the tools and production components, coupled with automated sequencing within operations. However, it is still important that humans remain in the loop. The largest single benefit in this area would stem from increased standardization of subsea equipment; unfortunately this seems to be very far away.

Application of Available Technology

Our industry lags behind other industries in the application of technology that has existed for some time: we know how to apply it to aid or benefit all of the three tasks just described but the requirement must stem from the operator. It may be that there is too little dialogue between ROV designers—and I do mean designers, not kit-assembly men—and the engineers from the operators. I believe that this time lag is also partly due to the awareness of financial risk which leads to continued reliance on tried and tested methods, because they are safe in a financial sense, long after they could have been cost-effectively replaced.

It's not all that long ago that an executive from a North Sea operator told me that I'd never see an ROV carrying out his drill support. His company has paid for a significant amount of ROV hire this last season.

A more recent example of our industry's conservative nature occurred this year when a group of engineers working for an operator on a major project decided that umbilicals containing fibre optics could not be used in a heave-compensation mode. The cost of inserting them, even alongside coppers, would have been insignificant in relation to the overall project costs, and the results of the ensuing trials would benefit us all. As an aside we have found that while flexing under tension causes failures of coppers, as yet none of the fibre optic strands in our ten or so umbilicals in service has failed.

Risk is a complex subject and I will not dwell on it further except to say that prior to judgement one must get sufficient information to be able to quantify the risk.

MARKET FORCES AND EFFECTS ON DEVELOPMENT

Until the end of 1985 the dramatically rapid evolution of the ROV—from being underpowered, unreliable, non-automated vehicles to today's powerful, and part-automated work-capable systems—has almost entirely depended upon the offshore oil and gas industry and that industry's drive to reduce operating costs: in addition, some limited governmental assistance has been available.

Today, when ROV hire-rates available from the operators are less than depreciation costs, there is no funding available from the oil industry's service companies; similarly ROV/RIV manufacturers have had to cut margins to levels where they can no longer sustain R. and D. expenditure at anywhere near the 1983 to 1985 levels (in those three years my own company spent some £6.15 million on R. and D.).

Thus, the majority of development in ROV/RIV automation will have to be funded by the operators in partnership with government, or it is unlikely to proceed.

There is the possibility that the defence ROV market may grow sufficiently to allow such worthwhile development that may be partly applicable to the offshore scene.

CONCLUSIONS

The provision of increased automation will largely depend on the perceived financial incentive for the operators, and perhaps to a lesser degree for the service companies.

The removal of the human being from the control loop totally will be expensive and it is difficult to see justification where operations are conducted in close proximity to a seabed structure or a fixed platform. However, there is much that some parts of our industry—the ROV manufacturing industry—have already achieved, there is more that we are completing now, and there is far more that we are capable of doing given the cash.

It is important for the offshore oil and gas industry to appreciate that we lag behind other industries in the application of available technology. This is probably due to the offshore industry's understandable but innate conservatism which all too often militates against advance.

The case for the viability of a small number of commercial autonomous vehicles will be argued by others. I would merely say that they are now technically feasible and could significantly reduce costs in survey operations, and perhaps in difficult areas such as under-ice operations. However, I suspect that such developments will first arise in other markets: perhaps deep-ocean mineral-mining survey or telecommunications.

To sum up: I believe our industry can invent a 'better mousetrap' but this is pointless unless the end client—the operator—will accept it and actively wants it, so that he will 'beat a path' to the manufacturers' doors, or make his service companies do so.

29

Operational Automatic Hyperbaric Welding

J. Blight and G. Hutt,
Comex Group, France

Manual hyperbaric welding in a dry hyperbaric environment is now a well established technique for underwater pipeline tie-ins and repairs. Over 600 hyperbaric tie-ins have been performed to date, and manual procedures have been successfully qualified down to simulated depths of 450 msw.

Despite the obvious success of manual welding, considerable resources have been invested in the development of automatic hyperbaric welding systems. The advantages of automatic welding will offer the following technical and economic advantages:

- improved weld quality and properties
- improved assurance of weld quality during offshore operations through increased process control and monitoring
- reduced costs for qualifying procedures and personnel
- lower costs for diving personnel
- improved working conditions for the habitat personnel

This chapter describes the Thor-1 system developed by the Comex Group and used on the North Alwyn Export Pipeline in August 1986 to perform the first commercial automatic weld on a submarine pipeline. The chapter concludes with a brief review of the future of automatic hyperbaric welding and the development of the second generation Thor-2 system which will permit diverless tie-ins.

DEVELOPING THOR-1

Thor-1 was designed as an autonomous 'stand alone' system which can be used with any suitable welding spread.[1] The system is suitable for deep water work and therefore the welding power source, weld head motors, and camera container were developed for use at 500 msw.

Important features incorporated into the design of Thor-1 include the following considerations.

- The entire welding procedure can be pre-programmed so that the offshore

Advances in Underwater Technology, Ocean Science and Offshore Engineering, Volume 14: Submersible Technology

Fig. 1 Thor-1 module being lowered into position on the seabed.

welds are made with parameters identical to those used for welds made onshore during the welding procedure qualification tests
- The welding operation can be initiated, monitored and supervised from the surface and the surface operators can intervene at any moment during the cycle to make modifications, etc. as required within pre-qualified/approved parameters ranges
- The habitat personnel are responsible for installing and servicing the equipment under the supervision and at the request of the surface operators, but they do not intervene during the welding cycle itself

The system used on the North Alwyn Pipeline tie-in consisted of:

- A lightweight track which is readily installed on the pipe by the habitat personnel
- A modular orbital weld head which is rapidly assembled on the track and connected to a short umbilical from the subsea module
- The weld head consisting of a motorized module (or 'engine') carrying the torch, etc. which pushes/pulls around the track a number of unmotorized modules (or 'wagons') carrying ancillary equipment such as wire feeders. This system permits ready modification of the weld head to include new equipment, for example NDT head
- A simple pendant control which permits the habitat personnel to perform function tests for the weld head under the supervision of the surface personnel
- A subsea module, installed beside the habitat and which houses the welding power source, weld head drive cards, teletransmission terminals, power transformers, etc.
- A surface container on the deck of the DSV which houses the modular remote control desk, the teletransmission terminals, etc. and which is

linked to the subsea module by a 70 mm diameter, 260 m long umbilical.
- A weld pool observation system which uses two-colour CCD cameras installed on the weld head to give the surface operator close-up views of the weld, both before and after the arc.
- A spare parts container and workshop.

Development work started in January 1985 and the principal components were delivered to Comex in January 1986. System testing and commissioning consisted of the following phases

- Factory testing of individual sub-systems, e.g. weld head motors, weld head, power source, teletransmission, cameras
- Integration of all the equipment in the Comex Hyperbaric Test Centre and tests at 50 bars (function tests and welding)
- Modification of equipment by the Comex technicians, e.g. camera focusing, lighting, wire feed, torch
- Welding parameter development

It can be seen that the use of an automatic welding system of this type has a major effect on the way in which the operation is conducted in so far as all the decisions and actions which affect quality are taken on surface under the directions of the supervising engineers.

The programming facility incorporates sufficient capacity (software design, memory, etc.) to optimize the pre-programming and to minimize the amount of 'real time' modifications which the surface operators may be required to perform. Up to 50 passes, each consisting of up to 8 separate sectors, can be pre-programmed and it is possible to pre-programme the transverse position of the torch to allow irregularities in track alignment to be compensated for before the start of welding.

QUALITY ASSURANCE

Particular attention was paid to defining the quality assurance functions which are now built into the system. These can be summarized as follows:

- The qualified procedure is recorded on cassette during the procedure qualification tests. The client keeps the cassette and gives it to the operator offshore, just before the start of welding
- The complete procedure is printed out before the start of welding and can be verified against the print-out obtained during procedure qualifications
- The permitted parameter modifications, e.g. qualified current level—10%, are pre-programmed such that the operator cannot exceed the approved range during welding (i.e. real time modifications have pre-programmed limits)
- All parameter modifications are automatically printed out during welding (positions, type of parameter, new value, method by which the modification was made, etc.)
- There is a permanent video display of the pre-programmed parameter values and the actual values (taking into account the real time modifications). This display is connected to a printer and the client's representative can take copies whenever he wishes

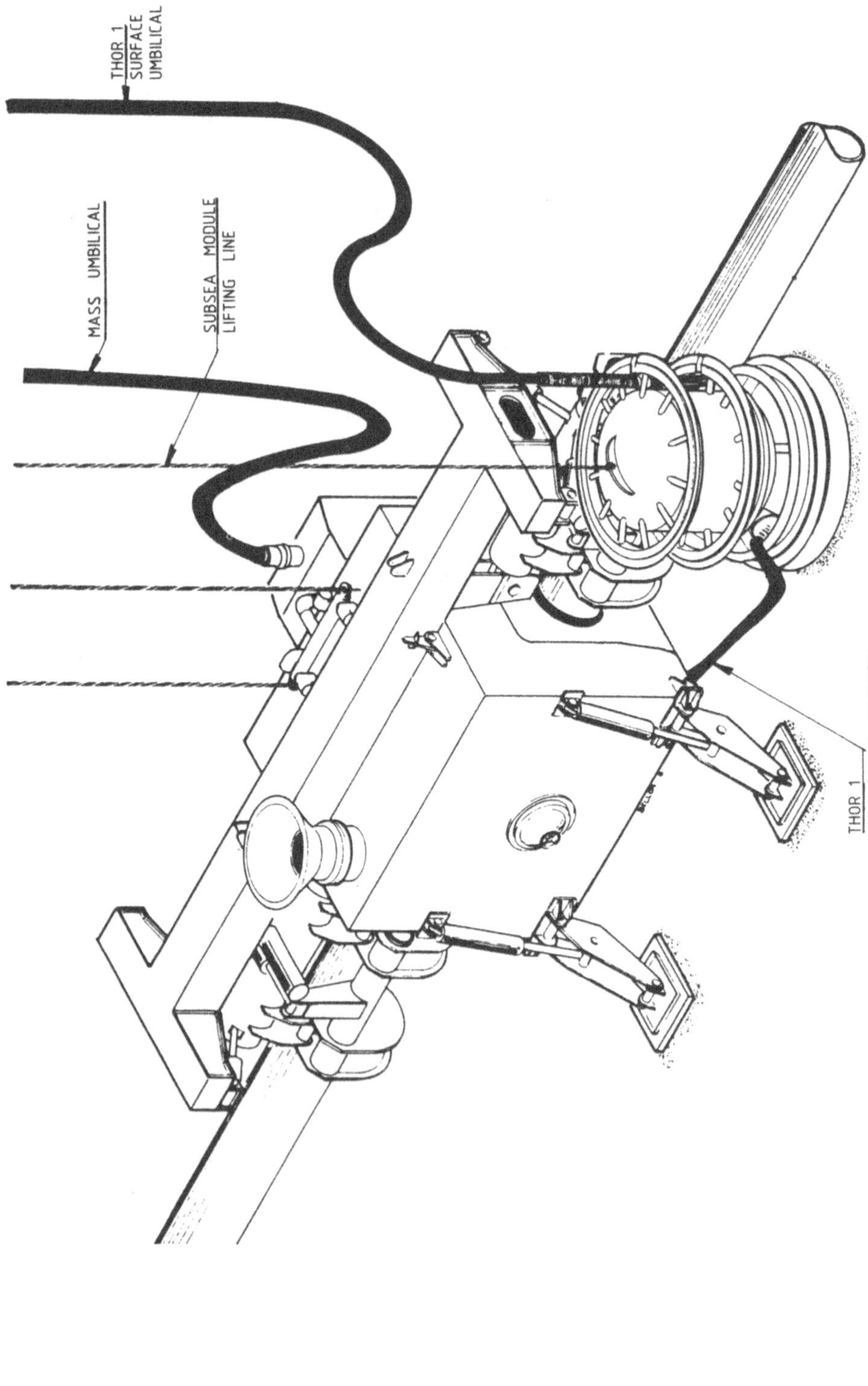

Fig. 2 Thor-1 subsea module and habitat in working position.

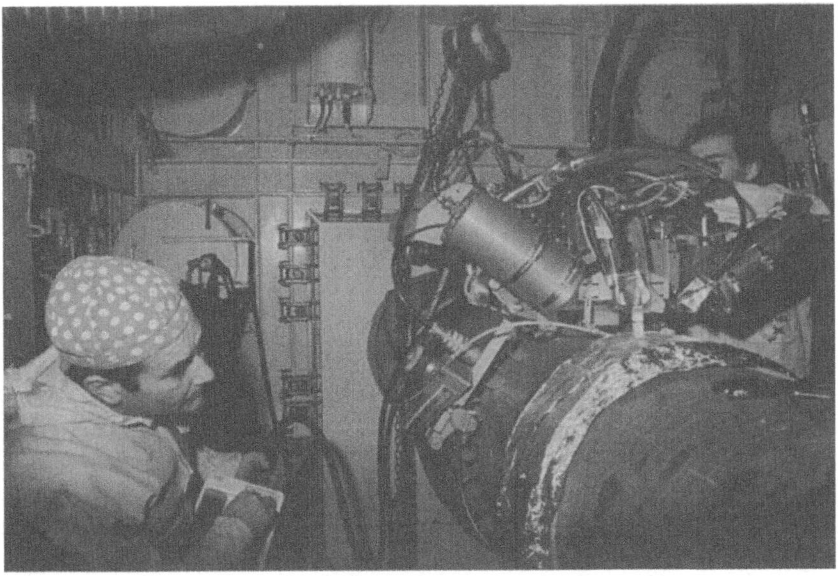

Fig. 3 Thor-1 weld head and track installed on pipe inside subsea habitat. Diver performing function test.

- There is a full analog display of essential variables as measured directly at the power source and weld head (current, voltage, wire feed speed, rotational speed, etc.)

OFFSHORE WELDING OPERATIONS

The offshore programme consisted of two separate phases. The first of these was a welding trial carried out by Comex/CHDL and the second was the actual production tie-in weld for Tomec. In preparation for the latter, a formal qualification trial was required which consisted of:

- Qualifying the system in terms of equipment and quality assurance
- Qualifying the welding procedures (main procedure, through thickness repair procedure and mid-thickness repair procedure)
- Qualifying the surface operators and demonstrating the competence of the habitat personnel as 'machine attendants'

Because this qualification programme was the first of its kind, considerable attention was paid to the quality assurance aspects and to clearly defining the roles and responsibilites of the surface and habitat personnel.

Three identical welds were made at a simulated depth of 130 msw and tested to the requirements of BS 4515 (1984).

The final stage of preparation of the system involved a shallow water test with equipment integrated into the subsea module and surface container. The offshore operations were performed on the Alwyn North Field using the semi-submersible support vessel *Uncle John*.

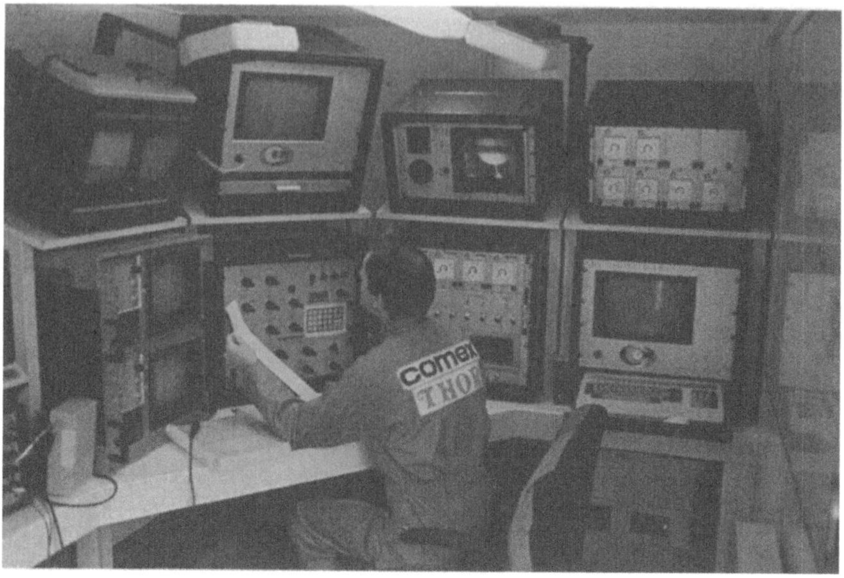

Fig. 4 Module remote control desk in surface container on deck of DSV.

The first step was to assemble the welding system on the deck of the DSV to ensure correct functioning and for calibration. This was followed by the welding test programme at 130 msw using pre-prepared pipe coupons (20 inch diameter × 17 mm API 5LX X60 pipe) which consisted of the following welds:

(*a*) Four identical root pass welds to demonstrate the validity of the 'onshore' developed parameters and their reproducibility
(*b*) Three root passes made with parameters outside the normal ranges to determine tolerances when operating in offshore conditions
(*c*) Completion of the fill and cap portions of two of the welds from part (*a*) using full weave and split weave techniques.

Thor-1 was then used to perform the last tie-in weld on the 110 km long pipeline using the previously qualified procedure. At this location, the pipeline consisted of 24 inch diameter, 17 mm wall thickness API LX X65 line pipe. The results of visual and radiographic examination were very satisfactory, and the production weld was accepted in August 1986, this being the first tie-in on a commercial pipeline for which an automatic welding system had been used.

One of the completed offshore test welds was destructively tested to BS 4515 (1984) under Lloyds and DNV supervision.

The weld metal Charpy test results were particularly good: they were characterised by a relatively narrow scatter band with an average value of greater than 90 J at −40°C. These results clearly confirmed that the same high level of weld quality found during the onshore weld procedure qualifications is automatically reproduced during offshore operations.

The success of the offshore operations clearly demonstrated that the Thor-1 system and personnel were fully operational and automatic hyperbaric welding is now a practical alternative to manual hyperbaric welding.

Fig. 5 Thor-1 welding on Alwyn North Pipeline, July 1986.

SUBSEQUENT DEVELOPMENTS

The experience gained during the offshore operation has been utilized in the further development of the system. An extensive data bank has now been developed which relates depth, pipe dimensions, bevel shape, root face thickness, gap and misalignment to weld parameters which will ensure a successful root pass. This data bank will serve as the basis for a future computer-based expert system for generating root pass parameters as a function of the metrology of the actual groove to be welded.

Comex has developed pipe preparation procedures and equipment based on a subsea lathe and a special tool box which ensure that the required J bevel shape and dimensions including the critical root face thickness are reproduced with the minimum of 'pipefitter intervention'.

A hydraulic re-forming clamp (HRC) has been designed which positions itself perpendicularly on the pipe, locks itself and reduces the pipe ovality automatically under simple 'push button' control by the 'pipefitter'.

A local remote control function has been developed for the alignment frame which facilitates the work of the pipefitter. The HRC is fitted with a series of fingers which are used to align and support the pup piece and which can be withdrawn sequentially to allow the passage of the weld head during root pass welding. As an alternative, Comex has developed an internal pup piece support which maintains the pup piece in position after alignment without the aid of external fingers or arms. The use of this pig, known as the stupid pig in order to differentiate it from the Houlder offshore intelligent pig and clamp (IPAC) has the advantage of reducing the 'overcrowding' of equipment around the weld joint area.

Another important factor in reducing this overcrowding has been the

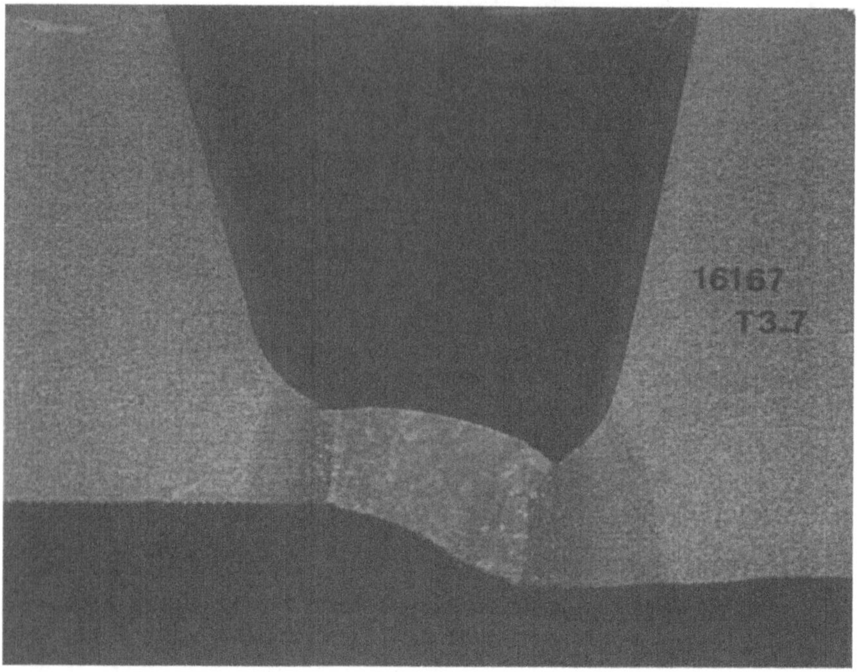

Fig. 6 Root pass made with Thor-1 at a simulated depth of 130 msw.

- Dissimilar root face thicknesses (1.8–2.3 mm)
- Root gap of 1.5 mm
- Root misalignment of 2 mm
- Position 4 o'clock
- Pipe API 5LX X60 20 in. OD × 17 mm WT

development of induction pre-heating for hyperbaric welding. Not only are the inductors smaller and less fragile than the traditional resistance pre-heating elements, the induction heating technique is also energetically more efficient and can be used for postweld heat treatment or hot tapping operations as was demonstrated earlier this year.[3]

Automatic root pass welding is also dependent on removing or nullifying the effects of remanent magnetism in the weld groove during welding. An automatic demagnetizing system has been incorporated into the Thor-1 system. Hall effect probes installed on the weld head provide readings of magnetism in the groove, and these values are used by the Thor-1 telemetry system to calculate the currents to be supplied to the two demagnetizing coils set each side of the weld. The system is now fully operational and can permit 'real time' correction of remanent magnetic fields which can vary in direction and magnitude up to 1000 gauss around the circumference of the pipe.

A further aim of the development has been to increase the welding speed. Although synergic MIG or plasma may eventually prove more rapid than GTAW, neither process is yet sufficiently developed for operational use. Comex has studied different variants of the GTAW process including hot wire TIG, welding with high deposition rate synthetic wires or large diameter solid wires. It was concluded, however, that the most efficient manner for increasing effective deposition rates, while maintaining 'operational tolerance' and high quality levels, is to weld with two weldheads simultaneously. A considerable number of tests have been made in this configuration, and both operating procedures and equipment layouts have now been prepared such that future Thor systems can be built as 'two weld head' systems.

It is obvious that during offshore operations the operators must be capable of maintaining and eventually repairing the welding equipment without the assistance of specialists and others. Particular attention has been paid to simplifying the equipment to aid inspection and maintenance and training the operators to perform these functions. Moreover a computer-based 'expert system' is being built up which aids the operators in identifying the causes of malfunctions and in replacing or repairing the defective components.

Future versions of the expert system will be interfaced directly with the welding equipment to allow equipment monitoring and the detection of abnormal functioning and so on.

CONCLUSIONS

August 1986 saw the first operational use of automatic hyperbaric welding (Thor-1). A second automatic welding system was also used operationally in autumn 1986. In autumn 1987, the Oseberg Automatic Welding Spread was delivered to Norsk-Hydro by Norcem Comex Subsea A/S.[2] Thus, offshore operators now have the choice between three operational automatic welding systems which, although differing in configuration and equipment design, all offer the basic advantages of automatic hyperbaric welding. Despite an extremely depressed market, automatic hyperbaric welding will gradually and ineluctably replace manual welding above all for tie-ins in the deeper waters (Norwegian Sector of North Sea) and for Emergency Pipeline Repair Services, where the use of automatic welding represents a major saving in terms of personnel qualification and salary costs.

The three automatic welding systems that are now operational all share one characteristic: the use of habitat personnel to either install, service or operate parts of the equipment.

The Comex Group is committed to developing the technology for deep water interventions, including welding. Although special diving techniques have been developed for example using hydrogen–heliox gas mixes, Comex believes that diver interventions should be limited as far as possible to 'back up' operations for otherwise diverless systems. Despite the depressed state of the market, Comex believes that there will be a need for diverless hyperbaric welding systems for deep water interventions in the foreseeable future. Given the technical success of Thor-1, it has been decided to start the development of Thor-2—a system that could be used either as an advanced diver-assisted system or as a diverless system.[4]

The 'Thor-2' development programme is being coordinated with the on-going IRDS diverless welding spread development such that a joint diverless hyperbaric tie-in trial will be made offshore in autumn 1988. The 'Thor-2' development is described in detail elsewhere, but it should be noted here that diverless equipment and procedures are being developed for all the wet and dry tasks associated with pipeline tie-ins or repairs, including pipe preparation, alignment, and metrology.

ACKNOWLEDGEMENT

The authors wish to thank the directors of Comex Services and CHDL for permission to present this paper and acknowledge the contributions of Polysoude, Cybernetix and Hytec to the 'Thor-1' development programme.

REFERENCES

1. Blight, J. 'Automation in hyperbaric welding', The Welding Institute Conference on 'Computer Technology in Welding', June 1986, London, Paper 10.
2. Knagenhjelm, H. O., Morris, A. W., Pinches, C. A., Bellis, G. S. and Gjermundsen K. 'The development of a mechanised welding system for deep waters'. The Welding Institute Conference on Advanced welding systems, November 1985, London, Paper 43.
3. Comex Group Hot Tapping Trials, December 1986–January 1987.
4. Blight, J. and Baylot, M. 'Thor-2' Diverless Welding System. The Second International Conference on Developments in Automated and Robotic Welding, 17–19 November 1987, London, Paper 12.

Part VI

Diversification and Innovation: Adapting to Change

30

Resources from the Sea Programme: a Technical Challenge

D. M. Hoddinott and A. H. Ruyten,
Department of Trade and Industry, UK

Marine resources have been exploited throughout history but, up until the advent of the recovery of hydrocarbon fuels, the technology needed developed comparatively slowly. In a number of cases the technology developed for other requirements was adapted for use in the marine environment.

Exploitation of the traditional resources, including offshore oil and gas, continued with little or no consideration being given to other forms of energy, minerals or biological resources. However, more recently, there has been an upsurge in interest in the longer term effect of the depletion of non-renewable on-shore resources and offshore hydrocarbons. This has given an impetus to the consideration of other sources of minerals and energy and interest focused on the oceans of the world. It is now apparent that with the use of existing sophisticated equipment and techniques and, perhaps even more significantly, with the expected developments of technologies and novel equipment, this vast source of the oceans can be tapped.

The exclusive economic zones (EEZs) defined by the Law of the Sea Convention extend for 200 miles from the coast of maritime nations. This gives those nations which do not have the benefit of an extensive continental shelf, as does the UK, rights to the exploitation of resources within this 200-mile limit. It is envisaged that over the next few decades, the nations that declare an EEZ, or have extensive continental shelf rights, will want or, perhaps more importantly, need to explore these areas with a view to exploiting the very extensive resources that lie within them. The technology and financial capabilities to carry out sophisticated surveying operations and, eventually, commercial recovery of the available resources at present lie within the advanced nations. The UK marine equipment industry is among the world's leaders in a rapidly moving and intensely competitive international market and, as such, is able to take advantage of the potential export market for the supply of equipment and

Advances in Underwater Technology, Ocean Science and Offshore Engineering, Volume 14: Submersible Technology

services for exploration and exploitation of resources. The rate of advance in technology world-wide requires that to sustain its position the UK marine industry will need to maintain a high level and quality of innovation and develop the underlying technological capability required for the future. It is envisaged that major commercial opportunities will arise in the provision of the equipment and services which contribute towards the exploration and exploitation of the potential resources. Companies with the appropriate technological capabilities will be in the best position to take advantage of these opportunities.

The Department of Trade and Industry's (DTI) Maritime Technology Committee (MTC) saw the concept of marine resources as a challenge which, if faced and overcome, could have far-reaching commercial benefits for the diverse UK industries concerned with the maritime sector. With this in mind an Advisory Committee on Resources of the Sea (RSC) was set up to 'review the current activities directed towards the exploration and exploitation of the marine resources and to develop an overall strategy and programme'.

STRATEGY AND PROGRAMME

The RSC, which included members of the Society for Underwater Technology, identified areas of interest for a wide range of industries as well as for other Government Departments such as the Ministry of Agriculture, Fisheries and Food and the Ministry of Defence and for Research Councils such as the Natural Environment Research Council (NERC) and the Science and Engineering Research Council (SERC).

RSC suggested a programme aimed at resources other than the well-established areas of fishing, sand and gravel, and offshore oil and gas. The principal target resources identified were:

- Biological: plankton, seaweed, mariculture
- Mineral: manganese nodules, polymetallic sulphides, mineral deposits on the seabed, minerals in solution
- Energy: tides, currents, wind, waves, thermal

Other commercially exploitable resources seem likely to come to light in the future, perhaps from the biocultures now known to exist around the submarine volcanoes known as 'black smokers'. New approaches to exploitation of traditional resources will no doubt also emerge to present new commercial opportunities for industry.

Based on RSC's recommendation the Resources from the Sea Programme (RSP) was launched in January 1986 to assist and encourage British industry to collaborate in building upon the sound existing basis of marine technology and undertake the development of advanced sensing, control, communication, robotic, power and materials technologies which will be required in the hostile environment of the deep ocean. The programme divides broadly into two main areas of exploration and exploitation with each having elements relating to the development of instrumentation and equipment to operational and analysis techniques.

It was apparent during the formative stages that the initial opportunities for industry lay in the exploration and surveying of the oceans where early interest, particularly from overseas countries, may result from a realization of the opportunities that are likely to exist within the limits of continental shelves and

EEZs. Exploitation of identified resources logically follows as a later phase but, nevertheless, early consideration needs to be given to this phase to ensure that the enabling technologies are available when the time is right for exploitation activity.

IMPLEMENTATION OF THE PROGRAMME

Following launch of the programme, it rapidly became clear that within this enormous sector, there was a need to identify those areas likely to provide the most promising commercial opportunities and upon which industry should thus concentrate their efforts. The DTI thus commissioned a number of feasibility studies to assess the state of present and future technologies in certain areas and to assess the world-wide market opportunities for systems and services based on them. The Department consulted widely among industry, as well as consultants, academia and research organizations in determining the areas of study and in the final selection of the consultants to carry them out. The studies, relating to the exploration phase, and the respective consultants for each were as follows:

Surveying and Exploration of the Seabed to study the technology and market opportunities of the subject area, including, but not limited to, acoustic and seismic surveying techniques and seabed sampling methods. (The Institute of Offshore Engineering at Heriot-Watt University.)

Marine Environmental Measurement, which covers the measurement technology and techniques associated with parameters such as waves, currents, tides, wind, temperatures and chemical and biological levels. (Metocean Consultancy Ltd.)

Manned Subsea Operations, which is not directed so much at manned submersibles as the deep sea operating environments and life support systems, sensing and manipulation technology and, where appropriate, propulsion. (The Underwater Engineering Group (UEG) of the Construction Industry Research and Information Association.)

Surface Support for Subsea Operations, which covers, among other aspects, the safety, power, communication, telemetry, logistics, control, data handling, maintenance and positioning requirements for both manned and unmanned subsea operations. (Marine Technology Centre of the University of Strathclyde.)

Technologies for Exploitation of Marine Resources, which covers the enabling technologies required for the future commercial exploitation of marine mineral, energy and biological resources. (General Technology Systems Ltd.)

A further study has been commissioned more recently into the technology of mariculture, being carried out by Aberdeen University Marine Studies Limited. This study covers the technologies associated with the 'farming' of fish or other biological resources in the marine environment.

The terms of reference for the studies are similar and provide for the assessment of world market opportunities and prospects for the commercial exploitation of equipment, systems and technique in each area together with the likely timescale; identification of the technologies required to enable these market opportunities and prospects to be exploited fully; assessment of the current or prospective availability of such technologies; and an assessment of costs, timescales and prospects for commercial exploitation, taking into account likely international competition.

There are two other studies related to the RSP. Within the Department's Advanced Robotics (AR) Programme, mobile autonomous robots for hostile environments are being investigated. An underwater application area has been addressed by a feasibility study undertaken by UEG supported by Hollobone Hibbert and Associates and Slingsby Engineering Ltd. (Other AR application areas addressed to date include tunnelling, nuclear, firefighting and rescue, construction, space, and medical and health care.) The second study related to the RSP is by the Foreign and Commonwealth Office who are preparing a study into market opportunities for exploration in the South Pacific oceans.

RESULTS OF THE FEASIBILITY STUDIES

A number of areas within the sub-sea sector have been identified by the feasibility studies as being ones that can be commercially viable to UK industry in the near, medium and long term, in response to the perceived markets in novel marine resources. Within the surveying and exploration phase of the RSP, the major opportunities are seen to be data handling and processing activities, both in real time and post data acquisition, in geophysical surveys and in the development of sampling technology. The relatively new and certainly high technology requirement of marine environmental measurement offers good potential growth in the short- to mid-term for small to medium-sized companies and it provides an area of application for UK skills in engineering and innovation. The development areas identified within these sectors are set out in the Appendix. Surface support activities comprise a number of inter-related basic elements, and each needs to be considered in its relationship to the overall system, for a particular application, to ascertain opportunities and constraints of the individual elements. The example of the sub-sea operation of seabed surveying highlights a number of areas which present potential opportunities, some of which will, of course, be applicable to other support operations. Opportunities for innovative equipment and systems include:

- Improved sub-sea vehicle hydrodynamics and control systems; integrated vehicle/task design; control packages and sensors and autonomous ROVs
- Application of artificial intelligence techniques are applicable in the development of hardware; interfacing; task programming and collision avoidance
- Navigation and control systems; inertial navigation; transponders and subsea use of global navigation systems are all of interest in the operational aspects of sub-sea vehicles and surveying devices
- Power source technology particularly closed diesel and sterling engines; inductive power transmission and energy storage systems
- Signal conditioning and telemetry systems for the transmission of data and control signals cover the areas of multiplexing systems; through water acoustics; very low frequency transmission and optical-fibre umbilicals

Exploitation of resources have, in general, longer term prospects, although some of the potential resources present opportunities in the shorter term. Typically, except for a few individual opportunities, the exploitation of deep sea minerals is not seen as being economic in the near-term, and recovery of mineral deposits in international waters, as distinct from in EEZs, is seen as being even more distant. However, the enabling technologies which will be needed to undertake the recovery of minerals from the deep ocean must be developed if the UK is going to be in a position to take advantage of these longer term opportunities.

Renewable energy sources present a real prospect for the shorter term in the harnessing of tidal energy, but this needs massive capital investment in the necessary civil and power conversion engineering. Ocean thermal energy conversion (OTEC) has longer term prospects and has limited application potential due to the need for a high temperature differential between the deep and surface waters.

The major growth area of mariculture provides a large future market with short-term prospects in the supply of equipment and services and, in the longer term, opportunities in products and services based on biosciences.

Among technology sectors of secondary interest, but for which opportunities could arise in the main market areas, are small-scale wave energy units; artificial islands, for ice-area operations and other off-shore civil engineering projects; and specific site investigation equipment.

The extensive list given in this chapter is, as already mentioned, taken from consultants' advice to the DTI and will be augmented by the mariculture and other studies. It is in the process of being refined, with the aid of industry and other interested parties to give a handful of particularly rewarding targets that can be concentrated on by the marine orientated industries and by research organization of the UK and by others looking for investment opportunities. It is the Department's intention to give priority to projects that fall within this grouping. However, this is not to say that proposals that do not fit directly into this short list of identified 'targets' will not also be considered.

DTI SUPPORT FOR R AND D PROJECTS

The Department of Trade and Industry's activities in the area of science and technology are geared to support the Department's policies and objectives, among which the competitiveness of UK industry is a key element. The forms of assistance which are provided are aimed at achieving the greatest benefit for UK industry at minimum cost to the taxpayer. They are based upon the Department's long and close involvement with the innovation process and experience of the problems of industry. In particular DTI concentrates increasingly upon those activities that are tailored to improve the capacity of wide ranges of companies to exploit technological opportunities. The importance that is attached to this work may be judged from the fact that science and technology expenditure was budgeted to account for 28.6% of the Department's total spend in the 1986/87 financial year. In cash terms this is 2.2 times as much as was spent in 1980/81, when the expenditure on science and technology was 6.5% of the total.

The RSP is a programme through which some of this expenditure on science and technology can be channelled to support projects proposed by industry that can be shown to obtain worthwhile benefit for the UK which would not otherwise be likely to arise. DTI has allocated £3 million the next three years to the RSP. This is by no means a large sum but if industry does not participate in the programme by putting forward projects then the demands of other marine technology and, indeed, of other sectors of industry, will have priority. The more that good projects are proposed and supported the stronger the case for more funds.

There are three particular schemes through which support for R and D can be made available and which can be used as a mechanism for such support. Each of the three schemes outlined below highlights the Department's policy that available funds should be used increasingly to support collaborative R and D projects as such projects enable companies to pool technical and commercial knowledge, share the costs and technical risk, avoid duplication of R and D effort and gain through the wider dissemination of the results.

Support for Innovation (SFI)

The Department of Trade and Industry offers financial support to encourage firms to undertake innovative research and development projects which, if successful, promise high returns. Applications from companies, partnerships and other organizations, and from individuals are welcome.

Projects must be innovative and result in the product or process which represents a significant advance for the industry or sector concerned. In addition the organization must have the technical, financial and managerial capability to carry them through. There should also be a clear appreciation of the commercial potential of the project including overseas markets where appropriate.

The main form of support is a grant towards the eligible cost. For companies applying individually the maximum rate of grant is 25%. For collaborative projects which involve a minimum of three independent partners including two in the private sector, the maximum rate for longer-term applied research projects is 50%.

All directly attributable costs properly incurred in carrying out the project up to the point of commercial production are eligible. These may include:

- The appropriate portion of the salaries of personnel working directly on the project
- The cost of materials consumed in the course of the project
- The cost of capital equipment purchased or constructed for the project, less the estimated value to the business of the equipment at the end of the project. In certain circumstances the cost of employing capital equipment that is already owned may be supported
- Licensing fees paid to third parties for acquiring new technology
- Sub-contract charges and consultancy fees paid, fees for trials and testing and preparation of technical manuals
- Project management costs, such as travel and office space, that are additional to those involved in the normal routine
- Training that is specific to the project including that of members of management
- Market assessment to ensure during the development phase that the project or process will satisfy the identified market need
- Cost of patenting, particularly for small firms
- An allowance for overheads which should not exceed their direct labour costs

EUREKA

An important role is also seen for international cooperation, and the EUREKA initiative is a further mechanism for tapping the budget allocated to the RSP for the support for collaborative projects. It is a framework for promoting collaborative projects in fields of advanced technology which was agreed between 19 European countries and the European Commission. Member countries are Austria, Belgium, Denmark, Federal Republic of Germany, Finland, France, Greece, Iceland, Ireland, Italy, Luxembourg, Netherlands, Norway, Portugal, Spain, Sweden, Switzerland, Turkey, and the UK. The aim of the initiative is to improve European competitiveness in world markets in the civil applications of new technology by encouraging industrial and technological collaboration within Europe. This initiative is seen as being industry led and is aimed at the joint development and exploitation of high technology products and has marine technology as one of its priority themes.

There has been a rapid growth in EUREKA since its inception in late 1985, and there are now in excess of 100 approved projects, with a total value of over £2.5bn, and with UK involvement in some 40 of them. Within the marine area of interest there is the EUROMAR project, which was first proposed by West Germany, in which the UK is participating. The DTI is coordinating UK interests in EUREKA and hope that many more marine technology projects, including some related to RSP, will come into it.

Support for UK participants in EUREKA projects is available under the SFI scheme, but in this case only one UK organization needs to be involved to qualify for assistance. The level of support that can be offered is up to 50% of the UK participant's share of the development costs. To achieve EUREKA status projects should comply with the following criteria:

- Cooperation between participants in more than one EUREKA country
- An identified benefit from pursuing the project on an international cooperative basis
- The use of advanced technologies
- The aim of securing a significant technological advance
- Adequate financial commitment by the participants
- Development work carried out in EUREKA countries and its results exploited to the benefit of EUREKA countries generally

LINK

LINK is another mechanism for the support of collaborative projects but, in this case, between industry and the scientific community within the UK. It has four main objectives:

- Fostering strategic areas of scientific research directed towards the development of innovative products processes and services by industry
- Stimulating a real increase in industry's own investment in R and D
- Aiding industry in the exploitation of developments in science and enhancing the awareness of researchers and scientists to the needs of industry
- The development of technologies that cross the boundaries of industrial sectors and scientific disciplines

To this end LINK will help to bridge the gap between science and the market by focusing on advances in the UK science and engineering base which need further exploration. This will in turn enhance the future exploitation of new processes, products and services by industry.

All Government Departments with a major spend in research and development are taking part in this initiative, with the aim of generating extra expenditure of at least £400 million over the next five years. Resources will be directed at a series of programmes in specific areas of science and technology each with particular promise. One of the areas considered to be of importance is ocean engineering, and the RSP has been suggested as a programme within this context. Each programme will consist of a series of individual collaborative projects, and it is envisaged that each portfolio will grow as more proposals are selected for inclusion within LINK.

Collaborative projects involving higher education and/or other research establishments together with commercial or industrial companies will be considered for LINK, the minimum level of participation being that the partnership comprises at least one research and one commercial organisation. The Government's contribution to suitable collaborative projects will be up to 50% of the eligible costs, and in exceptional circumstances a higher level of support may be available to aid initial expenditure. The normal Research Council practice for eligible costs will be followed in the case of higher education while SFI rules will apply to industry.

Other DTI Support

The DTI's support for the Resources from the Sea Programme is not limited to that of making finance available for research and development. Equally important is the coordinating role that the Department can assume in helping to develop links between related projects. It can assist in the identification of collaborative partners for organizations which have proposals suitable for support and wish to involve others in the project. It plays an active role in ensuring effective communication between the wide range of Government Departments and Research Councils involved in marine science and technology. Although they each need to pursue their own particular interests and responsibilities, many of which have little or no direct relevance to industry, individual programmes can lead to commercially exploitable products and can provide both support and opportunities to industry.

APPENDIX

Areas Offering Scope for Development within the Surveying and Exploration Phase of RSP

- The development of directional magnetometers for defining geological structures through the measurement of variations in the earth's magnetic fields.
- Electrical techniques, particularly induced polarization and spectral induced polarization methods to detect the presence of and identify specific minerals.
- Optical imaging for ground truthing and for navigation purposes provides development opportunities in wide angle video and still photography, improvements in lighting, for instance range-gating and laser scanning, and high resolution video transmission using fibre-optics and acoustics.

- Sonar systems such as combined side scan and bathymetry, to remove ambiguity from sonar signals; high speed scan sonar for pipe inspection and platform routing; higher acoustic survey speed and improved umbilicals using fibre-optic data telemetry.
- Sediment sampling equipment for soft and hard sediments, particularly grabs with washout resistances; multiple cores; precision corers; bulk samplers for cobalt crusts and massive sulphides; other specialist samplers capable of providing undisturbed cores and quantitative coarse sediment samples; and the ability to carry out the drilling of the deeper rocks without the use of drill ships.
- *In situ* testing requires developments in instruments for measuring sediment erosion, sediment strength and pressure among other parameters.
- Development systems and sensor platforms for the effective operation of towed platforms; towed ground contact sleds, low cost ROVs; 2000 m plus ROV and ancillary equipment such as high speed winching for deep towed platforms; acoustic navigation and hovering capability for detailed inspection.
- Data telemetry at high speed and with high resolution through fibre optics and acoustic methods.
- Techniques for data handling using on-line processing by interpretive software; real time processing; data compression and economical and accessible digital storage.

Some R and D Prospects in Instrumentation and Measurement Techniques

- Measuring waves using remote sensing techniques and systems based on satellites, on land and on vessels under way, and low cost wave direction sensors for use in coastal regions.
- Current meters with fast response vector averaging capability and with threshold speed, practical near sea surface current sensors and instruments that can provide a current profile from near surface and over a large part if not all of the water column in the ocean.
- Water quality samplers that can rapidly determine chemical substances in sea water; automatic samplers based on the ion exchange principle and monitoring equipment for coastal and estuarine use. Instruments to measure suspended organic materials, sediment density in fluid beds, drift and coastal sediment dynamics and suspended sediment concentrations *in situ*.
- Numerical modelling of the coastal and estuarine hydrodynamics distribution of pollutants, dynamics of continental shelf, shelf edge and deep ocean and the design of sea outfalls.

31

The Return of the Manned Submersible: a New Dawn?

D. J. Hampson, OSEL Group Limited, UK

At the first Subtech Conference in 1983, I chaired a seminar on manned submersibles. To a question on the length of the future of manned submersibles, Frank Busby replied, 'About ten minutes'. Four years later that prediction remains stubbornly unfulfilled, and companies, agencies and authorities still purchase manned submersibles.

It is, however, fair to state in the case of offshore work, that competing technologies outsell manned submersibles by about ten to one.

At one time it could be supposed that this was due to the reduced risk in using ROVs instead of manned submersibles, but the continued use of saturation diving, with far higher physiological risks, surely indicates that this is not the case. It is most probable that economic considerations in the overall sense are paramount, including costs of surface platforms, safety back-ups, limited-duration dives, and battery-charging times.

CHANGING PATTERNS OF USAGE

Since the late 1970s, changes in the use of manned submersibles have occurred, as the trends indicate below:

(1) Survey work

$x\%$ = Market share

	Manned submersibles	*Divers*	*Others*
1975	60% large, autonomous battery powered	40%	none—ROVs only experimental
1980	20% manned microsubs/ADS 10% autonomous subs	20%	ROVs 50%
1985	5% manned microsubs/ROV	10%	75 ROVs 10% controllable towed vehicles

Advances in Underwater Technology, Ocean Science and Offshore Engineering, Volume 14: Submersible Technology

1990 5% manned microsubs/ROV	5%	40% ROVs 20% controllable towed vehicles 10% intelligent tow fish

(2) Inspection work

Manned submersibles	*Divers*	*Others*
1975 10% autonomous	90%	nil
1980 10% microsubs/ADS	80%	10% ROV
1985 5% ROV/microsubs	75%	20% ROV
1990 5% ROV/microsubs	60%	35% ROV

(3) Salvage work

Manned submersibles	*Divers*	*Others*
1975 10% autonomous	80%	10% ROV
1980 10% microsubs/ADS	80%	10% ROV
1985 10% ROV/microsubs	70%	20% ROV
1990 10% ROV/microsubs	60%	30% ROV

(4) Rig support

Manned submersibles	*Divers*	*Others*
1975 10% autonomous	90%	nil
1980 5% autonomous 20% microsubs/ADS	60%	15% ROV
1985 20% ROV/microsub	10%	65% ROV 5% guidewire deployed tools
1990 15% ROV/microsub	5%	60% ROV 20% alternative technologies

These figures indicate a reducing usage of manned autonomous submersibles, a steady usage of ROV microsubs, which may be used in an ROV-only mode much of the time, and an increase in cost-effective alternatives to ROVs such as the patented Intelligent Tow Fish (OSEL Group 1984) and guidewire-deployed manipulators.

NEW WORK AREAS

Subsea completions

The rapid fluctuations in oil prices have concentrated the attitudes of oil companies to consider overall value for money very much more than previously, and to disregard technical 'snobbery'. In 1980 no self-respecting underwater engineer would consider designing a subsea completion to include diver intervention. Today, the most cost-effective system is chosen without high-technology bias, and many of the prophets of diverless technology have been converted to a less messianic, more unbiased approach, or have departed from the business altogether.

There is a great degree of difficulty inherent in predicting the optimum technology for any particular field. This is due to the very high sensitivity to parameters such as reservoir geology, predicted price of product, water depth, location, adjacent facilities, etc. For instance, it is generally accepted that diver intervention is limited to depths shallower than 300 m; deeper than 300 m, alternative technologies must be used.

It is also accepted that for the most complex field structures some form of atmospheric manifold will be used, requiring manned intervention using atmospheric, mobile, tethered bells or personnel lock-out submersibles. These may be deployed from surface platforms or have sufficient endurance to depart from their home base using on-board, long-endurance power sources.

The type of surface platform used to support subsea field production has a major impact on field economics. The most sophisticated field-support vessels (FSVs) may cost four to six times the day rate of the older-generation, simple vessels. A subsea completion design using the lowest-cost, surface support-platform will reduce field-maintenance costs by 50% or more, even if a sophisticated FSV is needed for one or two interventions per year.

An alternative to surface-deployed submersibles is the large, autonomous commercial submersibles, but power sources continue to be a problem.

A field-support submersible (FSS) will be equipped with inspection ROV, personnel transfer systems and perhaps one-man submersibles. On-site work-time at the full operating depth will be upwards of one week or more. The major perceived drawbacks of the FSS are the difficulty of crew changes on location or the time taken to return to port to change crews.

Alternative operational strategies and technologies will be discussed below.

Other Offshore Work Areas

Other work areas in the established offshore business include pipe and cable burial and subsea completion installation, support and repair. These work areas are dominated by ROV, divers or towed systems. There are few signs of any manned systems except ROV–microsubs finding application here.

Scientific Submersibles

Submersibles have been used for scientific research for the past 35 years. In one sense, their use will continue indefinitely because the initial investment was so great, and the operations each year so limited, that writing off the capital invested would be unacceptable. There is no doubt that a great deal of the work can be performed using alternative technologies, and in many cases this is done.

One novel approach would be to replace a surface platform with an autonomous commercial submarine, but the restricted facilities available make this unacceptable to most users.

The United States, Japan, France and the USSR are all developing submersibles capable of operating in depths of up to 6000 m, in addition to their existing submersibles.

The high initial and through-life costs of deep submergence systems, and the long build-time, makes it almost certain that deepwater commercial exploration and exploitation will initially be carried out using scientific submersibles.

Commercial Systems

During the exploratory phase of any deep-sea activity, it is most likely that scientific submersibles will be used. It is generally accepted that below 3000 m, the problems of cable handling through the trade-off between umbilical drag and the power required to overcome it means that autonomous submersibles, either manned or ROV, will be required. Epaulard demonstrated the capability of autonomous ROV in deepwater surveys. The technical problems of using autonomous vehicles, either manned or ROV, are similar. Assuming that the instrumentation payloads carried by each are the same, the difference between the two lies in the size and weight of the life-support capsule. This will necessitate an increase of probably 30% in total size. However, the extra flexibility and productivity of a man at the worksite and the improvement in achieving a number of differing or unforeseen tasks quickly, offset the weight penalty. The motive power requirement is much the same, and the size and cost of the surface platform is also of the same order. The major commercial applications, excluding oil and gas, are deep-sea mineral-resources exploitation and toxic-waste disposal.

Leisure and Recreation

This is to be a major growth area, and there is a great deal of activity at present. Improvements in technology have resulted in simpler, safer operations. A number of ex-commercial submersibles are being used for this purpose with some success.

It is unlikely that one-man units will find a place in this market, except perhaps the ROV–microsub types. As the usage matures it is likely that more automated approaches to ensure a high throughput of passengers will develop in order to maximize return on investment.

Salvage

The use of large, manned, autonomous submersibles for salvage work is considered unlikely; microsubs, ADS and ROV-microsubs have all been used in the past with some success and will continue to be used in the future. The salvage business has no innate technology bias either for or against any particular type of system, being concerned solely with the most cost-effective methods of salvage.

Military Uses

Various military uses of manned submersibles are planned or in existence. Apart from security-sensitive tasks, salvage, survey, personnel rescue, light work, and deep exploration have been performed. There is very little military diver lock-out work. Naval requirements using wet submersibles and devices such as the sub–skimmers are not discussed in open meetings to any great extent, but there are clearly applications for manned submersibles.

Where Are They Now?

As with ageing film stars, it is fascinating to discover the fate of old submersibles. Unlike many first-generation ROVs, surprisingly few submersibles or ADS are actually scrapped. Some are in museums, but many are in use for research, leisure, military use, oil and gas use, training and other areas. As the capital investments are written off, systems can be hired out at very low day rates. The use of ADS for research compares very favourably with divers in both cost and performance terms.

TECHNOLOGY INNOVATIONS

No new basic technology is needed for manned submersibles. However, the

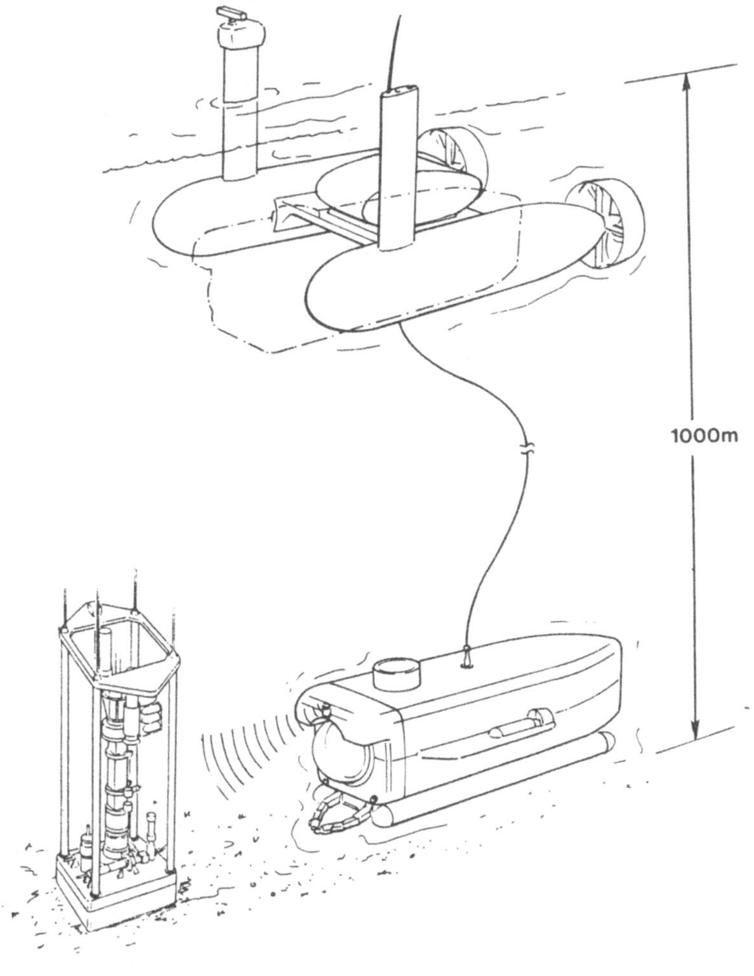

Fig. 1 The principle of TRANSUB.

enabling technologies now available can improve performance dramatically, and reduce operating costs.

Power sources dominate any discussion of autonomous submarines. At present the old-fashioned lead–acid battery offers the best compromise between initial cost, in-service costs, volume, weight and efficiency. The alternatives are usually expensive, are fragile, or primary cells only, or combinations of all these. Chemical or heat engines such as the Cosworth diesel or the Stirling engine from Sub-power AS have a great deal of merit, but are also expensive and very complex. Rather more intimidating is nuclear power, where civilian users are likely to be discouraged by preconceptions about cost and especially safety, despite the evident technical advantages of nuclear power.

In my opinion, the most likely candidate for a field-support submarine will be a conventionally aspirated (via a snorkel) diesel-powered submarine using lead–acid batteries for subsea power. The advent of extremely efficient DC–AC inverters means that thruster reliability, efficient hydraulic power sources and optimum battery-power management are obtainable in a way which was impossible with first-generation submersibles and submarines.

One of the major cost elements is the surface platform. Figure 1 shows TRANSUB, a typical manned submersible which is controlling its own surface platform. By the use of a semisubmersible catamaran, the technology risk is reduced dramatically. Figure 2 shows the elements of the system, including those which involve technical risk, and indicates relative costs in operation and purchase for a particular intervention requirement. A major benefit, as with ROVs, is that real-time data are transmitted back to base. The position-keeping problem for the submersible is much less severe than it is for a dynamically positioned DSV controlling a diving bell; control strategies to hold the catamaran within a 300-m radius of the submersible are readily available.

Commercial considerations for TRANSUB

	Conventional		*Autonomous, self-contained*	*TRANSUB submersible catamaran*	
Capital cost	support-vessel	10	15	catamaran and cable	4
	submersible	2			
				submersible	3
	total	12	15		7
Running costs	total spread	30	20		10
Technology risk		Nil	50%		20%

The use of ROV technology on manned submersibles brings a number of benefits. Systems designed for ROVs are usually smaller and lighter, the technology is in continuous development, and the use of automatic systems to regulate life support, altitude, depth, heading, and thrust monitoring reduces the pilot's workload and may eliminate one crew member.

Power management using modern inverters yields tremendous benefits in excluding battery endurance, battery life and instrumentation flexibility.

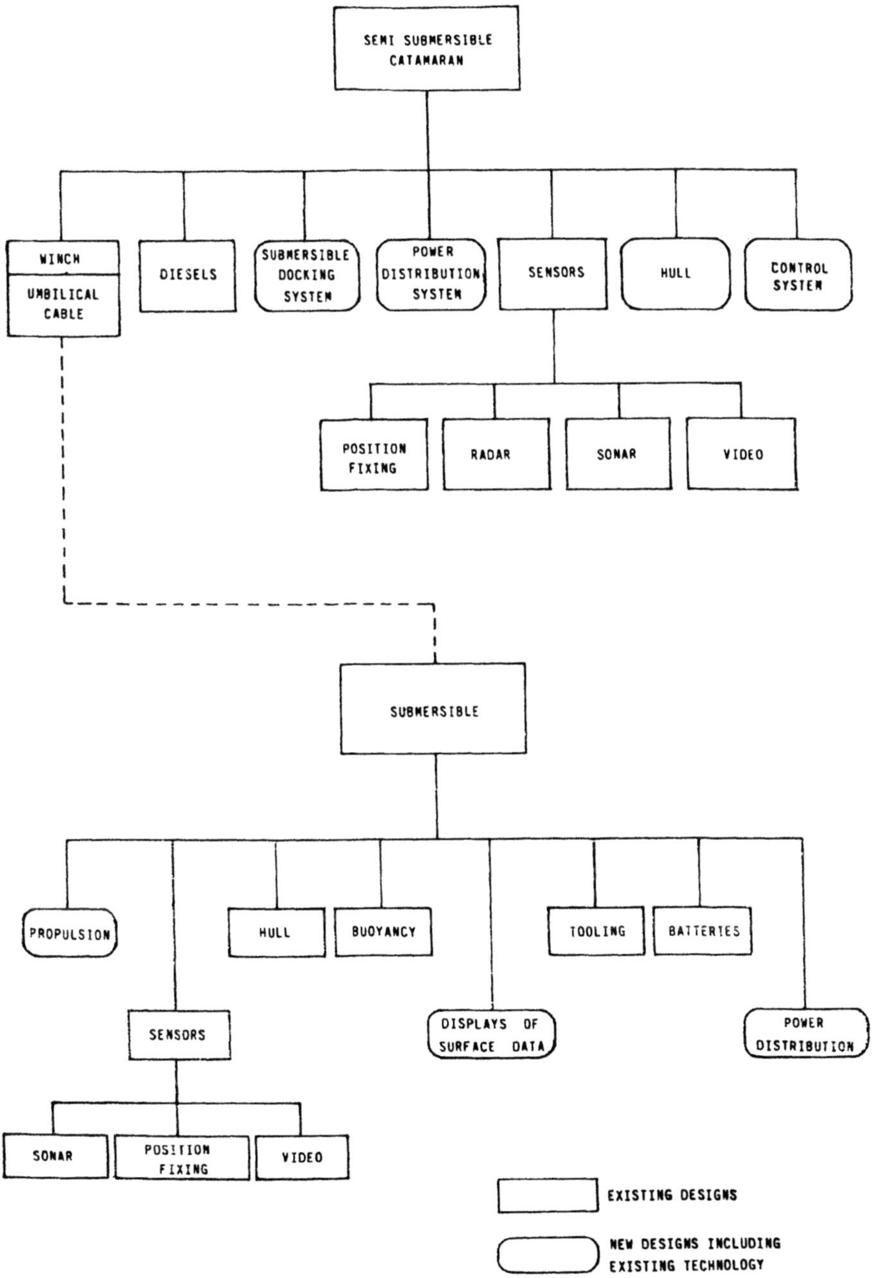

Fig. 2 Flow diagram showing the TRANSUB system.

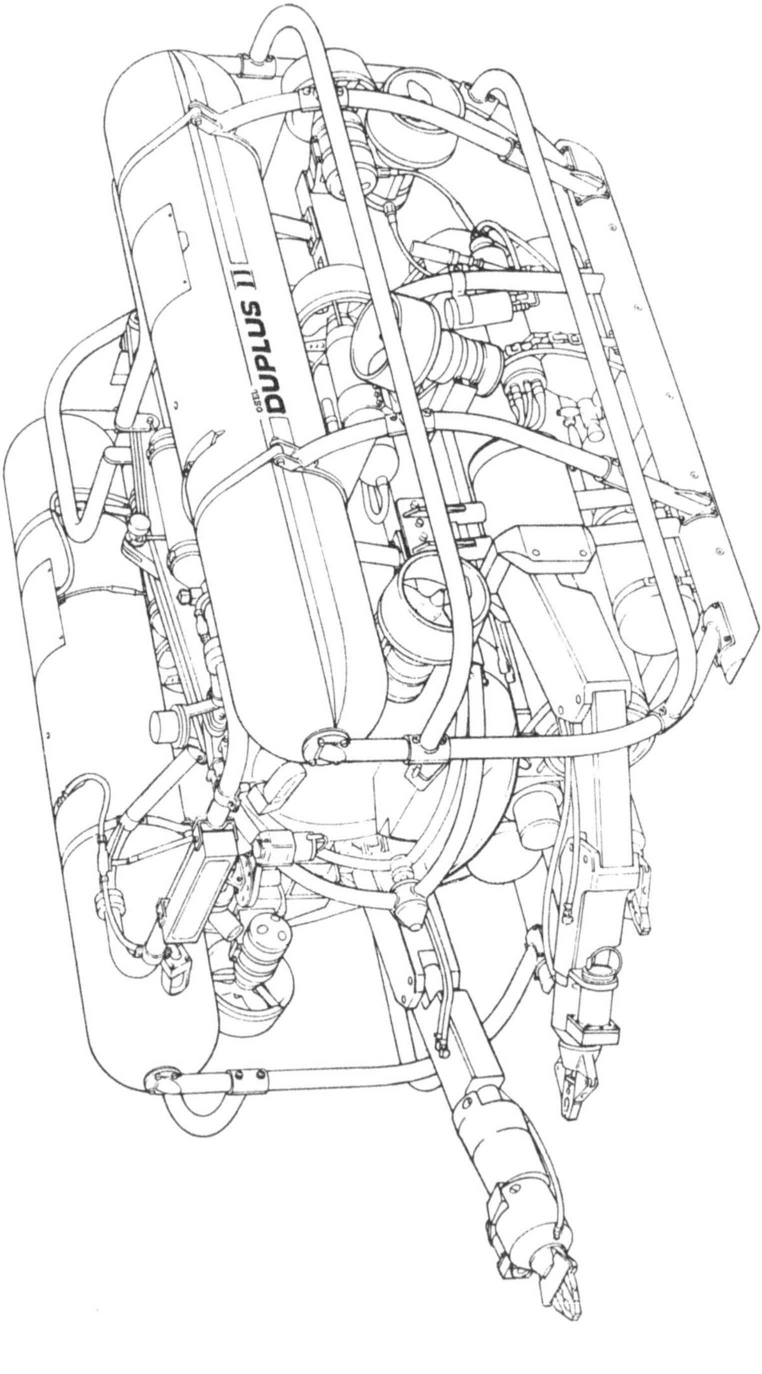

(a)

(b)

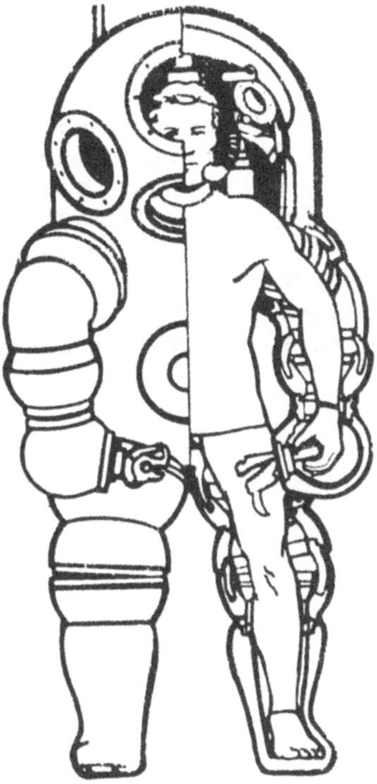

(c)

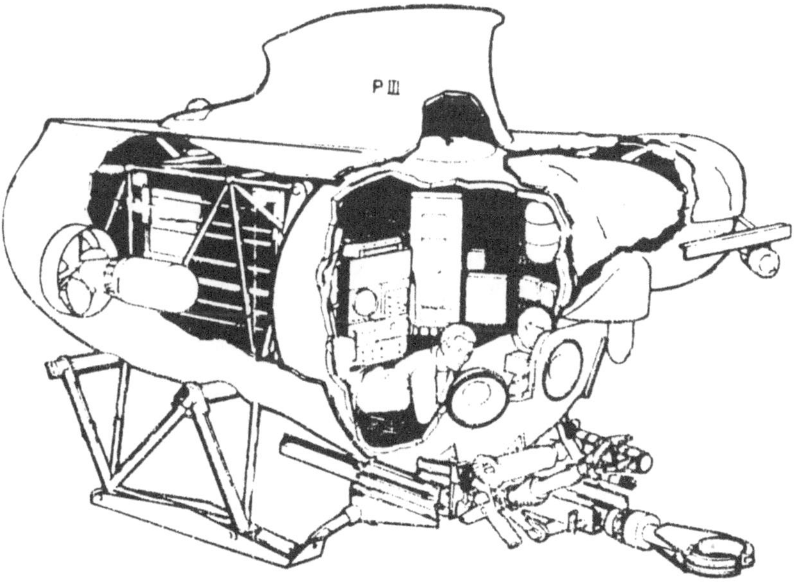

Fig. 3 (a) DUPLUS II ROV–microsub. (b) Jim ADS. (c) Pisces III autonomous manned sub.

CONCLUSIONS

It has been shown that the use of manned submersibles continues without signs of terminal contraction. Future use in the oil and gas sectors depends more on innovative techniques rather than on major technology advances. The other uses will continue to grow in areas where an unbiassed view of the systems prevails, and the enviable safety record is maintained.

In deep-water work the costs of long-endurance autonomous ROVs and manned submersibles are shown to be broadly similar, allowing the operator greater flexibility. The only sad aspect of the sector is the lack of UK government investment in an area where British technology once led the world.

32

Seabed Contact Vehicles

J. S. Wood, UDI Group Limited, UK

Since the early 1970s a new industry, consisting of the design, manufacture and operation of remotely operated vehicles (ROVs), has emerged. Most of the ROVs which have been developed are in the tethered free-swimming category and have been designed as multi-functional vehicles capable of carrying out tasks which previously had been undertaken by divers.

Whilst this market was expanding, a few companies instead concentrated their efforts in the development of seabed contact vehicles able to accomplish tasks which were outwith the capability of the diver. Vehicles in this category are primarily task specific in their design. The main function for which these vehicles have been developed is in the trenching/backfilling of pipelines and in the installation and burial of flexible flowlines and umbilical cables. Unlike the free-swimming ROVs most seabed contact vehicles are operated by the same company responsible for their construction.

There is no general configuration to which seabed contact vehicles tend to follow; however, the one aspect common to all vehicles is that they are large and require a highly skilled crew and an extensive range of support equipment. Launch, recovery, and maintenance of a vehicle of this type require a surface support vessel which has to satisfy a wide range of criteria.

Most operators of these vehicles therefore do not operate on the principle of using a vessel of opportunity; instead they try to operate from a single dedicated vessel throughout the working season. With such high overheads and falling day rates due to the downturn in the industry, vehicle operators have been compelled to be more innovative in their approach in an effort to increase their share of the declining market.

PRINCIPLE OF OPERATION

Whilst the different trenching techniques used by the few vehicles on the market vary dramatically, the principal mode of operation is the same.

The vehicle, as illustrated in Figure 1, usually consists of a rigid framework on to which is mounted a pair of hydraulically driven caterpillar tracks. These tracks are similar to those found on any land-based construction vehicle or military tank.

Seabed contact vehicles in general require more accurate positioning than

Advances in Underwater Technology, Ocean Science and Offshore Engineering, Volume 14: Submersible Technology

Fig. 1 Seabed contact vehicle.

their land-based counterpart and therefore usually have independent drive to each pressure compensated track.

Hydraulic power to the tracks is provided either through a series of umbilicals from the support vessel or more commonly by means of electro-hydraulic pumps mounted on the vehicle chassis. In this case power is transmitted to the vehicle using a single umbilical cable. UDI have found that the use of a single umbilical, not only to transmit power, telemetry data, TV and sonar signals, but also to launch and recover the vehicle is advantageous.[1]

Because of the vehicle's size and complexity of tooling, a wide selection of video and sonar systems is necessary to provide adequate visual coverage of the vehicle and its surroundings. A variety of control functions and monitoring equipment provides the operator on the support vessel with sufficient data to monitor the vehicle's progress and performance.

With a vehicle weight in air of anything up to 130 tons an extensive range of equipment is required on the support vessel to enable the safe launch/recovery and operation (Fig. 2).

Typically the support system will consist of:

- 'A' frame or large heave-compensated crane for launch/recovery
- Control umbilical, storage and traction winches
- Power generators
- Hydraulic power packages
- Control console
- Workshops/spares

To accommodate all this equipment a support vessel with a large deck area, dynamic positioning capability, and sufficient quarters for vessel and vehicle crew is essential. If the vehicle is to be used for installation of flexible flowlines or umbilical cables, a comprehensive range of storage and handling equipment has to be provided.

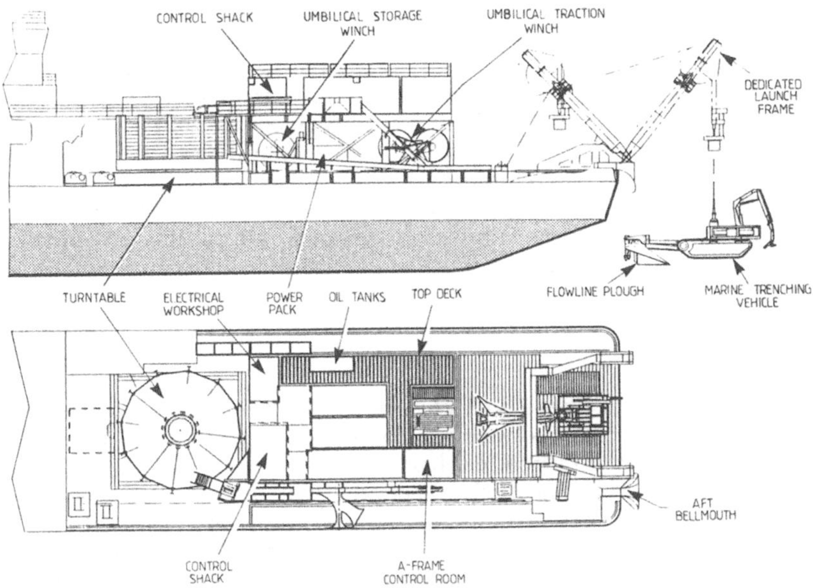

Fig. 2 Vehicle support equipment.

In some cases diver assistance is required during the launch/recovery operation.

OPERATING CRITERIA

Seabed contact vehicles have been used extensively in a wide spectrum of environmental conditions where their inherent weight and bottom contact have been advantageous. Vehicles of this type have been used both through the surf zone on shoreline approaches and in water depths of up to 200 m. Vehicle ground speed varied in accordance with the nature of the work task and the soil conditions.

In the pipeline trenching mode, it is the capability to move large quantities of soil rather than vehicle speed that is important. Whilst it is difficult to compare vehicle trenching rates because depth of cut/soil conditions have a substantial effect on speed, an average trenching rate of 300 m per hour is not unrealistic.

In a simple inspection mode, ground speeds in excess of 1 km per hour are feasible, dependent on the type of bottom, the slope gradient and the ability of the bottom to support the weight of the vehicle.

The nature of the bottom will determine whether the vehicle can work at all.

These type of vehicles are unsuitable for operation on a seabed consisting of either: (*a*) continuous cover of large boulders over 0.5 m in diameter or (*b*) in areas with several metres of liquid silt interface between water and seabed material of measurable strength.

Seabed contact vehicles have been known to operate in silty clays with shear strengths of between 2–6 kPa with assistance of buoyance tanks and wide tracks to reduce the vehicle's ground bearing pressure.

One area in which the use of seabed contact vehicles has achieved a greater bottom time than divers or even the heavy duty tethered free swimming ROV is in the Southern North Sea Gas Fields. This area is notorious for its high tidal currents, making the use of divers and ROVs very restrictive. With the correct support vessel, seabed contact vehicles can work continuously in this area and even the 1:5 sand waves gradient which can be found have proven not to be a major problem to this type of vehicle.

Seabed contact vehicles have proven to be very reliable in their operation. Bottom time varies with each vehicle; however, single vehicle dives in excess of 6 days have been recorded.

APPLICATION

As previously stated, these vehicles have been used extensively for the trenching/backfilling of pipelines (up to 36 inch NB) and the installation of flexible flowlines and umbilical cables.

In the pipeline trenching mode these vehicles are normally only used on infield flowlines where the vehicle's ability to manoeuvre close into the subsea structure and to other submarine pipelines has been an advantage. The ability to operate from a single dynamic positioning vessel close to existing platforms without the use of anchors is also a major advantage.

Surface towed ploughs tend to be favoured for the trenching of longer pipelines because of the higher trenching rates achieved by this method.

Unlike vehicles where the trenching tool is an integral part of the vehicle, e.g. Heerema's Eager Beaver and Allsea's Digging Donald, UDI pioneered the development of a seabed contact vehicle that was capable of deploying a variety of tool packages. This has enabled this particular vehicle to be used not only for pipeline trenching, where it has over a 6-year period successfully trenched 17 pipelines, but also to install umbilical cables and flexible flowlines. A total of 36 umbilical cables have been installed by the simultaneous lay/burial method using this vehicle.

This vehicle has also proven to provide a stable and powerful platform on to which a wide range of tooling can be mounted.

In addition to the range of ploughs available which can be attached to a common tool mounting point on the vehicle, UDI recently developed an umbilical/pipe dredge tool. This tool (Fig. 3) was designed and used during the 1986 season for post burial of a section of umbilical cables for Maersk Olie og Gas in the Rolf Field and for British Petroleum on the South East Forties Project. In both cases the use of this diverless tool package, attached to the vehicle to bury short (approximately 200 m) lengths of umbilical cable close to the platform structure, proved to be an economical alternative to diver-held dredge units, installation of protection mattresses and rock dumping.

Both these jobs, although short in duration, helped to improve the utilization of this vehicle and contributed to the overall profitability of the system during the working season.

Another vehicle operator who has maximized the use of its trench vehicle is Coflexip, who along with Colus developed a new method of protecting and stabilizing submarine pipelines. The COPS system involves the installation, by the trenching vehicle, of a continuous fabric mattress which is spread on the seabed over the pipeline. The mattress is simultaneously filled with cement,

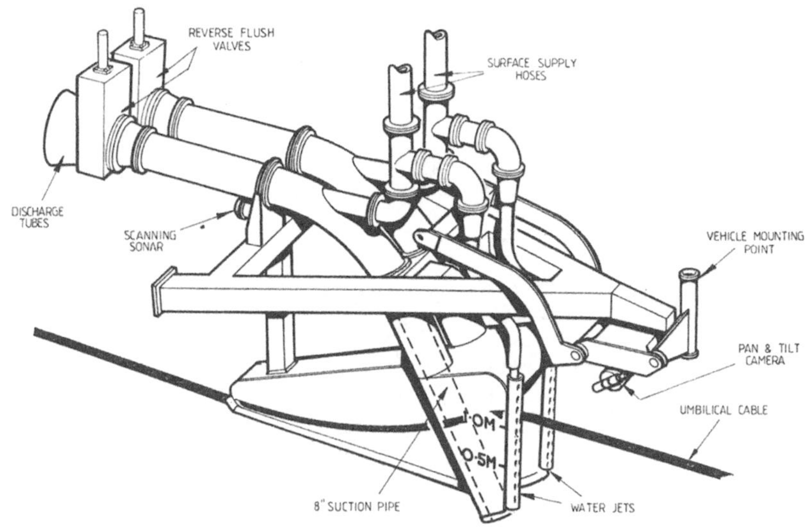

Fig. 3 Cable dredge system.

which allows the installation of continuous lengths of up to 500 metres without diver intervention.

SEABED LEVELLING OPERATION

During the spring of 1987, UDI carried out a seabed levelling operation where they moved 4000–5000 cubic metres of material, of up to 1.7 m deep, from around the Gullfaks 'A' platform. A new innovative tool package developed for use with their trenching vehicle was utilized for this operation. This soil-removal tool, designed by UDI, was built specifically for this project between contract award from Norwegian Contractors in December 1986 and start of work in March 1987.

The problem facing the client was that grout which had been pumped down the platform structure during installation had overflowed to form a carpet of hard crust material preventing pipelaying and other subsea operations. The grout material is a cement/sodium silicate composition which when mixed with seawater, forms a solid material with a density of 1.15 tonne/cubic metre. It had been established that the grout carpet extended over an area of approximately 7500 square metres. Previous attempts to remove this material using divers and ROV mounted dredges had proven to have limited success and to be very time consuming.

The system had to be designed such that the grout was able to be removed to original seabed level leaving a substantially level surface. It also had to operate below and in between the anchor chains of the *Polycastle* flotel, which was located adjacent to the platform. It was also very desirable for this operation to be diverless.

To meet the above criteria the trenching vehicle was fitted with a multi-purpose bucket (Fig. 4) incorporating an articulated jaw, a pair of rotating

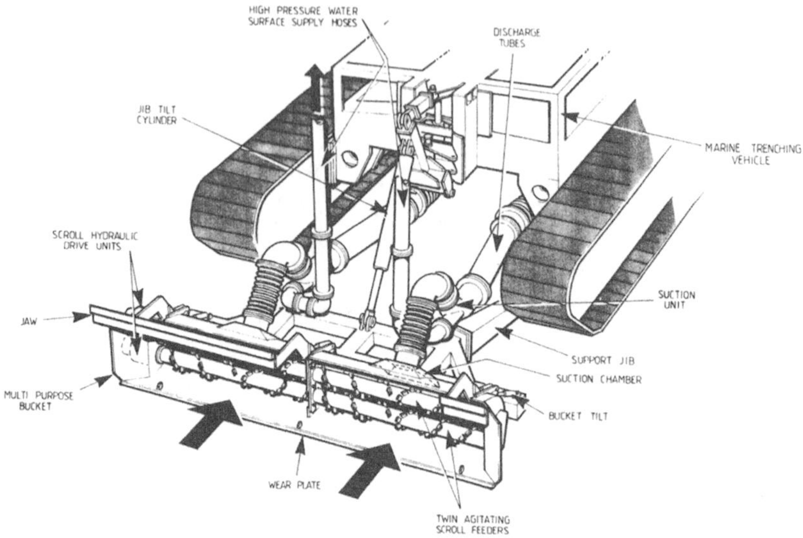

Fig. 4 Grout removal tool.

aggressive scrolls, and a pair of suction dredge tubes connected to discharge hoses.

To enable the vehicle to drive along the channel cleared through the grout by the bucket, an overall width of 5.6 m, which slightly exceeded the width of the vehicle's tracks, was chosen. Although the type of bucket arrangement proposed is similar to that used by the land-based construction industry, the widest bucket commercially suitable was only 2.6 m, therefore a special unit had to be made. The bucket was fabricated as one piece with a central division providing two compartments. The bucket was fitted with a single hydraulically operated jaw. Its purpose was to break off manageable sections of grout and guide them into the bucket for disposal.

Two parallel rotating scrolls were assembled inside the bucket. Each scroll was fitted with steel plates arranged to form two pairs of opposing screws, one pair for each bucket compartment. The dual purpose of the opposing hydraulically driven scrolls was to draw the grout towards the suction heads at the centre of each bucket compartment while breaking up the grout.

The suction for the tool was provided by two Alluvial Mining venturi pumps which were fitted below the chassis of the vehicle. Alluvial Mining also supplied the two surface water pumps and hoses on board the support vessel.

The total weight of the final tool including support jib was approximately 5.6 tonnes. The seabed contact vehicle was used to provide all hydraulic power for the tool, and full use was made of the extensive range of instrumentation, video and sonar systems which are a standard feature of the UDI trenching machine.

To determine the depth of removal and to conduct a seabed survey after completion of work a bathymetric system was mounted on the vehicle. Using the vehicle to carry out the post-operative survey eliminated the need for an ROV on board the vessel to monitor the tool performance.

An interesting feature of the operation was the need to assemble the two 12 inch NB discharge hoses to the vehicle subsea using the vehicle's manipulator.

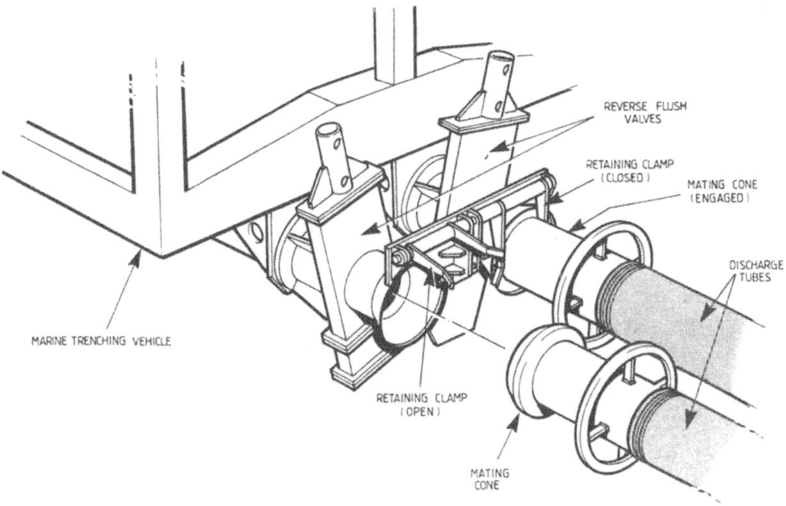

Fig. 5 Remote discharge connections.

Figure 5 illustrates the easy-fit connections that were developed. UDI found that these hose connections could be remotely coupled in less than 20 minutes.

It should be noted that the manipulator fitted to this vehicle is not of the light-weight articulated type currently available on the ROV equipment market; instead it is a robust 3-function manipulator capable of manoeuvring equipment of up to 1 tonne in weight. The manipulator is fitted on the end of an 8.5 m reach crane, with an 11 tonne/m capability, which is mounted as a standard unit on the vehicle chassis.

In addition to carrying out this remote connection task, the manipulator/crane has proved to be an indispensable tool for other tasks, especially clearing debris found during the grout removal operation.

After the equipment was built, shallow water tests were conducted off the support vessel *Northern Installer* in the Moray Firth before the vessel made passage to the Gullfaks Field at the beginning of March.

Once the navigational equipment calibration was completed in the field, the grout removal operation began.

First, the two 25 m lengths of discharge hose were deployed to the seabed in a location some distance away from the edge of the grout carpet.

The vehicle was deployed, and an offset between vessel and vehicle was established, thus ensuring a good umbilical catenary existed. The vehicle was driven towards the discharge hose ends, and the connection was made. The vehicle was manoeuvred along a predetermined route selected to minimise the risk of vehicle operations under the *Polycastle* anchor chains. The vehicle was driven into the grout in steps with the bucket jaw raised. The jaw would then be fully lowered breaking off and guiding a section of grout into the bucket. The rotating scrolls would break up the grout further within each compartment before guiding it towards the suction heads in the centre of each bucket section. The grout seawater mix would be transported along the discharge hoses.

Once the grout was broken up, the silicone bond would break down and the

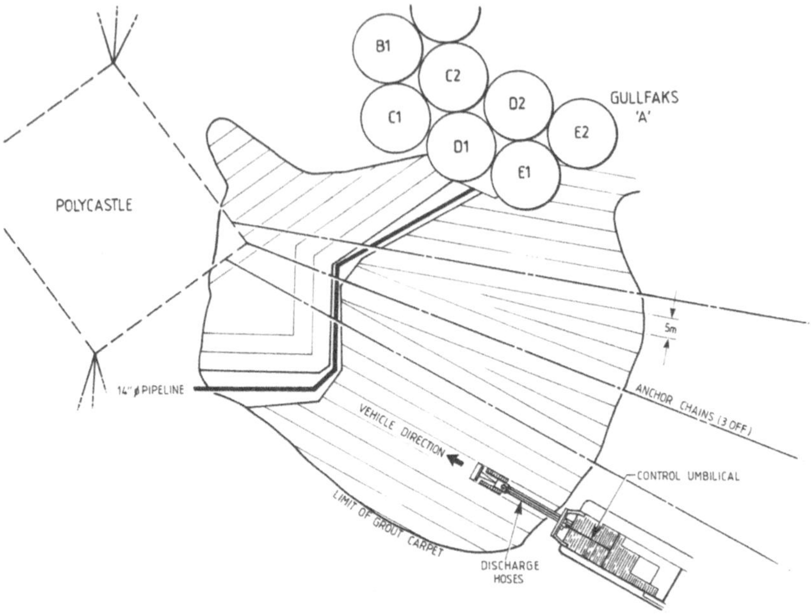

Fig. 6 Seabed levelling operation.

material would reduce to a tenth of its original volume. The cement particles were then able to disperse in the sea current.

The vehicle was able to clear a 5 m wide corridor in every run starting from the edge of the carpet working towards the existing pipeline as illustrated in Fig. 6.

Each run was approximately 45 m in length and was cleared in about 1 hour. Grout removal operations were ceased approximately 5 metres from the pipeline and platform. A single dredge tool was later fitted to the vehicle to clear these areas at the end of the operation.

The weather on location, even for the time of year, was very poor, and considerable weather downtime was incurred. The system, however, was only required to operate on the seabed for a total of 10 days, and in that time it moved over 4000 tonnes of material and left a flat seabed to the satisfaction of both the client and operator.

CONCLUSION

Seabed contact vehicles have proven that they can perform useful work for the offshore industry. With the correct support from the oil operators, this type of vehicle can be adapted for a wider range of tasks than they have been used for in the past.

As the offshore oil industry is forced into the development of oil reservoirs that have already been located in water depths beyond the capability of existing diving technology, seabed contact vehicles equipped with the correct hardware could offer an economical alternative to the use of expensive construction rigs and guidewire systems for a number of subsea tie-in and maintenance operations.

REFERENCE

1. Cameron, G. M. Trenching of Flowlines and Installation of Umbilicals to Subsea Completion. ASME, Dallas, TX, 1985.

33

Diversification and Innovation: Adapting to Change—the View of One Small Business

J. Turner, Camera Alive Group, UK

The most surprising aspect of the collapse in oil prices was the speed in which service companies were affected. In the case of Camera Alive, revenue for January and February 1986 was running about 20% ahead of budget and capital spending and expansion plans were being implemented according to budget. Everything looked good, and like most other companies we considered the small drop in the price of oil to be a short term fluctuation.

In March, April and May, revenue fell to about 15% of that budgeted. The panic button had been pressed. The offshore oil industry was not spending money. If we had been Japanese, we would have been sharpening the ceremonial swords. June showed a sudden recovery, with much of the lost revenue recovered, and the recovery extended into August, when the season came to an early halt. Looking back on the year, revenue was perhaps 25% down on what had been budgeted.

Will the industry ever recover? Will the industry ever be the same again? What can small companies do to protect their future?

THE CAMERA ALIVE GROUP

In early 1986 the Camera Alive Group was made up of four companies active in four main areas. Camera Alive Limited provided photogrammetric services and electronics-based products. Camera Alive (Underwater Video) Limited was active in underwater video equipment. Camera Alive Marine Equipment Limited manufactured underwater photographic cameras. Turner Round Limited was active outside the oil industry. From entering the oil industry in 1976, with one company offering consultancy services, the group had expanded into the provision of specialist services and was concentrating on expanding its manufacturing capability.

The Initial Reaction

The initial reaction to the reduction in revenue was to stop all unnecessary expenditure, and that was defined as any expenditure which was not a necessity

Advances in Underwater Technology, Ocean Science and Offshore Engineering, Volume 14: Submersible Technology

for continued day-to-day trading. The secondary action was to review all activities to ascertain which should be allowed to continue and where cost savings could be made. Both actions were made especially urgent by the initial reactions of the banks to reduce borrowing limits overnight.

Having made a determined effort to keep costs at minimum levels the 'hatches were then battened down to weather the storm', which was initially envisaged to be short lived. It was soon realized that this was not to be the case, so long term plans were formulated. These can be broadly defined within the categories of cooperation, consolidation and investigation.

The Action

The first action was to negotiate an agreement with Osprey Electronics Limited rather than continue in competition. The product lines were compared and rationalized, and it was agreed that Osprey would concentrate on the video products while Camera Alive would concentrate on the photographic products. The revised product lines would then be marketed alongside one another as a total British Underwater Visual Imaging package. This decision resulted in the winding up of Camera Alive (Underwater Video) Limited.

The main reasoning behind this agreement was that it would maximize exposure of Camera Alive's products outside the North Sea virtually overnight and, from Osprey's point of view, would remove a competitor capable of snapping at their heels and biting into their reducing profits. Many other companies have since followed a similar policy with some progressing even further to carry out a full merger.

On a more long term basis it was decided that vertical diversification would hold more success than horizontal diversification on the basis that it was better to hold a larger share of a market that one was familiar with than to blunder into completely new markets when cash was in short supply and where success in overcoming the barriers to entry would not be cheap. It was also decided that the highly segmented offshore market could be better exploited by maximizing the niche strategy already pursued. This meant analysing areas in which there had been a monopoly supply situation, or in which no strong competition from a British company was evident, and providing a product that was better, cheaper and had high British content.

The third strategy was to investigate the market outside the offshore oil industry for similar products.

The Results

A year has passed from the implementation of these strategies. The fact that the Camera Alive Group is still in existence when many other small companies have floundered is possibly some testiment to their success; but what has been achieved and what has been learnt?

Cooperation

The Cooperation with Osprey has undoubtedly been a success. Their marketing muscle, in cooperation with our technical expertise, has resulted in photographic camera sales all over the world. The production that was budgeted to go into the North Sea has been diverted to America, Brazil, Scandinavia, and even

Japan. This has allowed development to continue and hence has allowed increased pressure to be directed on to the products of our two American competitors in the underwater photographic market, one of which had been allowed to make the North Sea a 'home' market by lack of UK competition until the formation of Camera Alive Marine Equipment Limited in 1985. It is extremely important in a world wide recession to pursue competitors in their home markets.

This arrangement has resulted in a considerable saving on up front marketing costs and has hence helped cash flow extensively.

Consolidation

Whilst many others decided to 'bail-out' of the offshore market, the Camera Alive Group decided to increase its presence. It has been evident over the past year that companies that had markets outside of Aberdeen have expended their efforts on those markets, thus leaving a partial vacuum that had to be filled. It is also evident that the total market has contracted considerably, but our policy was to take a larger share of this reduced market and so to maintain our turnover. An example of the implementation of this policy is the market for underwater electrical connectors.

In 1986 our turnover in sales of underwater electrical connectors was maintained at 1985 levels, despite the smaller market, but in 1987 it has more than doubled. Within this period we have introduced alternatives to most of the well known branded connectors, most of which have been more competitively priced, all of which have been of similar or better quality, and many of which are now the first choice of the user. The introductions were backed up by good stock levels maintained in Aberdeen, thus allowing the users to reduce their inventory. This, together with the reduction in price, has allowed the users some compensation for the squeeze in rates that they have experienced.

In other areas a determined market lead development strategy has led to the identification of market segments from which products are missing.

The paramount interest in the offshore oil industry at present is that of cost savings. Any technique that can result in a cost saving must be evaluated. In many cases techniques that can save offshore underwater time can produce such a saving even though they are themselves more expensive than a similar technique, or a technique that is used for the same purpose.

Camera Alive was one of the first companies to promote the use of photogrammetry for underwater inspection. This photographic technique allows three-dimensional measurements of great accuracy to be taken from a pair of photographs and has resulted in ROVs being used to carry out inspection tasks that were previously the domain of divers. The technique has also been used extensively to take the measurements required for engineering tasks underwater, so reducing the total underwater time as well as generally increasing the accuracy of the resulting measurements.

Camera Alive has now introduced videogrammetry to the North Sea. In association with Marseilles-based Travocean, 1987 has seen the introduction of a videogrammetric system that uses video cameras to obtain the images that were previously produced by photographic cameras. This has the major advantage that measurements may be taken in 'real time'. A subject need only to be imaged by the video cameras, with real time quality control ensuring that the image is satisfactory, and relayed to the surface using normal umbilicals. It can then be processed to obtain measurements in as little as 60 seconds from receipt.

Fig. 1 The introduction of new, cost effective, products. Videogrammetric cameras configured in the close-up mode.

Fig. 2 The videogrammetric analysis container for offshore measurement in three dimensions.

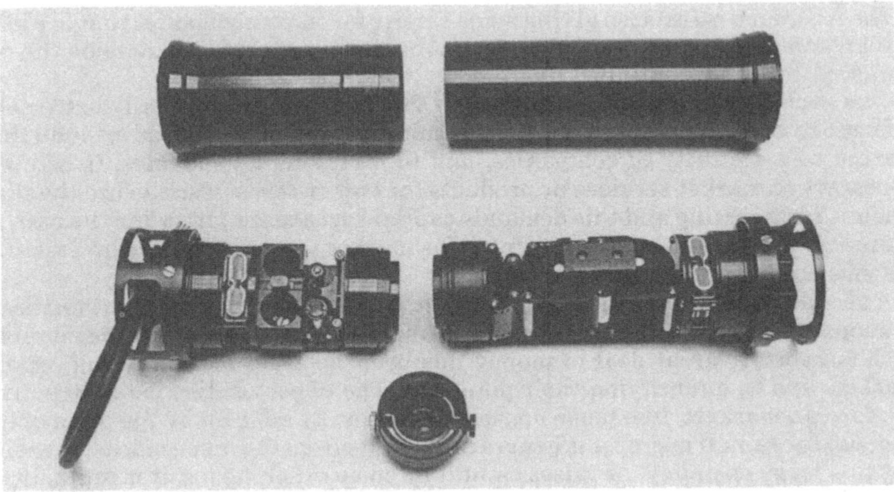

Fig. 3 Product rationalisation. A 36 exposure camera (left) and a 250 exposure camera (right) built from common parts.

This innovative approach will overcome one of the inherent disadvantages of photogrammetry, that of the time taken to process the film and to produce the measurements, a job that is normally carried out onshore. Although videogrammetry does not currently have the accuracy of photogrammetry, it currently can give accuracies of 1:200 in the subject plane against the 1:5000 commonly obtained by photogrammetry; it can still give an accuracy of 0.5 mm in close-up. Videogrammetry will allow ROVs to be used for more inspection work, thus producing overall cost savings to the oil companies.

One important act of consolidation was to look at the costs of production and to try to reduce or at least rationalize them. Introducing the concept of design modules, essentially building blocks that could be transferred from one camera design to another, enabled other products to be designed using existing components. The CI36 35-mm 36-exposure camera was built from components used for the CI250 35-mm 250-exposure camera, with only the film chamber and housing length requiring to be changed.

A further market lead diversification is that into internal pipeline inspection. Various oil companies had indicated that they required to inspect the inside of risers and other pipe systems and that systems were only available to do this at ambient pressure. Camera Alive produced a solution to allow this to be carried out at depth, by the use of a tracked crawler fitted with video and photographic systems. This solution now needs to be developed further to see whether it can be developed into a worthwhile product or permanent service. It is a technique worth further development as it has markets outside of the offshore oil industry.

Investigation

The Camera Alive Group is fortunate in that much of its work has traditionally come from outside Aberdeen. The recession is hitting companies that depend on Aberdeen for their livelihood extremely hard. It is forcing other companies out of the Aberdeen market, so giving some respite for those companies that are left, but all companies must look outside of Aberdeen and certainly outside the oil industry, for their continued livelihood.

One such market is defence, although this has been traditionally extremely difficult to break into. Barriers such as quality control standards exclude this market to a majority of companies, and to most small companies. It is often necessary to market services or products for two or three years before the first order is forthcoming and this demands capital investment just when the cash is reducing. The best form of entry into this market is on the back of an existing defence contractor, in our case Osprey.

A second possibility is the export market. Government agencies are very keen to support a determined push into foreign markets, and many offer considerable cash support. A great deal of money has been invested in identifying export markets and in quantifying their potential. The opportunities look attractive but foreign markets, like home markets, are equally affected by the price of oil and, unless a small business is generating profits from the home market, or sells out to a large company, or raises capital by some other means, it is unlikely to have the initial capital to market its wares and to develop its export contacts to turn them into sales.

In any case it is likely that only those companies that have specialist products which are of a relatively small capital value or are so specialised as not to warrant local production under licence or joint venture will ultimately succeed. Those companies offering services are unlikely even to gain an entry.

Exporting still provides the icing on the cake but the home market must still provide the bread and butter. Possibly government should do more to stimulate the home market for the home producers rather than pour money into export drives that benefit only a few.

On the home front, adopting the strategies as outlined has led to a resistance in branching out into new ventures that are completely outside the range of experience. New markets for similar products to that already produced have been sought and developed.

One such market is that for oceanographic cameras. The oceanographic market is experiencing a new resurgence, possibly following recent media interest in the environment or in such expeditions as the *Titanic*. This interest has resulted in a slight increase in the demand for oceanographic cameras which has been stimulated by the development of improved and new cameras based on technology incorporated into offshore cameras.

For our CI6036 seabed photography system we introduced acoustic technology to replace the use of a clump weight to trigger the camera. The acoustic ranger had been developed initially as an auto-focus device for a video camera, but its use was adapted to overcome the problems of the rope and clump weight becoming tangled in the lowering frame, fouling on the seabed, or even disappearing into a soft seabed, in all cases not allowing the required photographs to be taken. In the CI6036 system, which is rated to 6000 m, the acoustic ranger triggers an acoustic pinger at a distance of 8 m from the seabed. This warns the operators that the equipment package is nearing the seabed. At a pre-set distance of between 0.3 metres and 7.9 metres, the ranger will cause a photograph to be taken and, simultaneously, double the rate of output from the pinger. This indicates to the operators that a photograph has been taken and that the frame should be raised to re-set the camera. When the frame has been raised above the pre-set photographic distance the ranger causes the camera to

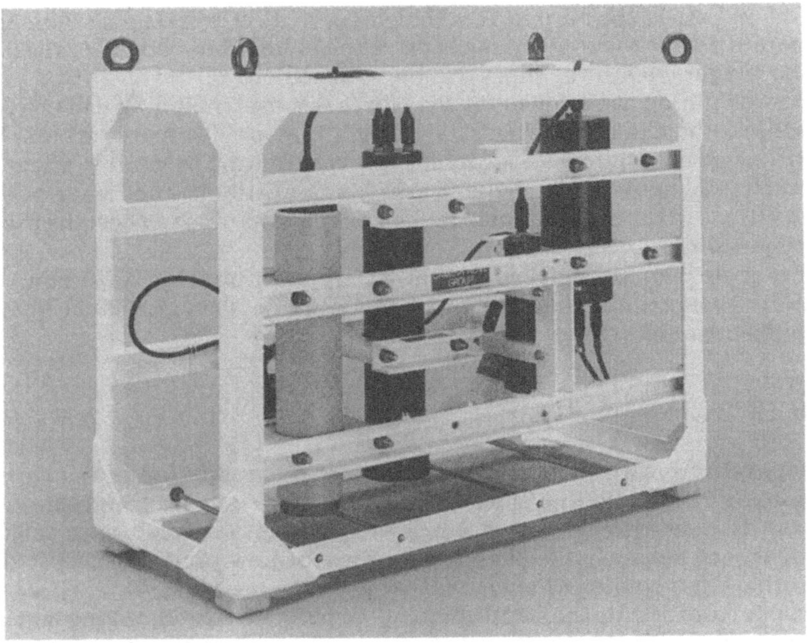

Fig. 4 Technical innovation. The CI6036 seabed photographic system with camera, acoustic rangefinder, acoustic pinger and flash unit.

Fig. 5 Diversification. A 400 exposure 35 mm camera for oceanographic and deep sea mining applications to a depth of 9000 metres.

be re-set and alters the output from the pinger to its original level, indicating to the operators that the system has been re-set. When the system is raised above 8 m, the ranger causes the pinger to be switched off.

Other technology developed for the North Sea use, including camera control electronics, film chamber design, miniaturization and flash unit design, has all been transferred to the oceanographic market, resulting in smaller, cheaper and more technologically advanced cameras being available and hence offering a serious threat to the previous market leader who was offering designs that were many years old.

Another developing market is that of deep sea mining. A 70 mm camera originally designed for offshore was adapted to be used at 6000 m to support mining operations.

CONCLUSIONS

It is apparent from all the market studies that oil production in the North Sea has passed its peak and that the overall market for equipment and services will contract. It must also be true that a shift in market emphasis must mean that some markets, for example the construction of new platforms, must decline while others, for example the removal of old platforms, must grow. There will be many opportunities for the small company to pursue in the changing market but it is becoming evident that all companies must look outside of the oil industry for their long term survival.

In the short term the industry will recover. It will not be the same again, because investments will not be made when a return on capital is in doubt.

It is believed that the industry is now more professional as a result of the recession, as the poorly managed companies and the 'cowboys' have long disappeared. It has been proven that anyone can manage a company when it is expanding but that few can manage one that is contracting. It is evident that the well managed companies, those that have survived and will continue to prosper, will be more guarded about developing equipment for the industry. In general its use will have to be written off over a shorter period as working seasons contract and utilization lessens. This must mean that rates will soon rise well above the level of inflation for certain equipment.

Some oil companies have assisted small companies through the bad patches by giving them much-needed work, even though that work was not a strict requirement, in order to protect services or products that they had identified as being required in the future. Others have introduced a more cavalier attitude by indicating that their money would spawn new companies to offer those same services and products in the future, hence expecting a repeat of the inflationary pricing spiral that characterized the early days of the industry. They believe that this will be cheaper in cash terms. Time only will indicate which of these approaches is correct.

However, it is apparent that many skilled and experienced people are being lost to the industry, never to return as they are of such an age where job security ranks above short term earning potential. It is also apparent that, where success has been achieved outside the offshore industry other factors make corporate life more pleasant. Deadlines are not as critical, operations rarely are round-the-clock, busy times are normally not in the summer, allowing holidays to be taken with the family when it is warm, payment arrives on time, specifications are accurately laid down and adhered to, planning allows requirements to be foreseen well ahead, reducing last minute panics. Successful companies may be tempted to develop in these markets, reducing competition in the offshore markets and allowing prices to rise.

The future of the industry is uncertain but, while cars are still on the road and industry still needs fuel, there will still be an offshore industry. It will not be an industry for the weak hearted and will certainly favour those companies that are innovative and can react rapidly to the many, and often sudden, changes demanded on them.

34

University/Industry/Governmental Interfaces—Joint R & D Projects—Do They Work?

G. Russell,
Heriot-Watt University, UK

INTRODUCTION

New technologies, changing markets, emerging nations, population, employment, disposal of waste, diminishing known resources, wars and rumours of war have for time immemorial created great social and economic pressures for the industrial world. Today is no exception as we experience a rapid rate of change both nationally and internationally. The need and ability to adapt to technical, economic and political change have never been more prevalent both in our nation and in our volatile industry.

With this background the role of research and development has received a new focus. The rate of gaining new knowledge to bring to bear on increasingly complex problems, within stringent financial constraints, has brought a new pattern of research programme. These emerging directed programmes cross the university/industry/government interfaces and attempt to manage and/or steer the available resources in a collaborative rather than competitive manner. Many benefits can accrue from this collaborative spirit. These may include a better mutual understanding of research motivation, access to valuable data or equipment and an easier transfer of new knowledge into development and production. Some pitfalls can also present themselves. Overmanagement may stifle innovation, arguments over joint ownership of intellectual property may sour relationships, and time-scales for research may be too short.

Nevertheless, these parameters define the way in which R. and D. programmes are being structured to meet the goals of cost-effective research.

The whole topic of research funding and exploitation of results is immense. This brief paper can only touch on some aspects of university/industry/governmental interfaces in an attempt to define models for joint R. and D. projects. The question 'Do they work?' can only be answered as existing projects

Advances in Underwater Technology, Ocean Science and Offshore Engineering, Volume 14: Submersible Technology

begin to deliver results and the participants analyse their performance. The question of 'how we make them work' must, however, be more important at this point in time.

In 1981 the UK university sector was drastically cut back and has been steadily eroded each year since that time. Engineering and Science faculties, although receiving some preferential treatment, have not escaped the intense pressures experienced over the last eight years. The North Sea oil and gas industry has, however, experienced a tremendous growth and success over this time, and only in the last year or so have we seen the results throughout the industry of a turn down in oil price. The manufacturing and information technology industries have not faired well over this time against other European, North American and Japanese competition. Many of these industries have suffered from short-term investment policy together with skill shortage in many areas of new technology.

The enforced impoverishment of our education system in concert with government department initiatives to encourage industry, the eventual users, to invest in university research has induced change. This has drawn the technologically based university disciplines closer to those industries seeking innovative development of their products or processes.

Seeking to improve the quality of research in terms of achieving defined aims within a well managed framework of awareness, relevance and timeliness has always been the goal of many university groups. The current climate has further encouraged these practices to spread into many collaborative ventures.

THE INTERFACES

The demand for research activity is often perceived differently by the three domains of university, industry and government. The common boundaries between the domains, the interfaces, can be shown diagrammatically in Fig. 1. Each sector—university, industry or government—has its own activities in areas of research for perhaps different reasons but for similar effects, each seeking to establish new knowledge. A simple view is that universities seek to discover and publish new knowledge, gradually raising the nation's base line of awareness and understanding. Industry seeks to convert new knowledge into exploitable products or processes and government needs to 'know' about new knowledge for procurement and national policy decisions.

The University Grants Committee and Research Council bridge the government/university interface, while the Department of Trade and Industry, the Department of Energy and the Ministry of Defence provide the bridge for joint government/industry activity. Direct industry/university activities are usually formed through individual groups involved in specific contract research projects.

The area of growing importance for universities is that involving all three domains in collaborative programmes initiated by SERC Directorates such as the Marine Technology Directorate (MTD) and the Application of Computers to Manufacturing Engineering (ACME) Directorate, or the Alvey Programme in Information Technology and the DTI initiatives in Advanced Robotics and Resources from the Sea.

Of course, the collaborative research model is widespread in European

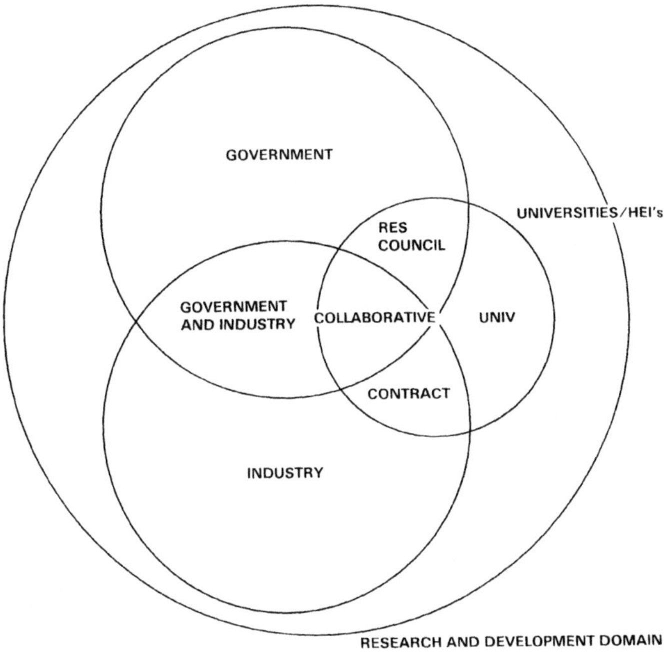

Fig. 1 The interfaces.

Economic Community programmes such as ESPRIT, BRITE, etc. and other European-based programmes such as EUREKA.

The mechanisms and opportunities for establishing collaborative programmes across the domain interfaces are many and varied. Arranging the partnerships and defining the roles, contributions and interactions between the participants is often a complex operation which can only be achieved through a great deal of effort and goodwill. Perhaps a limiting factor—or is it the driving force for collaboration?—is the shortage of skilled and experienced people to initiate and undertake the required scope of innovative research.

The research and development domain can have a wide interpretation. The industry and government participants can be considered to include charitable bodies and overseas organizations. Coupled with the research core in universities, there are often postgraduate or undergraduate courses that can act as sources of income by way of 'overseas fees'.

A 29% increase in money for research awarded by charitable organizations and from organizations overseas is reported in the University Grants Committee statistics published in August. Also reported is an increase by 23% to over £58 m from contracts with industry. The UGC noted a record level of 23% increase to £85 m of overall income from research grants, contracts and other services. On the negative side, however, there were nearly 800 fewer full-time undergraduates paying 'home fees', indicating a future reduction in output of graduates. Postgraduate home fee numbers did rise by 1350 and over 4000 additional students paid 'overseas fees', showing that many universities are marketing their courses abroad to meet projected financial deficits.

Industry often views academe with suspicion, believing that they, the academics, are too far removed from 'the real world'. Unfortunately this is sometimes true and university researchers must work doubly hard to keep abreast of technology and state-of-the-art application. On the other hand, it is apparent that within many companies different departments or sections often engage in developing the same product. Communication and planning is very poor and staff are overstretched and lack a long-term vision. European and American manufacturers often have quite a different philosophy to that of the Japanese competition. This can be illustrated by an example related to quality control. Conformance to specification is typically cited as the goal for quality control efforts, leading to the development of quality assurance and inspection departments within a company. The delivered products are of good quality, but the rejected products are expensive both to the manufacturer and the customer.

The Japanese idea is to set the goal of eliminating inspection procedures by designing products that are proven to be robust to conditions of normal use and material wear. Combined with the production process, the Japanese quality control effort is aimed at reducing variability around a target value. Resulting consistency of product quality increases, warranty service costs decrease and customer complaints decrease, thus increasing the market share. British industry has in many instances been reluctant to adopt scientific methods and indeed new technologies. Working more closely with research establishments in universities and elsewhere can improve the transfer of new ideas and bring new ways of approaching problems.

GOVERNMENT R & D FUNDING

Research funding from government is shown briefly in Figs 2 and 3.

The Research Council component of 12% amounts to about £600 m, divided between:

(i)	Science and Engineering Research Council (SERC)	—	£290 m
(ii)	Agriculture and Food Research Council (AFRC)	—	£99 m
(iii)	Natural Environment Research Council (NERC)	—	£94 m
(iv)	Medical Research Council (MRC)	—	£120 m

The SERC funding is further divided into:

(i)	Engineering	—	25%
(ii)	Science	—	25%
(iii)	Nuclear Physics	—	20%
(iv)	Astronomy and Space Radio	—	20%
(v)	Central Support	—	10%

The Engineering Budget is split into:

(i) Specially promoted programmes
(ii) Marine Technology Directorate
(iii) Information Technology
(iv) Biotechnology
(v) Application of Computers to Manufacturing Engineering
(vi) Teaching Companies

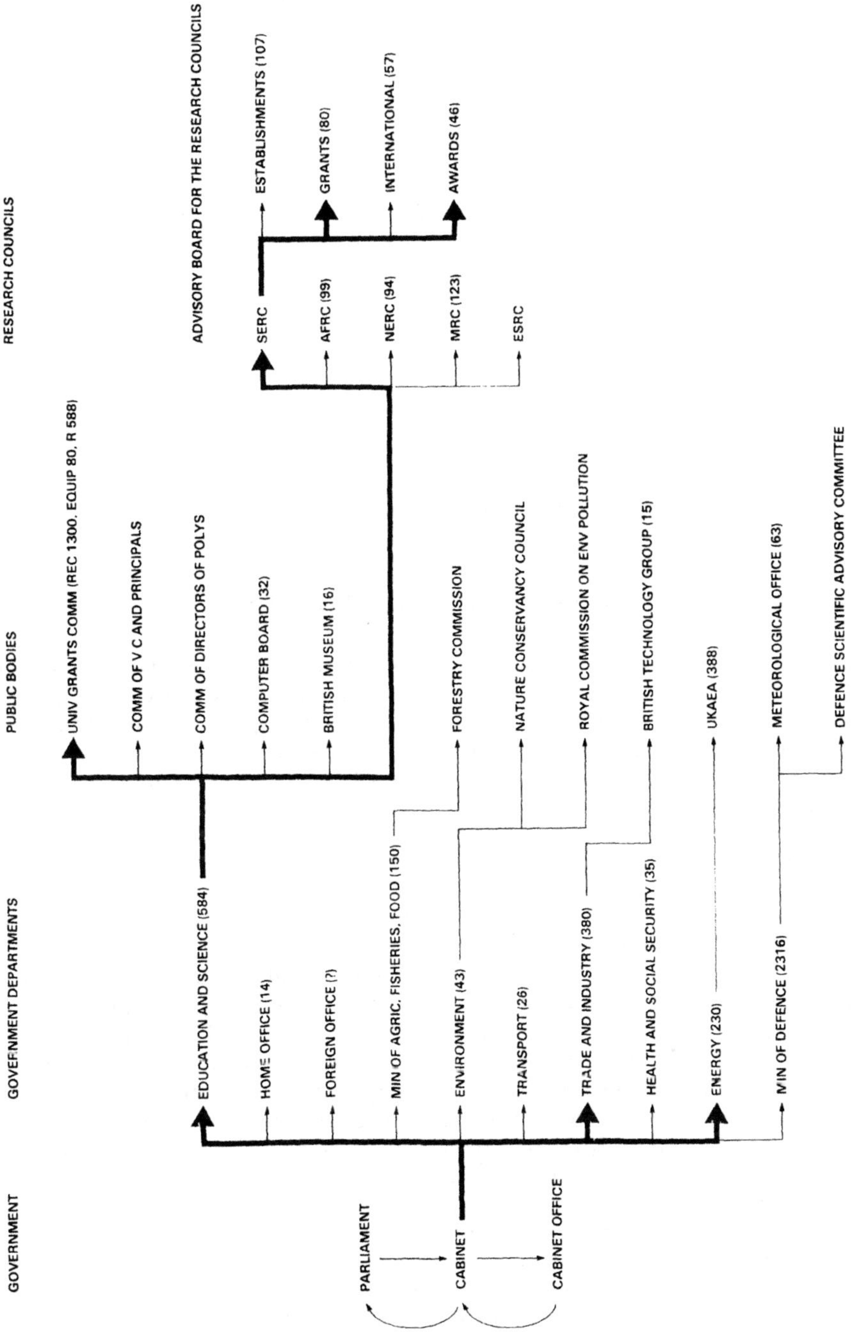

Fig. 2 Research funding (£m).

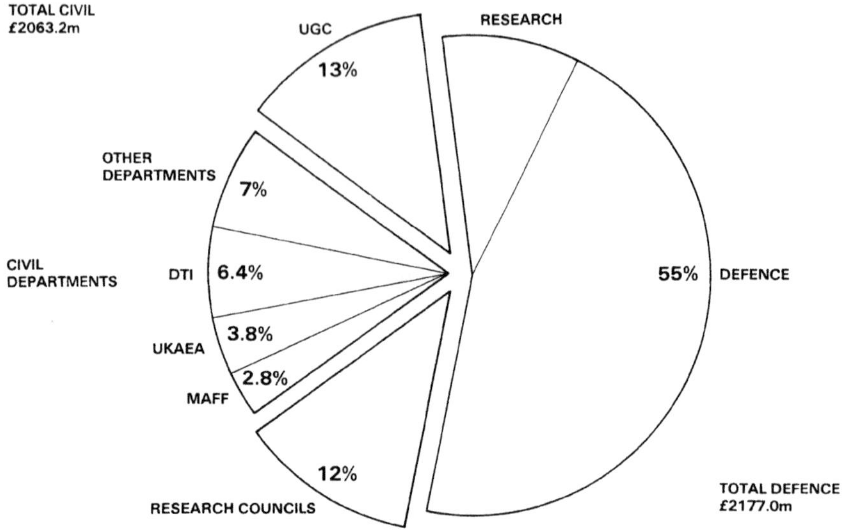

Fig. 3 Government research budget.

A RESEARCH AND DEVELOPMENT MODEL

A simplified systematic R. & D. model to show the interaction and interdependence of university, industry and government is postulated in Fig. 4.

Research programmes within the university system produce new knowledge and trained staff. The changing 'demand' is compared with existing knowledge and further research attempts to generate new knowledge to satisfy the demand. Excessive delays in this loop due to underfunding, lack of competent staff or inadequate equipment, means that the difference between the demand and knowledge output can grow to a point beyond recovery.

A similar loop exists in industry where the demand is for new products and development (and/or design) plays the major role matching the output and demand. Delays in this loop can also have a devastating effect, leading to products that do not satisfy the demand. The demand can be generated by the commercial part of industry with the customer accepting the products and feeding-back *cash* to maintain the growth of the company. The commercial part of industry requires *investment* to generate *profit, demand* and *revenue* to *government* which can in turn provide funding to the university sector and help with research investment to industry. The interdependence of research and development can be achieved through the flow of new knowledge and experienced staff to improve the effectiveness of the development process. Similarly, new product and market-awareness can steer research methodology to improve the cost effectiveness of the research process.

Figure 4 demonstrates the systematic flow through and across the interfaces between university, industry and government.

A more detailed model shown in Fig. 5 gives a structure for a collaborative research programme. Government departments and industry (or industries) provide a steering committee for the planned research programme. The

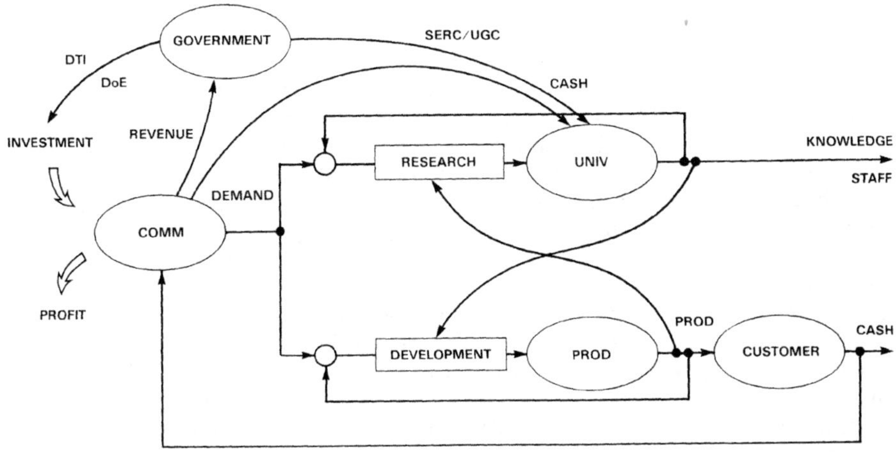

Fig. 4 Simplified R. and D. model.

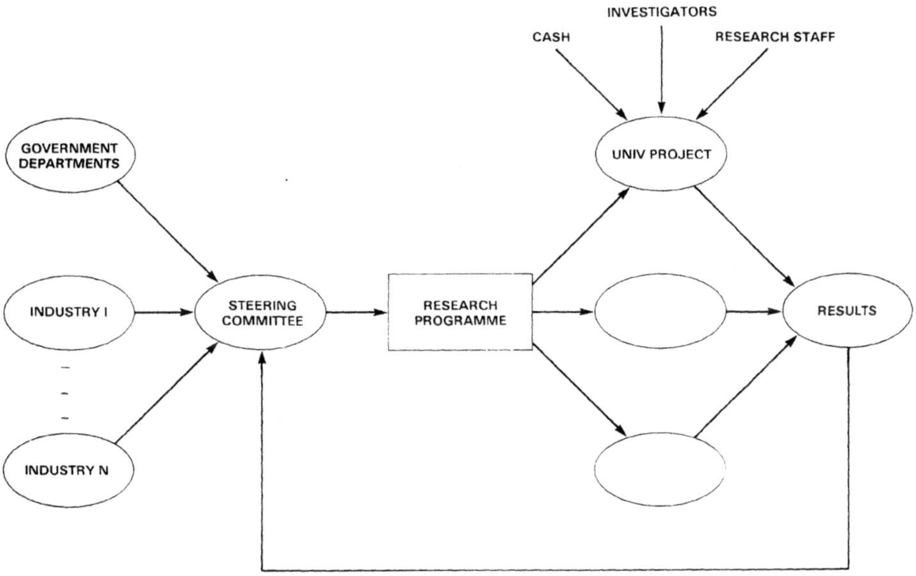

Fig. 5 Collaborative model.

component parts of the programme are partitioned into specific projects matching available resources in university departments to the requirements of the programme aims. Coordinated results are fed-back to the steering committee to aid the direction and management of the research projects.

SUMMARY

To maintain quality in research output, providing the necessary new knowledge for our developing industries must be of prime importance to our community. This is an extremely difficult task when viewed in the context of the rapid rate of change being experienced within our national industrial scene.

Collaborative research programmes designed to constantly monitor and improve the quality of output, through regular reporting and assessment of achieved goals against defined targets, can help the growth of our industry and benefit the rest of society.

The government has established initiatives to encourage collaboration and the transfer of technology between the research and development domains. There are many signs that industry and the universities are making a concerted effort to adapt their working practices and relationships to meet the rapidly changing industrial environment. There are still many shortfalls and areas for improvement. Some of the extremes of government educational policy can, and will, seriously impair the evolution of, and benefits gained from, the collaboration initiatives. These must be resisted by academe and industry alike if we are to make new advances against foreign competition.

Joint R. and D. projects do work, and can be made to work. To achieve the steady growth of healthy and fruitful relationships they must, however, be given a good environment and a nourishing diet.

REFERENCES

1. *Alvey Programme Annual Report*(s), Alvey Directorate, Millbank Tower, Millbank, London, SW1P 4QU.
2. *Annual Review of Government Funded R. & D. (1985).* (From the Cabinet Office), Her Majesty's Stationery Office.
3. Burgam, P. M. Design of Experiments—The Taguchi Way, *Manufacturing Engineering*, May 1985.
4. Report on a study of Support for Engineering, Science and Engineering Research Council, May 1986.
5. *A Strategy for the Science Base*, A discussion document prepared for the Secretary of State for Education and Science.
6. University statistics 1985–86, Vol. 3, Finance. (Published by University Statistical Record on behalf of the UGC).

GPSR Compliance

The European Union's (EU) General Product Safety Regulation (GPSR) is a set of rules that requires consumer products to be safe and our obligations to ensure this.

If you have any concerns about our products, you can contact us on ProductSafety@springernature.com

In case Publisher is established outside the EU, the EU authorized representative is:

Springer Nature Customer Service Center GmbH
Europaplatz 3
69115 Heidelberg, Germany

Batch number: 10607212

Printed by Printforce, the Netherlands